STUDENT SOLUTIONS MANUAL

Susan H. Porter and Laurel Technical Services

to accompany

TECHNICAL MATHEMATICS

Fourth Edition

TECHNICAL MATHEMATICS WITH CALCULUS

Fourth Edition

Paul A. Calter
Professor Emeritus
Vermont Technical College

Michael A. Calter, Ph.D.
Associate Professor
University of Rochester

John Wiley & Sons, Inc.
New York ♦ *Chichester* ♦ *Weinheim* ♦ *Brisbane* ♦ *Singapore* ♦ *Toronto*

To order books or for customer service call 1-800-CALL-WILEY (225-5945).

ISBN 0-471-37345-1

Printed in the United States of America

10 9 8 7 6 5 4 3 2

Printed and bound by Hamilton Printing Company

CONTENTS

CHAPTER 1: NUMERICAL COMPUTATION

Exercise 1: The Real Numbers

1. $7 < 10$

5. $\frac{3}{4} = 0.75$

9. $-|3 - 9| - |5 - 11| + |21 + 4|$
    $= -|-6| - |-6| + |25|$
    $= -6 - 6 + 25 = 13$

13. 2

17. 5

21. 55.86

25. 278.38

29. 0.5

33. 0.8

37. 274,800

41. 484,000

45. 8370

Exercise 2: Addition and Subtraction

1. -1090

5. 105,233

9. $-576 - 553 = -1129$

13. $1123 + 704 = 1827$

17. 593.44

21. Area difference $= 237,321 - 158,933$
    $= 78,388 \text{ mi}^2$

25. Total resistance $= 27.3 + 4.0155 + 9.75$
    $= 41.1 \text{ ohms (rounded)}$

Exercise 3: Multiplication

1. $73\overline{0}0$

5. $-17,800$

9. Cost $= 52.5 \text{ tons} \times \$63.25/\text{ton} = \$3320$

13. Cost $= 274 \text{ km} \times \$5,723/\text{km} = \$1,570,000$

17. 385.84 in. $\times 2.54 \text{ cm/in.} = 980.03 \text{ cm}$

Exercise 4: Division

1. 163

5. 0.7062

9. 17

13. Total work needed $= 5 \text{ masons} \times 8 \text{ days}$
    $= 40 \text{ mason days}$
    No. of masons required
    $= 40 \text{ mason days} \div 4 \text{ days}$
    $= 10 \text{ masons}$

17. -0.00000253

21. 0.2003

25. 0.9930

29. $\sin \theta = \frac{1}{3.58} = 0.279$

Exercise 5: Powers and Roots

1. 8

5. 1

9. 1

13. 100

17. 1000

21. 10,000

25. 1.035

29. 0.0146

33. 59.8

37. $P = (0.5855)^2 (365) = (0.3428)(365)$
$\quad\quad = 125$ watts

41. 5

45. -2

49. 4.45

53. -7.28

57. $Z = \sqrt{(3540)^2 + (2750)^2} = \sqrt{20{,}094{,}100}$
$\quad\quad = 4480\,\Omega$

Exercise 6: Combined Operations

1. 3340

5. 5

9. 121

13. 27

17. 12

21. 30

25. 978

29. 0.160

33. 55.8

37. 7.17

41. 0.871

Exercise 7: Scientific Notation

1. 10^2

5. 10^8

9. 0.1

13. 2.5742×10^4, 25.742×10^3

17. 2850

21. 0.003667

25. 10^3

29. 10^7

33. $(5 \times 10^4)(8 \times 10^{-3}) = 40 \times 10 = 4 \times 10^2 = 400$

37. $(6 \times 10^4)/0.03 = 2 \times 10^6$

41. $49{,}000/(7.0 \times 10^{-2}) = 7.0 \times 10^5$

45. $0.037 - (6.0 \times 10^{-3})$
$\quad\quad = (3.7 \times 10^{-2}) - (0.60 \times 10^{-2})$
$\quad\quad = 3.1 \times 10^{-2}$

49. $(3.87 \times 10^{-2})(5.44 \times 10^5) = 2.11 \times 10^4$

53. $R = (4.98 \times 10^5) + (2.47 \times 10^4) + (9.27 \times 10^6)$
$\quad\quad = (0.498 \times 10^6) + (0.0247 \times 10^6)$
$\quad\quad\quad\quad + (9.27 \times 10^6)$
$\quad\quad = 9.79 \times 10^6 \Omega$

57. $C = 8.26 \times 10^{-6} + 1.38 \times 10^{-7} + 5.93 \times 10^{-5} = 0.826 \times 10^{-5} + 0.0138 \times 10^{-5} + 5.93 \times 10^{-5}$
$= 6.77 \times 10^{-5}$ F, rounded to 2 decimal places.

Exercise 8: Units of Measurement

1. $152 \text{ in.} = 152 \text{ in.} \left(\frac{1 \text{ ft}}{12 \text{ in.}} \right) = 12.7 \text{ ft}$

5. $29 \text{ tons} = 29 \text{ tons} \left(\frac{2000 \text{ lb}}{1 \text{ ton}} \right) = 58,000 \text{ lb}$

9. $364,000 \text{ m} = 364,000 \text{ m} \left(\frac{1 \text{ km}}{1000 \text{ m}} \right) = 364 \text{ km}$

13. $6.2 \times 10^9 \text{ ohms} = 6.2 \times 10^9 \text{ ohms} \left(\frac{1 \text{ megohm}}{10^6 \text{ ohms}} \right)$
$= 6.2 \times 10^3 \text{ megohms}$

17. $364.0 \text{ m} = 364.0 \text{ m} \left(\frac{3.281 \text{ ft}}{1 \text{ m}} \right) = 1194 \text{ ft}$

21. $4.66 \text{ gal} = 4.66 \text{ gal} \left(\frac{3.785 \text{ l}}{1 \text{ gal}} \right) = 17.6 \text{ l}$

25. $2840 \text{ yd}^2 = 2840 \text{ yd}^2 \left(\frac{9 \text{ ft}^2}{1 \text{ yd}^2} \right) \left(\frac{1 \text{ acre}}{43,560 \text{ ft}^2} \right)$
$= 0.587 \text{ acre}$

29. $0.982 \text{ km}^2 = 0.982 \text{ km}^2 \left(\frac{247.1 \text{ acre}}{1 \text{ km}^2} \right) = 243 \text{ acres}$

33. $73.8 \text{ yd}^3 = 73.8 \text{ yd}^3 \left(\frac{1 \text{ m}}{1.094 \text{ yd}} \right)^3 = 56.4 \text{ m}^3$

37. $4.86 \text{ ft/s} = \frac{4.86 \text{ ft}}{\text{s}} \left(\frac{3600 \text{s}}{\text{h}} \right) \left(\frac{1 \text{ mi}}{5280 \text{ ft}} \right) = 3.31 \text{ mi/h}$

41. $953 \text{ births/yr} = \frac{953 \text{ births}}{\text{yr}} \left(\frac{1 \text{ yr}}{52 \text{ wk}} \right)$
$= 18.3 \text{ births/week}$

45. $\$4720/\text{ton} = \frac{4720 \text{ dollars}}{\text{ton}} \left(\frac{100 \text{ cents}}{1 \text{ dollar}} \right) \left(\frac{1 \text{ ton}}{2000 \text{ lb}} \right)$
$= 236\text{¢/lb}$

49. $6.35 \text{ kg} = 6.35 \text{ kg} \left(\frac{2.205 \text{ lb}}{1 \text{ kg}} \right) = 14.0 \text{ lb}$

53. $8834 \text{ in.}^2 = 8834 \text{ in.}^2 \left(\frac{2.54 \text{ cm}}{1 \text{ in.}} \right)^2 \left(\frac{1 \text{ m}}{100 \text{ cm}} \right)^2$
$= 5.699 \text{ m}^2$

Exercise 9: Substituting into Equations and Formulas

1. $y = 5(3) + 2 = 15 + 2 = 17$

5. $y = 2(3) + 3(-4)^2 - 5(2)^3 = 6 + 48 - 40 = 14$

9. $y = 8 - (-8.49) + 3(-8.49)^2$
$= 8 + 8.49 + 3(72.1) = 233$

13. $y = 3000[1 + 0.065(5)] = \$3975$

17. $y = \$9570(1 + 0.0675)^5 = \$13,266$

Exercise 10: Percentage

1. 372%

5. 40.0%

9. 0.23

13. $\frac{3}{8}$

17. $(0.411)(255) = 105 \text{ tons}$

21. $12\overline{0}$ liters

25. Tax credit $= (\$1100)(0.42)$
$+ (\$5500 - \$1100)(0.25)$
$= \$1562$

29. 65.6

33. Let y = range on earlier electric vehicles
$y + 0.495y = 161 \text{ km}$
$1.495y = 161 \text{ km}$
$y = 161 \text{ km}/1.495 = 108 \text{ km}$

37. $12.3/26.8 = 0.459 = 45.9\%$

41. 33.4%

45. Ripple percentage $= (0.75 \text{ V}/51 \text{ V})(100)$
$= 1.5\%$

49. % change $= [(298 - 227)/227] (100) = 31.3\%$

53. $\left(\dfrac{37.625 - 35.5}{35.5}\right)(100) = 5.99\%$

57. $\left(\dfrac{1\ hp}{550\ ft\ lb/sec}\right)\left[\dfrac{(72\ ft)\ (10,000\ lb/h)\ (h/3600\ sec)}{0.50\ hp}\right](100)$
$= 73\%$

61. Low voltage $= 125.0V - (0.10)(125.0V)$
$= 112.5\ V$
High voltage $= 125.0\ V + (1.50)(125.0V)$
$= 312.5\ V$

65. $0.055(455\ L) = 25\ L$

53. 109

57. 14.7

61. $3^2 - 9 + 2 = 2$

65. 36.82 in.(2.54 cm/in.) $= 93.52$ cm

69. $\left(\dfrac{746}{992}\right) \times 100\% = 75.2\%$

73. 7.239

77. $\dfrac{1}{0.825} = 1.21$

Chapter 1 Review Problems

1. 83.35

5. 0.346

9. 5.46

13. $y = 3(-2.88)^2 - 2(-2.88)$
$= 24.88 + 5.76 = 30.6$

17. (a) 179   (b) 1.08   (c) 4.86   (d) 45,700

21. 3.63×10^6

25. 15 gal (3.785 l/gal) $= 56.8\ l$ thus
$\left(\dfrac{2.0\ l}{2.0\ l + 56.8\ l}\right)100 = 3.4\%$

29. $-39.2 \div -0.003826 = 10,200$

33. 2.42

37. $0.492(4827) = 2370$

41. $\dfrac{405}{628} = 64.5\%$ of former consumption

45. $\left(\dfrac{284\ m}{1\ min}\right)(5.25\ s)\left(\dfrac{1\ min}{60\ s}\right) + \frac{1}{2}\left(\dfrac{9.807\ m}{s^2}\right)(5.25\ s)^2$
$= 160\ m$
$160\ m\left(\dfrac{3.281\ ft}{m}\right) = 525\ ft$

49. $pop + 0.125\ pop = 8118$
$1.125\ pop = 8118$
$pop = 8118 \div 1.125 = 7216$

CHAPTER 2: INTRODUCTION TO ALGEBRA

Exercise 1: Algebraic Expressions

1. 2

5. 5

9. $\frac{3}{2a}$

Exercise 2: Addition and Subtraction of Algebraic Expressions

1. $7x + 5x = 12x$

5.
$$
\begin{array}{rrrrr}
7a & - & 3b & + & m \\
-7a & + & 3b & + & m & - & c \\
\hline
0 & + & 0 & + & 2m & - & c \\
= 2m & - & c
\end{array}
$$

9.
$$
\begin{array}{rrrrrr}
& 8ax & + & 2(x+a) & + & 3b \\
& 9ax & + & 6(x+a) & - & 9b \\
11x & - & 7ax & - & 8(x+a) & + & 6b \\
\hline
11x & + & 10ax & + & 0 & + & 0 \\
= 11x & + & 10ax
\end{array}
$$

13.
$$
\begin{array}{l}
x - y - z \\
-x + y + z \\
\hline
0 + 0 + 0 \\
= 0
\end{array}
$$

17.
$$
\begin{array}{l}
9b^2 - 3ac + d \\
4b^2 - 4ac + 7d \\
-4b^2 + 6ac + 3d \\
5b^2 - 2ac - 12d \\
4b^2 \qquad - d \\
\hline
18b^2 - 3ac - 2d
\end{array}
$$

21. $3.52(a+b) + 4.15(a+b) - 1.84(a+b)$
$= 5.83(a+b)$

25. $6.4 - 1.8x - 7.5 - 2.6x = -1.1 - 4.4x$

29. $a + b - m - (m - a - b)$
$= a + b - m - m + a + b$
$= 2a + 2b - 2m$

33. $(w + 2z) - (-3w - 5x) + (x + z) - (z - w)$
$= w + 2z + 3w + 5x + x + z - z + w$
$= 5w + 2z + 6x$

37. $2[2w^2 + 3w^2 + 6w^2] = 2[11w^2] = 22w^2$

Exercise 3: Integral Exponents

1. $10^2 \cdot 10^3 = 10^{2+3} = 10^5$

5. $y^3 \cdot y^4 = y^{3+4} = y^7$

9. $w \cdot w^2 \cdot w^3 = w^{1+2+3} = w^6$

13. $\frac{y^5}{y^2} = y^{5-2} = y^3$

17. $\frac{10^{x+5}}{10^{x+3}} = 10^{[x+5-(x+3)]} = 10^2 = 100$

21. $(x^3)^4 = x^{3 \cdot 4} = x^{12}$

25. $(x^{a+1})^2 = x^{2(a+1)} = x^{2a+2}$

29. $(3abc)^3 = 3^3 \cdot a^3 \cdot b^3 \cdot c^3 = 27a^3b^3c^3$

33. $\left(\frac{2a}{3b^2}\right)^3 = \frac{(2a)^3}{(3b^2)^3} = \frac{2^3 \cdot a^3}{3^3 \cdot (b^2)^3} = \frac{8a^3}{27b^6}$

37. $\left(\frac{3}{y}\right)^{-3} = \frac{1}{(3/y)^3} = \frac{1}{(3^3/y^3)} = \frac{y^3}{27}$

41. $2x^{-2} + 3y^{-3} = 2\left(\frac{1}{x^2}\right) + 3\left(\frac{1}{y^3}\right) = \frac{2}{x^2} + \frac{3}{y^3}$

45. $\frac{x^2}{y^2} = x^2 \cdot y^{-2} = x^2 y^{-2}$

49. $(a + b + c)^0 = 1$

53. $\frac{x^{2n} \cdot x^3}{x^{3+2n}} = \frac{x^{2n+3}}{x^{2n+3}} = x^{(2n+3)-(2n+3)} = x^0 = 1$

57. $\left(\frac{i}{3}\right)^2 R = \left(\frac{i^2}{9}\right)R = \frac{i^2 R}{9}$

Exercise 4: Multiplication of Algebraic Expressions

1. $x^3(x^2) = x^{3+2} = x^5$

5. $(6ab)(2a^2b)(3a^3b^3)$
$= 6 \cdot a \cdot b \cdot 2 \cdot a^2 \cdot b \cdot 3 \cdot a^3 \cdot b^3$
$= 36a^6b^5$

9. $2a(a - 5) = (2a)a + (2a)(-5) = 2a^2 - 10a$

13. $\quad -4pq(3p^2 + 2pq - 3q^2) = (-4pq)(3p^2) + (-4pq)(2pq) + (-4pq)(-3q^2) = -12p^3q - 8p^2q^2 + 12pq^3$

17. $\quad (a+b)(a+c) = a^2 + ac + ab + bc$

21. $\quad (m+n)(9m - 9n) = m(9m) + m(-9n) + n(9m) + n(-9n) = 9m^2 - 9mn + 9mn - 9n^2 = 9m^2 - 9n^2$

25. $\quad (x^2 + y^2)(x^2 - y^2) = (x^2)(x^2) + (x^2)(-y^2) + (y^2)(x^2) + (y^2)(-y^2) = x^4 - x^2y^2 + x^2y^2 - y^4 = x^4 - y^4$

29. $\quad (m^2 - 3m - 7)(m - 2) = m^3 + m^2(-2) + (-3m)m - 3m(-2) + (-7)m + (-7)(-2)$
$$= m^3 - 2m^2 - 3m^2 + 6m - 7m + 14 = m^3 - 5m^2 - m + 14$$

33. $\quad (1 - z)(z^2 - z + 2) = 1(z^2) + 1(-z) + 1(2) + (-z)(z^2) + (-z)(-z) + (-z)(2) = z^2 - z + 2 - z^3 + z^2 - 2z$
$$= -z^3 + 2z^2 - 3z + 2$$

37. $\quad (a^2 + ay + y^2)(a - y) = (a^2 \cdot a) + a^2(-y) + (ay)a + (ay)(-y) + (y^2 \cdot a) + y^2(-y)$
$$= a^3 - a^2y + a^2y - ay^2 + ay^2 - y^3 = a^3 - y^3$$

41. $\quad (m^4 + m^3 + m^2 + m + 1)(m - 1)$
$$= (m^4 \cdot m) + m^4(-1) + (m^3 \cdot m) + m^3(-1) + (m^2 \cdot m) + m^2(-1) + (m \cdot m) + m(-1)$$
$$+ 1(m) + 1(-1)$$
$$= m^5 - m^4 + m^4 - m^3 + m^3 - m^2 + m^2 - m + m - 1 = m^5 - 1$$

45. $\quad (x+1)(x+1)(x-2) = (x-2)[(x \cdot x) + x(1) + 1(x) + 1(1)] = (x-2)(x^2 + 2x + 1)$
$$= (x \cdot x^2) + x(2x) + x(1) + (-2)(x^2) + (-2)(2x) + (-2)(1)$$
$$= x^3 + 2x^2 + x - 2x^2 - 4x - 2$$
$$= x^3 - 3x - 2$$

49. $\quad (1+c)(1+c)(1-c)(1+c^2) = (1+c)(1-c)(1+c)(1+c^2) = (1 - c^2)[(1)(1) + 1(c^2) + c(1) + (c \cdot c^3)]$
$$= (1 - c^2)(1 + c^2 + c + c^3)$$
$$= (1)(1) + 1(c^2) + 1(c) + 1(c^3) + (-c^2)(1) + (-c^2)(c^2) + (-c^2)c + (-c^2)c^3$$
$$= -c^5 - c^4 + c + 1$$

53. $\quad (a+b+c)(a-b+c)(a+b-c) = (a^2 - ab + ac + ab - b^2 + bc + ac - bc + c^2)(a+b-c)$
$$= (a^2 - b^2 + 2ac + c^2)(a+b-c)$$
$$= a^3 + a^2b - a^2c - ab^2 - b^3 + b^2c + 2a^2c + 2abc - 2ac^2 + ac^2 + bc^2 - c^3$$
$$= a^3 + a^2c - ab^2 + b^2c - b^3 + 2abc - ac^2 + bc^2 - c^3 + a^2b$$

57. $\quad (a+c)^2 = a^2 + ac + ac + c^2 = a^2 + 2ac + c^2$

61. $\quad (A+B)^2 = A^2 + 2AB + B^2$

65. $\quad (2c - 3d)^2 = (2c)^2 - 2(2c)(3d) + (3d)^2 = 4c^2 - 12cd + 9d^2$

69. $\quad (y^2 - 20)^2 = (y^2)^2 - 2(y^2)(20) + 20^2 = y^4 - 40y^2 + 400$

73. $\quad (1 + x - y)(1 + x - y) = 1 + x - y + x + x^2 - xy - y - xy + y^2 = x^2 + 2x - 2xy - 2y + y^2 + 1$

77. $\quad (a+b)^3 = (a^2 + 2ab + b^2)(a+b) = a^3 + a^2b + 2a^2b + 2ab^2 + ab^2 + b^3 = a^3 + 3a^2b + 3ab^2 + b^3$

81. $\quad (a-b)^3 = (a^2 - 2ab + b^2)(a-b) = a^3 - a^2b - 2a^2b + 2ab^2 + ab^2 - b^3 = a^3 - 3a^2b + 3ab^2 - b^3$

85. $\quad (p-q) - [p - (p-q) - q] = p - q - [p - p + q - q] = p - q$

89. $\quad (x-1) - \{[x - (x-3)] - x\} = (x-1) - [(x - x + 3) - x] = (x-1) - (3 - x) = x - 1 - 3 + x = 2x - 4$

93. $A = (L + 4.0)(W - 3.0) = LW - 3.0L + 4.0W - 12$

97. $V = \frac{4}{3}\pi(r - 3.60)^3 = \frac{4}{3}\pi(r - 3.60)(r^2 - 7.20r + 12.96)$

$\quad = \frac{4}{3}\pi(r^3 - 10.8r^2 + 38.88r - 46.66)$

$\quad = 4.19r^3 - 45.2r^2 + 163r - 195$

Exercise 5: Division of Algebraic Expressions

1. $z^5 \div z^3 = z^{5-3} = z^2$

5. $\frac{30cd^2 f}{15cd^2} = \frac{30}{15} \cdot \frac{c}{c} \cdot \frac{d^2}{d^2} \cdot \frac{f}{1} = 2f$

9. $\frac{15axy^3}{-3ay} = \frac{15}{-3} \cdot \frac{a}{a} \cdot \frac{x}{1} \cdot \frac{y^3}{y} = -5xy^2$

13. $\frac{15ay^2}{-3ay} = \frac{15}{-3} \cdot \frac{a}{a} \cdot \frac{y^2}{y} = -5y$

17. $\frac{35m^2 nx}{5m^2 x} = \frac{35}{5} \cdot \frac{m^2}{m^2} \cdot \frac{n}{1} \cdot \frac{x}{x} = 7n$

21. $\frac{ab^2 c}{ac} = \frac{a}{a} \cdot \frac{b^2}{1} \cdot \frac{c}{c} = b^2$

25. $\frac{32r^2 s^2 q}{8r^2 sq} = \frac{32}{8} \cdot \frac{r^2}{r^2} \cdot \frac{s^2}{s} \cdot \frac{q}{q} = 4s$

29. $\frac{25xyz^2}{-5x^2 y^2 z} = \frac{25}{-5} \cdot \frac{x}{x^2} \cdot \frac{y}{y^2} \cdot \frac{z^2}{z} = \frac{-5z}{xy}$

33. $\frac{6abc}{2c} = 3ab$

37. $\frac{42xy}{xy^2} = \frac{42}{y}$

41. $\frac{2a^3 - a^2}{a} = \frac{2a^3}{a} - \frac{a^2}{a} = 2a^2 - a$

45. $\frac{27x^6 - 45x^4}{9x^2} = \frac{27x^6}{9x^2} - \frac{45x^4}{9x^2} = 3x^4 - 5x^2$

49. $\frac{-3a^2 - 6ac}{-3a} = \frac{-3a^2}{-3a} - \frac{6ac}{-3a} = a + 2c$

53. $\frac{4x^3 y^2 + 2x^2 y^3}{2x^2 y^2} = \frac{4x^3 y^2}{2x^2 y^2} + \frac{2x^2 y^3}{2x^2 y^2} = 2x + y$

81.

$$
\begin{array}{r}
x \;\; -2y \;\; -z \\
(x+2y+z)\overline{)x^2 \;\; -4y^2 \;\; -4yz \;\; -z^2} \\
\underline{x^2 \qquad\qquad +2xy \;\; +xz} \\
-4y^2 \;\; -4yz \;\; -z^2 \;\; -2xy \;\; -xz \\
\underline{-4y^2 \;\; -2yz \qquad\quad -2xy} \\
-2yz \;\; -z^2 \qquad\qquad -xz \\
\underline{-2yz \;\; -z^2 \qquad\qquad -xz} \\
0
\end{array}
$$

57. $\frac{x^2 y^2 - x^3 y - xy^3}{xy} = \frac{x^2 y^2}{xy} - \frac{x^3 y}{xy} - \frac{xy^3}{xy}$

$\quad = xy - x^2 - y^2$

61. $\frac{a^2 - 3ab + ac^2}{a} = \frac{a^2}{a} - \frac{3ab}{a} + \frac{ac^2}{a} = a - 3b + c^2$

65. $\frac{m^2 n + 2mn - 3m^2}{mn} = \frac{m^2 n}{mn} + \frac{2mn}{mn} - \frac{3m^2}{mn}$

$\quad = m + 2 - \frac{3m}{n}$

69.

$$
\begin{array}{r}
a \;\; +5 \\
(2a+1)\overline{)2a^2 \;\; +11a \;\; +5} \\
\underline{2a^2 \;\; + \;\; a} \\
10a \;\; +5 \\
\underline{10a \;\; +5} \\
0
\end{array}
$$

73.

$$
\begin{array}{r}
x \;\; -6 \\
(x+2)\overline{)x^2 \;\; -4x \;\; +3} \\
\underline{x^2 \;\; +2x} \\
-6x \;\; +3 \\
\underline{-6x \;\; -12} \\
\text{Remainder} = 15
\end{array}
$$

77.

$$
\begin{array}{r}
a \;\; -b \qquad +c \\
(a-b-c)\overline{)a^2 \;\; -2ab \;\; +b^2 \;\; -c^2} \\
\underline{a^2 \;\; - \;\; ab \qquad\qquad -ac} \\
-ab \;\; +b^2 \;\; -c^2 \;\; +ac \\
\underline{-ab \;\; +b^2 \qquad\qquad +bc} \\
-c^2 \;\; +ac \;\; -bc \\
\underline{-c^2 \;\; +ac \;\; -bc} \\
0
\end{array}
$$

1. $(b^4 + b^2x^3 + x^4)(b^2 - x^2) = b^6 - b^4x^2 + b^4x^3 - b^2x^5 + b^2x^4 - x^6$

5. $(3x - m)(3x - m)(x^2 + m^2) = (9x^2 - 3mx - 3mx + m^2)(x^2 + m^2) = (9x^2 - 6mx + m^2)(x^2 + m^2)$
   $= 9x^4 + 9m^2x^2 - 6mx^3 + 6m^3x + m^2x^2 + m^4 = 9x^4 - 6mx^3 - 6m^3x + 10m^2x^2 + m^4$

9. $(3ax^2)(2ax^3) = 6a^2x^5$

13. $(4a - 3b)(4a - 3b) = 16a^2 - 12ab - 12ab + 9b^2 = 16a^2 - 24ab + 9b^2$

17. $(2a - 3b)(2a - 3b) = 4a^2 - 6ab + 9b^2 = 4a^2 - 12ab + 9b^2$

21. $(2m - c)(2m + c)(4m^2 + c^2) = (4m^2 - c^2)(4m^2 + c^2) = 16m^4 - c^4$

25. $y - 3[y - 2(4 - y)] = y - 3(y - 8 + 2y) = y - 9y + 24 = 24 - 8y$

29. $-2[w - 3(2w - 1)] + 3w = -2(w - 6w + 3) + 3w = -2(-5w + 3) + 3w = 10w - 6 + 3w = 13w - 6$

33. $\dfrac{(a-c)^m}{(a-c)^2} = (a - c)^{m-2}$

37. $(b - 3)^3 = (b - 3)^2(b - 3) = (b^2 - 6b + 9)(b - 3) = b^3 - 6b^2 + 9b - 3b^2 + 18b - 27 = b^3 - 9b^2 + 27b - 27$

41. $(x - 2)(x + 4) = x^2 + 4x - 2x - 8 = x^2 + 2x - 8$

45. $\dfrac{6a^3x^2 - 15a^4x^2 + 30a^3x^3}{-3a^3x^2}$

$$= \frac{6a^3x^2}{-3a^3x^2} - \frac{15a^4x^2}{-3a^3x^2} + \frac{30a^3x^3}{-3a^3x^2}$$

$$= -2 + 5a - 10x$$

49. $(2xy^3)(5x^2y) = 10x^3y^4$

53. $(1.33 \times 10^4)^2 = (1.33)^2 \times (10^4)^2 = 1.77 \times 10^8$

57.
$$
\begin{array}{r}
x^4 + x + 1 \\
(x^4 - x)\overline{\big)\,x^8 + 0x^5 + x^4 + 0x^2 + 0x + 1} \\
\underline{x^8 - x^5} \\
x^5 + x^4 + 0x^2 + 0x + 1 \\
\underline{x^5 - x^2} \\
x^4 + x^2 + 0x + 1 \\
\underline{x^4 - x} \\
x^2 + x + 1
\end{array}
$$
$$= x^4 + x + 1 \quad R(x^2 + x + 1)$$

61. $16.1\left(\dfrac{t}{2}\right)^2 \text{ ft} = 16.1\left(\dfrac{t^2}{4}\right) \text{ ft} = 4.02t^2 \text{ ft}$

CHAPTER 3: SIMPLE EQUATIONS AND WORD PROBLEMS

Exercise 1: Solving First-Degree Equations

1. $x - 5 = 28$
   $x = 28 + 5$
   $x = 33$

5. $9x - 2x = 3x + 2$
   $9x - 2x - 3x = 2$
   $4x = 2$
   $x = \frac{2}{4}$
   $x = \frac{1}{2}$

9. $3x = x + 8$
   $3x - x = 8$
   $2x = 8$
   $x = \frac{8}{2}$
   $x = 4$

13. $2.80 - 1.30y = 4.60$
    $1.30y = 2.80 - 4.60 = -1.80$
    $y = \frac{-1.80}{1.30} = -1.38$

17. $p - 7 = -3 + 9p$
    $-4 = 8p$
    $p = -\frac{1}{2}$

41. $7r - 2r(2r - 3) - 2 = 2r^2 - (r - 2)(3 + 6r) - 8$
    $7r - 4r^2 + 6r - 2 = 2r^2 - (3r + 6r^2 - 6 - 12r) - 8$
    $13r - 4r^2 - 2 = 2r^2 + 9r - 6r^2 + 6 - 8$
    $13r - 2 = 9r - 2$
    $4r = 0$
    $r = 0$

45. $ax + 4 = 7$
    $ax = 3$
    $x = \frac{3}{a}$

49. $c - bx = a - 3b$
    $-bx = a - 3b - c$
    $x = -\frac{a - 3b - c}{b} = \frac{-a + 3b + c}{b}$

21. $8x + 7 = 4x + 27$
    $8x - 4x = 27 - 7$
    $4x = 20$
    $x = \frac{20}{4}$
    $x = 5$

25. $5x + 22 - 2x = 31$
    $5x - 2x = 31 - 22$
    $3x = 9$
    $x = 3$

29. $2x + 5(x - 4) = 6 + 3(2x + 3)$
    $2x + 5x - 20 = 6 + 6x + 9$
    $7x - 6x = 15 + 20$
    $x = 35$

33. $3(6 - x) + 2(x - 3) = 5 + 2(3x + 1) - x$
    $18 - 3x + 2x - 6 = 5 + 6x + 2 - x$
    $12 - x = 5x + 7$
    $6x = 5$
    $x = \frac{5}{6}$

37. $3 - 6(x - 1) = 9 - 2(1 + 3x) + 2x$
    $3 - 6x + 6 = 9 - 2 - 6x + 2x$
    $9 - 6x = 7 - 4x$
    $2 = 2x$
    $x = 1$

Exercise 2: Solving Word Problems

1. If x = the number, then $\mathbf{5x + 8}$ is the required expression.

5. If x = gallons of antifreeze, then $\mathbf{6 - x}$ = gallons of water.

9. $82x$ km

13. $y + (y + 1) + (y + 2) = 174$
$3y + 3 = 174$
$3y = 174 - 3$
$3y = 171$
$y = 57$
$y + 1 = 58$
$y + 2 = 59$

17. $\frac{n}{3n+2} = \frac{3}{10}$
$10(3n + 2)\left(\frac{n}{3n+2}\right) = 10(3n + 2)\left(\frac{3}{10}\right)$
$10n = 3(3n + 2)$
$10n = 9n + 6$
$10n - 9n = 6$
$n = 6$

21. $2x - 3 = 29$
$2x = 29 + 3$
$2x = 32$
$x = 16$
Then, $4 + x = 4 + 16 = 20$

25. $3(x - 6) = x + 144$
$3x - 18 = x + 144$
$3x - x = 144 + 18$
$2x = 162$
$x = 81$

Exercise 3: Financial Problems

1. Let x = number of technicians
$17 - x$ = number of helpers
$210x + 185(17 - x) = 3345$
$x = 8$ technicians

5. Let x = thousands of gallons to break even
$(1.95)x = (1.16)x + 45$
$x = 57$, or 57,000 gallons

9. Let y = cost of the house
$100,000 - y$ = amount invested
$\left(\frac{1}{3}\right)(0.06)(100,000 - y)$
$+ \left(\frac{2}{3}\right)(0.05)(100,000 - y)$
$= 320$
$y = \$94,000$

Exercise 4: Mixture Problems

1. Let x = gallons of 12% mixture needed
$(0.05)(250) + (0.12)(x) = (0.09)(x + 250)$
$x = 333$ gal

5. Let x = number of kg of 63% copper mix needed
$(0.63)x + (0.72)(1100) = (0.67)(1100 + x)$
$0.63x + 792 = 737 + 0.67x$
$55 = 0.04x$
$x = 1380$ kg

9. Let x = lb of water needed to be evaporated
$300(0.80) - x(1.00) = (300 - x)(0.75)$
$x = 60.0$ lb

Exercise 5: Statics Problems

1. Taking moments about the right end,
$14,500(18.0 - x) = 10,500(18.0)$
$18.0 - x = 13.03$
$x = 18.0 - 13.03 = 4.97$ ft

5. Let x = bar length, cm. Taking moments about the left end,
$341x = 624(185)$
$x = 339$ cm

Chapter 3 Review Problems

1. $2x - (3 + 4x - 3x + 5) = 4$
$2x - (8 + x) = 4$
$2x - 8 - x = 4$
$2x - x = 4 + 8$
$x = 12$

5. $(2x - 5) - (x - 4) + (x - 3) = x - 4$
$2x - 5 - x + 4 + x - 3 = x - 4$
$2x - x + x - x = 5 - 4 + 3 - 4$
$x = 0$

9. $3x + 4(3x - 5) = 12 - x$
$3x + 12x - 20 = 12 - x$
$3x + 12x + x = 12 + 20$
$16x = 32$
$x = 2$

13. $x^2 + 8x - (x^2 - x - 2) = 5(x + 3) + 3$
    $x^2 + 8x - x^2 + x + 2 = 5x + 15 + 3$
    $8x + x - 5x = 15 + 3 - 2$
    $4x = 16$
    $x = 4$

17. $3p + 2 = \frac{p}{5}$
    $15p + 10 = p$
    $14p = -10$
    $p = -\frac{10}{14}$
    $p = -\frac{5}{7}$

21. $4.50(x - 1.20) = 2.80(x + 3.70)$
    $4.50x - 5.40 = 2.80x + 10.36$
    $1.70x = 15.76$
    $x = 9.27$

25. $20 - x + 4(x - 1) - (x - 2) = 30$
    $20 - x + 4x - 4 - x + 2 = 30$
    $4x - x - x = 30 - 20 + 4 - 2$
    $2x = 12$
    $x = 6$

29. Let c = kg of tin required
    $(0.39)(55) + c(1.00) = (55 + c)(0.50)$
    $c = 12$ kg

33. Let z = number of revolutions of the rear wheel
    $z + 250$ = number of revolutions of the front
                    wheel
    $10(z + 250) = 12z$
    $10z + 2500 = 12z$
    $2z = 2500$
    $z = 1250$
    distance traveled $= 12z = 12(1250) = 15,000$ ft

37. $5(x - 1) - 3(x + 3) = 12$
    $5x - 5 - 3x - 9 = 12$
    $2x - 14 = 12$
    $2x = 26$
    $x = 13$

41. $2(x - 7) = 11 + 2(4x - 5)$
    $2x - 14 = 11 + 8x - 10$
    $2x - 14 = 8x + 1$
    $6x = -15$
    $x = -\frac{5}{2}$

45. $bx + 8 = 2$
    $bx = -6$
    $x = \frac{-6}{b}$

49. $8 - ax = c - 5a$
    $-ax = c - 5a - 8$
    $x = \frac{c - 5a - 8}{-a} = \frac{5a - c + 8}{a}$

53. x = number of masons
    $125(x) + 92(10 - x) = 1118$
    $125x + 920 - 92x = 1118$
    $33x = 198$
    $x = 6$ masons

57. x = gallons of oil in final mixture
    $x = 0.0315(235) + 0.0505(186)$
    $x = 16.8$ gallons of oil

61. x = price of computer
    $x + 1\frac{1}{2}x = 995$
    $2\frac{1}{2}x = 995$
    $x = \$398$ for the computer
    $1\frac{1}{2}(398) = \$597$ for the printer

65. x = \$ invested at 10.25%
    $0.1025(x) + 0.0425(125,815 - x) = 11,782$
    $x = \$107,248$ at 10.25%
    $\$125,815 - \$107,248 = \$18,567$ at 4.25%

CHAPTER 4: FUNCTIONS

Exercise 1: Functions and Relations

1. Is a function

5. Not a function

9. $y = x^3$

13. $y = \left(\frac{2}{3}\right)(x - 4)$

17. Twice the cube root of x.

21. $V = f(r) = \frac{4\pi r^3}{3}$

25. $H = f(t) = 125t - \frac{gt^2}{2}$

29. $x \geq 7, \quad y \geq 0$

33. $x \neq 9, \quad y \neq 1$

37. $x \geq 1, \quad y \geq 0$

Exercise 2: Functional Notation

1. Explicit

5. x is independent, y is dependent

9. x and y are independent, z is dependent

13. $x^2 + y = x - 2y + 3x^2$

    $3y = -x^2 + 3x^2 + x$

    $3y = 2x^2 + x$

    $y = \frac{2x^2 + x}{3}$

17. $E = \frac{PL}{ae}$

    $ae = \frac{PL}{E}$

    $e = \frac{PL}{aE}$

21. $f(x) = 5 - 13x$

    $f(2) = 5 - 13(2)$

    $\qquad = 5 - 26$

    $\qquad = -21$

25. $f(x) = x^2 - 9$

    $f(-2) = (-2)^2 - 9$

    $\qquad = -5$

29. $f(x) = 2x - x^2$

    $f(5) = 2(5) - 5^2 = -15$

    $f(2) = 2(2) - 2^2 = 0$

    $f(5) + 3f(2) = -15 + 3(0) = -15$

33. $f(x) = 2x^2 + 4$

    $f(a) = 2a^2 + 4$

37. $f(x) = x^2, \ g(x) = \frac{1}{x}$

    $f(3) = 3^2 = 9$

    $g(2) = \frac{1}{2}$

    $f(3) + g(2) = 9 + \frac{1}{2} = 9\frac{1}{2}$

41. $f(x, y) = 3x + 2y^2 - 4$

    $f(2, 3) = 3(2) + 2(3)^2 - 4$

    $\qquad = 6 + 18 - 4$

    $\qquad = 20$

45. $g(a, b, c) = 2b - a^2 + c$

    $g(1, 1, 1) = 2(1) - 1^2 + 1 = 2$

    $g(1, 2, 3) = 2(2) - 1^2 + 3 = 6$

    $g(2, 1, 3) = 2(1) - 2^2 + 3 = 1$

    $\frac{3g(1,1,1) + 2g(1,2,3)}{g(2,1,3)} = \frac{3(2) + (6)}{1} = 18$

49. $f(t) = R_0(1 + \alpha t)$

    $\qquad = 98\overline{0}0(1 + 0.00427t)$

    $f(20.0) = 98\overline{0}0(1.0854)$

    $\qquad\qquad = 1.06 \times 10^4 \quad \Omega$

    $f(25.0) = 98\overline{0}0(1.10675)$

    $\qquad\qquad = 1.08 \times 10^4 \quad \Omega$

    $f(30.0) = 98\overline{0}0(1.1281)$

    $\qquad\qquad = 1.11 \times 10^4 \quad \Omega$

Exercise 3: Composite Functions and Inverse Functions

1. If $g(x) = 2x + 3$, and $f(x) = x^2$, then

    $g[f(x)] = 2(x^2) + 3 = 2x^2 + 3$

   If $g(x) = x^3$ and $f(x) = 4 - 3x$, then

5. $f[g(x)] = 4 - 3(x^3) = 4 - 3x^3$

9. $y = 8 - 3x$, so $x = \frac{(8-y)}{3}$

   Interchanging variables, $y = \frac{(8-x)}{3}$

13. $y = 3x - (4x + 3)$
     $\quad = 3x - 4x - 3 = -x - 3$
     $x = -y - 3$
     Interchanging variables, $y = -x - 3$

17.
```
10  PRINT " X",   " Y"
20  REM f1(X) = FNA(X)f2(X)=FNB(X)
30  DEF FNA(X) = 2 * X - 5
40  DEF FNB(X) = 3 * X^2 + 2 * X - 3
50  FOR X = 1 TO 10
60  LET Y = (4 * FNA(X))/(FNB(X) + 3)
70  PRINT X, Y
80  NEXT X
90  END
OK
RUN
    X       Y
    1      -2.4
    2      -.25
    3       .121212
    4       .214286
    5       .235294
    6       .233333
    7       .223603
    8       .211538
    9       .199234
    10      .1875
```

21. $y = 3x + 4(2 - x)$
     $y = 3x + 8 - 4x = 8 - x$
     $x = 8 - y$
     Interchanging the variables, $y = 8 - x$

25. $h(p, q, r) = 5r + p + 7q$
     $h(1, 2, 3) = 5(3) + 1 + 7(2) = 30$
     $h(1, 1, 1) = 5(1) + 1 + 7(1) = 13$
     $h(2, 2, 2) = 5(2) + 2 + 7(2) = 26$
     $h(3, 3, 3) = 5(3) + 3 + 7(3) = 39$
     $h(1, 2, 3) + \dfrac{2h(1,1,1) - 3h(2,2,2)}{h(3,3,3)}$
     $\qquad = 30 + \dfrac{2(13) - 3(26)}{39} = 28\frac{2}{3}$

29. $y = 9x + (x + 5) = 10x + 5$
     $x = \frac{y-5}{10}$
     Interchanging variables, $y = \frac{x-5}{10}$

33. If $g(x) = 3x - 5$, and $f(x) = 7x^2$, then
     $f[g(x)] = 7(3x - 5)^2 = 63x^2 - 210x + 175$

Chapter 4 Review Problems

1. (a) Is a function
    (b) Not a function
    (c) Not a function

5. (a) Explicit, with y independent and w dependent.
    (b) Implicit

9. $y = 3x^2 + 2z, \ z = 2x^2$
    $y = 3x^2 + 2(2x^2)$
    $\quad = 3x^2 + 4x^2$
    $\quad = 7x^2$

13. $f(3, 2, 1) = 3^2 + 3(3)(2) + 1 = 28$

17. y is seven less than five times the cube of x.

CHAPTER 5: GRAPHS

Exercise 1: Rectangular Coordinates

1. Fourth

5. Fourth

9. $x = 7$

13. $E(-0.3, -1.3)$, $F(-1.1, -0.8)$, $G(-1.5, 1)$, $H(-0.9, 1.1)$

17.

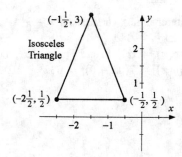

9.

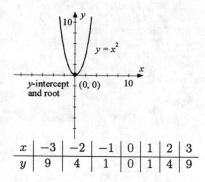

x	-3	-2	-1	0	1	2	3
y	9	4	1	0	1	4	9

13.

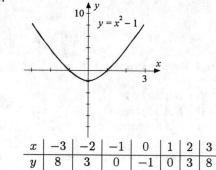

x	-3	-2	-1	0	1	2	3
y	8	3	0	-1	0	3	8

Exercise 2: The Graph of a Function

1.

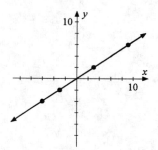

5.

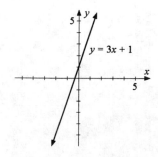

17.
```
10  ' XY-PAIRS
20  '
30  ' This program will compute y
40  ' for a function given in
50  ' line 110, for x values
60  ' specified in line 100, and
70  ' print the x,y pairs.
80  '
90  LPRINT " X", " Y"
100 FOR X = 0 TO 5
110    LET Y = 2 * X^2 - 3 * X + 5
120    LPRINT X, Y
130 NEXT X
140 END
    X              Y
    0              5
    1              4
    2              7
    3              14
    4              25
    5              40
```

Exercise 3: Graphing the Straight Line

1. Slope $= m = \dfrac{\text{rise}}{\text{run}} = \dfrac{4}{2} = 2$

5. $m = \dfrac{7-4}{5-2} = \dfrac{3}{3} = 1$

9. $y = mx + b$
    Since $m = \text{slope} = 4$
    and $b = y$ intercept $= -3$,
    $y = 4x - 3$

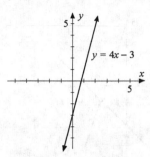

13. $m = 2.3,\ b = -1.5$
    Thus, $y = 2.3x - 1.5$

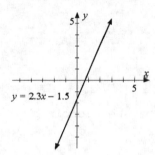

17. $y = -\dfrac{1}{2}x - \dfrac{1}{4}$
    Slope $= -\dfrac{1}{2}$
    y intercept $= -\dfrac{1}{4}$

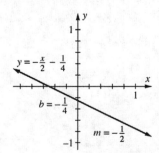

Exercise 4: The Graphics Calculator

Note: For Problems 1-8 graphs, please see Exercise 2,
Problems 9-16

9.

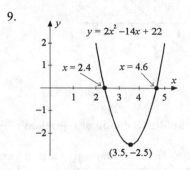

Exercise 5: Graphing Empirical Data, Formulas, and Parametric Equations

1.

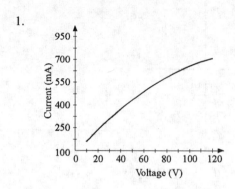

5.

N	100	200	300	400	500
f	45	90	135	180	225

N	600	700	800	900	1000
f	270	315	360	405	450

9.

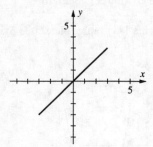

5.

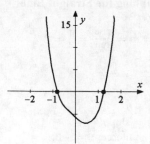

Exercise 6: Graphical Solution of Equations

For Problems 1-6, Note: You may get a mirror image about the *x*-axis if terms are transposed to the side of the equation other than what we did here.

9.

$\left(\frac{9}{4}, -\frac{7}{4}\right)$

1.

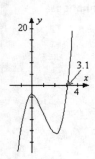

11.
```
10  ' MIDPOINT
20  '
30  ' This program finds the root of an equation by the midpoint
40  ' method.
50  ' Define the equation on line 80.
60  '
70  INPUT "WHAT ARE YOUR LEFT AND RIGHT LIMITS"; X1, X2
80  DEF FNA(X) = 2.4 * X^3 - 7.2 * X^2 - 3.3
90  Y1 = FNA(X1) : Y2 = FNA(X2)
100 IF Y1 < 0 AND Y2 > 0 THEN GOTO 140
110 IF Y2 < 0 AND Y1 > 0 THEN GOTO 140
120 PRINT "Your endpoints are not on different sides of the root."
130 GOTO 70
140 X3 = (X1 + X2) / 2
150 Y3 = FNA(X3)
160 IF Y3 > 0 AND Y1 > 0 THEN X1 = X3
170 IF Y3 < 0 AND Y1 < 0 THEN X1 = X3
180 IF Y3 > 0 AND Y2 > 0 THEN X2 = X3
190 IF Y3 < 0 AND Y2 < 0 THEN X2 = X3
200 IF ABS(Y3) < .00001 THEN GOTO 220
210 GOTO 90
220 PRINT "The root of this equation is "; X3
230 END
```

Chapter 5 Review Problems

1.

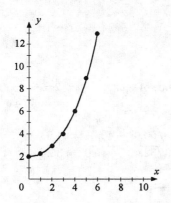

5.

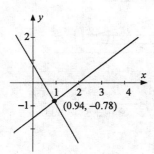

9.

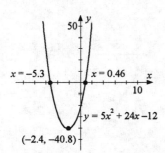

13.

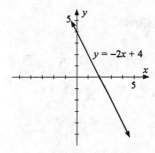

CHAPTER 6: GEOMETRY

Exercise 1: Straight Lines and Angles

1. (a) $\theta = 90° - 27.2° = 62.8°$
   (b) $\theta = 180° - 115.4° = 64.6°$
   (c) $\theta = 37.7°$

5. $A = 46.3°$, $B = 180° - 46.3° = 134°$
   $C = A = 46.3°$, $D = B = 134°$

9. $x = $ side of square, $\qquad 2x^2 = 2500$
   $x^2 = 1250,$ $\qquad\qquad x = 35.4\,\text{m}$

13. $d_1 = \sqrt{50.0^2 + 40.0^2} = 64.03$
    $d_2 = \sqrt{32.02^2 + 15.0^2} = 35.4\,\text{ft}$

17.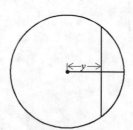

    $AC = AD - BE = \dfrac{23.4}{2} - \dfrac{15.8}{2} = 3.8\,\text{in.}$
    $AB = \sqrt{38.8^2 + 3.8^2} = 39.0\,\text{in.}$

21. $\left(\dfrac{x}{2.16\,\text{m}}\right) = \left(\dfrac{5.16\,\text{m}}{4.25\,\text{m}}\right) \qquad x = 2.62\,\text{m}$

Exercise 2: Triangles

1. $C = \left[\dfrac{(412')(828')}{2}\right]\left(\dfrac{\text{acre}}{43,560\,\text{ft}^2}\right)\left(\dfrac{\$1125}{\text{acre}}\right) = \$4405$

5. $l_{60} = \sqrt{45.0^2 + 60.0^2} = 75.0\,\text{ft}$
   $l_{108} = \sqrt{45.0^2 + 108.0^2} = 117\,\text{ft}$
   $l_{200} = \sqrt{45.0^2 + 200.0^2} = 205\,\text{ft}$

25.
```
10 INPUT "WHAT IS SIDE 1"; A
20 INPUT "WHAT IS SIDE 2"; B
30 INPUT "WHAT IS SIDE 3"; C
40 LET S = (A + B + C) / 2
50 LET AA = (S * (S - A) * (S - B) * (S - C))^.5
60 PRINT "SIDE 1", "SIDE 2", "SIDE 3", "AREA"
70 PRINT A, B, C, AA
80 END
```

Exercise 3: Quadrilaterals

1. $A = bh = (16.8)(25.2) = 423$ square units

5. $C = (312\,\text{ft})(6.5\,\text{ft})\left(\dfrac{\$13.50}{\text{yd}^2}\right)\left(\dfrac{\text{yd}^2}{9\,\text{ft}^2}\right) = \$3042$

9. $A_{\text{walls}} = [2(40\,\text{ft} + 36\,\text{ft})(22\,\text{ft})] - 1375\,\text{ft}^2$
   $\qquad\qquad = 1969\,\text{ft}^2$
   $A_{\text{ceiling}} = (40\,\text{ft})(36\,\text{ft}) = 1440\,\text{ft}^2$
   $A_{\text{total}} = 1969\,\text{ft}^2 + 1440\,\text{ft}^2 = 3409\,\text{ft}^2$
   $C = A\left(\dfrac{\$}{A}\right) = 3409\,\text{ft}^2\left(\dfrac{\$8.50}{\text{yd}^2}\right)\left(\dfrac{\text{yd}^2}{9\,\text{ft}^2}\right) = \$3220$

Exercise 4: The Circle

1. $d = \dfrac{C}{\pi} = \dfrac{22.5\,\text{m}}{\pi} = 7.16\,\text{m}$

5. $C = \pi d = \pi(33.0\,\text{m}) = 104\,\text{m}$

9.

$y = \sqrt{2.06^2 - 1.58^2} = 1.32$
$x = 4.12 - 2.06 - 1.32 = 0.74$ unit

13. $C = \pi d = \pi(15.0 \text{ cm}) = 47.0 \text{ cm}$

$h = \sqrt{48.0^2 + 52.0^2} = 70.8 \text{ cm}$

$L = \left[2\left(\frac{3}{8}\right) + \frac{1}{4}\right]C + 48.0 + 52.0 + h$

$\quad = 47.0 + 48.0 + 52.0 + 70.8 = 218 \text{ cm}$

17. $C = \pi(78.5) = 247 \text{ cm}$

13. The sum of the angles is $(5 - 2)180 = 540°$ so the missing angle is
$540 - 38 - 96 - 112 - 133 = 161°$.

17. Area $= \frac{38.4(53.8)}{2} = 1030 \text{ in.}^2$

21. Volume $= \pi(22.3)^2(56.2) = 87,800 \text{ cm}^3$

25. Volume $= 4.65(24.6) = 114 \text{ cm}^3$

Exercise 5: Volumes and Areas of Solids

1. $s = \frac{306 + 552 + 772}{2} = 815$

Base area
$= \sqrt{815(815 - 306)(815 - 552)(815 - 772)}$
$= 68,494 \text{ in.}^2$
Volume $= 68,494(925) = 6.34 \times 10^7 \text{ in.}^3$

5. altitude $= \sqrt{11.2^2 - 5.6^2} = 9.7 \text{ in.}$

$V = (18.2 \text{ in.}^2)(9.7 \text{ in.}) = 176 \text{ in.}^3$

9. $12 \text{ oz}\left(\frac{\text{lb}}{16 \text{ oz}}\right)\left(\frac{\text{ft}^3}{485 \text{ lb}}\right)\left(\frac{1728 \text{ in.}^3}{\text{ft}^3}\right) = 2.67 \text{ in.}^3$

$\frac{2.67 \text{ in.}^3}{3.50 \text{ in.}} = 0.763 \text{ in.}^2$

$d = \sqrt{\frac{4(0.763)}{\pi}} = 0.99 \text{ in.}$

13. $V = \frac{1}{3}\pi h(R^2 + r^2 + Rr)$

$\quad = \frac{1}{3}\pi 50\left[\frac{25}{4\pi^2} + \frac{9}{4\pi^2} + \frac{15}{4\pi^2}\right] = 65.0 \text{ ft}^3$

$W = 65.0 \text{ ft}^3\left(\frac{58.5 \text{ lb}}{\text{ft}^3}\right) = 3800 \text{ lb}$

17. $V = \frac{4}{3}\pi(744)^3 = 1.72 \times 10^9$

$SA = 4\pi(744)^2 = 6.96 \times 10^6$

21. $V = \frac{4}{3}\pi\left(\frac{2.50}{2}\right)^3 = 8.18 \text{ in.}^3$

$8.18 \text{ in.}^3\left(\frac{485 \text{ lb}}{\text{ft}^3}\right)\left(\frac{\text{ft}^3}{1728 \text{ in.}^3}\right)100 \text{ balls} = 23\overline{0} \text{ lb}$

Chapter 6 Review Problems

1. $\frac{12.0}{1} = \frac{x}{2}$, $x = 24.0 \text{ mi at } \frac{1}{60}\text{h}$
Speed $= 1440 \text{ mi/h}$

5. $w = \sqrt{12^2 + 5^2} = 13 \text{ m}$

9. $C = 8.00\pi$
Length $= 8.00\pi + 2(9.00) = 43.1 \text{ in.}$

CHAPTER 7: RIGHT TRIANGLES & VECTORS

Exercise 1: Angles and Their Measures

1. $27.8°\left(\frac{\pi}{180°}\right) = 0.485$ rad

5. 0.55 rev $\left(\frac{2\pi}{\text{rev}}\right) = 3.5$ rad

9. $68.8°\left(\frac{1 \text{ rev}}{360°}\right) = 0.191$ rev

13. 2.83 rad $\left(\frac{180°}{\pi \text{ rad}}\right) = 162°$

17. $29°27' = 29° + \frac{27'}{60'} = 29.45°$

21. 0.44975 rev $\left(\frac{360°}{\text{rev}}\right) = 161.91° = 161° + (0.91)(60') = 161°54.6' = 161°54' + (0.6)(60'') = 161°54'36''$

25.
```
10 ' DEG-MIN
20 '
30 ' This program accepts a decimal angle
40 ' and converts to deg, min, & sec.
50 '
60 INPUT "Enter the angle in decimal degrees"; A
70 LET B = A - INT (A)
80 LET B = B * 60
90 LET C = B - INT(B)
100 LET C = C * 60
110 PRINT INT(A); "DEG, "; INT(B); "MIN, "; C; "SEC"
120 END
```

Exercise 2: The Trigonometric Functions

1. $0.50, 0.87, 0.58, 1.73, 1.15, 2.00$

5. $\theta = 53°$

9. $h = \sqrt{196^2 + 122^2} = 231$

$\sin\theta = \frac{o}{h} = \frac{122}{231} = 0.528$

$\cos\theta = \frac{a}{h} = \frac{196}{231} = 0.849$

$\tan\theta = \frac{o}{a} = \frac{122}{196} = 0.622$

$\cot\theta = \frac{a}{o} = \frac{196}{122} = 1.61$

$\sec\theta = \frac{h}{a} = \frac{231}{196} = 1.18$

$\csc\theta = \frac{h}{o} = \frac{231}{122} = 1.89$

13. $r = \sqrt{2^2 + 3^2} = \sqrt{13}$

$\sin\theta = \frac{3}{\sqrt{13}} = 0.832$

$\cos\theta = \frac{2}{\sqrt{13}} = 0.555$

$\tan\theta = \frac{3}{2} = 1.50$

17. $\tan 44.4° = 0.9793$

21. $\cos 21.27° = 0.9319$

	sin	cos	tan
25.	0.955	0.296	3.230
29.	0.677	0.736	0.920

33. $3.72(\sin 28.3° + \cos 72.3°)$
$= 3.72(0.4741 + 0.3040) = 2.89$

Exercise 3: Finding the Angle When the Trigonometric Function is Given

1. $A = 30.0°$

5. $B = 28.9°$

9. $a = \sqrt{13^2 - 12^2} = 5$
    $\sin A = \frac{5}{13}$
    $\cos A = \frac{12}{13}$
    $\tan A = \frac{5}{12}$

13. $\cot^{-1} 1.17 = \tan^{-1} \frac{1}{1.17} = 40.5°$

Exercise 4: Solution of Right Triangles

1. $B = 90° - 42.9° = 47.1°$
    $\tan B = \frac{b}{155}$
    $b = 155(\tan 47.1°) = 167$
    $\sin A = \frac{155}{c}$
    $c = \frac{155}{\sin 42.9°} = 228$

5. $B = \frac{\pi}{2} - 1.13 = 0.441$ rad
    $\tan B = \frac{b}{284}$
    $b = 284(\tan 0.441) = 134$
    $\sin 1.13 = \frac{284}{c}$
    $c = \frac{284}{\sin 1.13} = 314$

9. $B = 90 - 31.4 = 58.6°$
    $\tan 31.4° = \frac{a}{82.4}$ $a = 82.4(\tan 31.4°) = 50.3$
    $\cos 31.4° = \frac{82.4}{c}$ $c = \frac{82.4}{\cos 31.4°} = 96.5$

13. $a = \sqrt{4.86^2 - 3.97^2} = 2.80$
    $A = \arccos\left(\frac{3.97}{4.86}\right) = \arccos 0.817 = 35.2°$
    $B = \arcsin\left(\frac{3.97}{4.86}\right) = \arcsin 0.817 = 54.8°$

17. $b = \sqrt{63.7^2 - 41.3^2} = 48.5$
    $\sin A = \frac{41.3}{63.7}$
    $A = 40.4°$
    $B = 90 - 40.4 = 49.6°$

21. $\sin 38° = \cos(90° - 38°) = \cos 52°$

25. $\cot 63.2° = \tan(90° - 63.2°) = \tan 26.8°$

29. $\sin 0.475 \text{ rad} = \cos\left(\frac{\pi}{2} - 0.475\right) = \cos 1.10 \text{ rad}$

Exercise 5: Applications of the Right Triangle

1.

$\tan 57.6° = \frac{h}{255\,\text{m}}$
$h = (255\,\text{m})(\tan 57.6°) = 402\,\text{m}$

5.

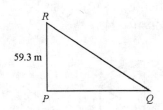

$\tan 47.6° = \frac{PQ}{59.3\,\text{m}}$
$PQ = (59.3\,\text{m})(\tan 47.6°) = 64.9\,\text{m}$

9.

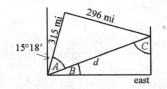

$\theta = 90° - 6.70° = 83.30°$
$\tan \theta = \frac{SS'}{15.0\,\text{m}}$
$SS' = (15.0\,\text{m})(\tan 83.30°) = 128\,\text{m}$

13.

$d = \sqrt{315^2 + 296^2} = 432 \text{ mi}$
$\tan A = \frac{296}{315}$
$A = 43°13'$
$B = 90° - 43°13' - 15°18' = 31°28'$
$C = 90° - B = 58°31' \quad \text{or} \quad \text{S } 58°31' \text{ W}$

17. Let x = length of horizontal beam at A

$$x = \sqrt{12.5^2 + 12.5^2} = 17.7 \text{ ft}$$
$$AB = \sqrt{10.3^2 + 17.7^2} = 20.5 \text{ ft}$$
$$\theta = \arctan \frac{10.3 \text{ ft}}{17.7 \text{ ft}} = 30.2°$$

21.

$$\sin 68.0° = \frac{a}{150}$$
$$a = (\sin 68.0°)(150) = 139$$
$$\cos 68.0° = \frac{(b/2)}{150}$$
$$\frac{b}{2} = (150)(\cos 68.0°) = 56.2$$
$$b = 2(56.2) = 112$$

25.

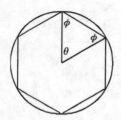

central angle $\theta = \frac{360°}{6} = 60°$
since the angles opposite the radii are equal
(isosceles triangle) they equal
$$\frac{180° - 60°}{2} = 60°$$
an equilateral triangle results, so the length
of one side of the hexagon = the radius = 125 cm

29.

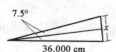

central angle $\theta = \frac{360°}{24} = 15°$
$$\frac{\theta}{2} = \frac{15°}{2} = 7.5°$$
$$\sin 7.5° = \frac{(x/2)}{36.000 \text{ cm}}$$
$$x = 2(\sin 7.5°)(36.000 \text{ cm}) = 9.3979 \text{ cm}$$

33.

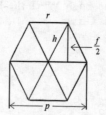

all central angles are 60.0°

$$\frac{f}{2} = \frac{0.750 \text{ cm}}{2} = 0.375 \text{ cm}$$
$$\sin 60.0° = \frac{0.375 \text{ cm}}{h}$$
$$h = \frac{0.375 \text{ cm}}{\sin 60.0°} = 0.433 \text{ cm}$$
$$p = 2h = 2(0.433 \text{ cm}) = 0.866 \text{ cm}$$
because small triangles are equilateral
$$r = h = 0.433 \text{ cm}$$

Exercise 6: Vectors

1.

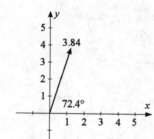

5.

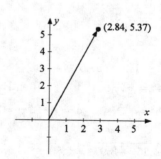

9. $V_x = (1.884)(\cos 58.24°) = 0.9917$
$V_y = (1.884)(\sin 58.24°) = 1.602$

13. $A = 7364, \ B = 4837$
$$R = \sqrt{7364^2 + 4837^2} = 8811$$
$$\theta = \arctan \frac{7364}{4837} = 56.70°$$

17. $A_x = 483, \ A_y = 382$
$$A = \sqrt{483^2 + 382^2} = 616$$
$$\theta = \arctan \frac{382}{483} = 38.3°$$

21.

$$\frac{x}{\begin{array}{r} 4.83\cos 18.3° = 4.586 \\ 5.99\cos 83.5° = 0.678 \end{array}}$$
$$R_x = 5.264$$
$$\frac{y}{\begin{array}{r} 4.83\sin 18.3° = 1.517 \\ 5.99\sin 83.5° = 5.951 \end{array}}$$
$$R_y = 7.468$$
$$R = \sqrt{5.264^2 + 7.468^2} = 9.14$$
$$\tan\theta = \frac{7.468}{5.264} \qquad \theta = 54.8°$$

25.

$$\frac{x}{\begin{array}{r} 1.38\cos 22.4° = 1.276 \\ 2.74\cos 49.5° = 1.779 \\ 3.32\cos 77.3° = 0.730 \end{array}}$$
$$R_x = 3.785$$
$$\frac{y}{\begin{array}{r} 1.38\sin 22.4° = 0.5259 \\ 2.74\sin 49.5° = 2.0835 \\ 3.32\sin 77.3° = 3.2388 \end{array}}$$
$$R_y = 5.8482$$
$$R = \sqrt{3.785^2 + 5.848^2} = 6.97$$
$$\tan\theta = \frac{5.848}{3.785} \qquad \theta = 57.1°$$

Exercise 7: Applications of Vectors

1. $W_y = (56.5\text{ N})(\sin 12.6°) = 12.3\text{ N}$

5. $\theta = \arcsin\frac{551}{1270} = 25.7°$

9. $V_W = \frac{(20.5\text{ km/h})}{(\tan 9°45')} = 119\text{ km/h}$

13. $V_s = \frac{(10.6\text{ mi/h})}{(\tan 5°15')} = 115\text{ mi/h}$

17. $Z = \sqrt{5.75^2 + 4.22^2} = 7.13\,\Omega$
$$\theta = \arctan\frac{5.75}{4.22} = 53.7°$$

Chapter 7 Review Problems

1. $38.2\text{ deg} = 38° + 0.2\text{ min} = 38°12'$
$$38.2\text{ deg}\left(\frac{\text{rad}}{180°}\right) = 0.667\text{ rad}$$
$$38.2\text{ deg}\left(\frac{1\text{ rev}}{360°}\right) = 0.106\text{ rev}$$

5. $r = \sqrt{3^2 + 7^2} = 7.62$
$$\sin\theta = \frac{o}{h} = \frac{7}{7.62} = 0.919$$
$$\cos\theta = \frac{a}{h} = \frac{3}{7.62} = 0.394$$
$$\tan\theta = \frac{o}{a} = \frac{7}{3} = 2.33$$
$$\cot\theta = \frac{1}{\tan\theta} = 4.29$$
$$\sec\theta = \frac{1}{\cos\theta} = 2.54$$
$$\csc\theta = \frac{1}{\sin\theta} = 1.09$$
$$\theta = \arcsin 0.919 = 66.8°$$

9. $\begin{array}{ll} \sin 72.9° = 0.9558 & \csc 72.9° = 1.0463 \\ \cos 72.9° = 0.2940 & \sec 72.9° = 3.4009 \\ \tan 72.9° = 3.2506 & \cot 72.9° = 0.3076 \end{array}$

13. $r = \arccos 0.824 = 34.5°$

17. $\text{arcsec } 2.447 = \arccos\left(\frac{1}{2.447}\right) = \arccos 0.4087$
$$= 65.88°$$

21. $V_y = 885(\sin 66.3°) = 810$
$V_x = 885(\cos 66.3°) = 356$

25.

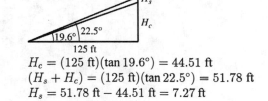

$$H_c = (125\text{ ft})(\tan 19.6°) = 44.51\text{ ft}$$
$$(H_s + H_c) = (125\text{ ft})(\tan 22.5°) = 51.78\text{ ft}$$
$$H_s = 51.78\text{ ft} - 44.51\text{ ft} = 7.27\text{ ft}$$

29. $\tan 52.8° = 1.3175$

33. $\theta = \arctan 1.7362 = 60.1°$

37. $\theta = \text{arccot } 0.9475 = 46.5°$

41. $9.26(\cos 88.3° + \sin 22.3°)$
$$= 9.26(0.02967 + 0.3795) = 3.79$$

45.

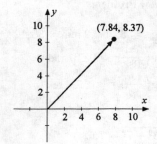

49. $AC = \sqrt{(158.2)^2 - (139.3)^2} = 74.98$ ft

$\cos B = \frac{139.3}{158.2}$

$B = 28.2935° = $ N 28.29° E or N 28°18' E

CHAPTER 8: FACTORS AND FACTORING

Exercise 1: Common Factors

1. $3y^2 + y^3 = y^2(3 + y)$

5. $3a + a^2 - 3a^3 = a(3 + a - 3a^2)$

9. $\frac{3}{x} + \frac{2}{x^2} - \frac{5}{x^3} = \frac{1}{x}\left(3 + \frac{2}{x} - \frac{5}{x^2}\right)$

13. $5a^2b + 6a^2c = a^2(5b + 6c)$

17. $3a^3y - 6a^2y^2 + 9ay^3 = 3ay(a^2 - 2ay + 3y^2)$

21. $8x^2y^2 + 12x^2z^2 = 4x^2(2y^2 + 3z^2)$

25. $L_0 + L_0\alpha t = L_0(1 + \alpha t)$

29. $v_0 t + \frac{at^2}{2} = t\left(v_0 + \frac{at}{2}\right)$

Exercise 2: Difference of Two Squares

1. $4 - x^2 = 2^2 - x^2 = (2 - x)(2 + x)$

5. $4x^2 - 4y^2 = 4(x^2 - y^2) = 4(x - y)(x + y)$

9. $9c^2 - 16d^2 = (3c)^2 - (4d)^2$
$= (3c + 4d)(3c - 4d)$

13. $(m^4 - n^4) = (m^2 - n^2)(m^2 + n^2)$
$= (m - n)(m + n)(m^2 + n^2)$

17. $a^{16} - b^8 = (a^8 - b^4)(a^8 + b^4)$
$= (a^4 - b^2)(a^4 + b^2)(a^8 + b^4)$
$= (a^2 - b)(a^2 + b)(a^4 + b^2)(a^8 + b^4)$

21. $16a^4 - 121 = (4a^2 + 11)(4a^2 - 11)$

25. $\pi r_2^2 - \pi r_1^2 = \pi(r_2^2 - r_1^2) = \pi(r_2 - r_1)(r_2 + r_1)$

29. $\frac{1}{2}mv_1^2 - \frac{1}{2}mv_2^2 = \frac{m}{2}(v_1^2 - v_2^2)$
$= \frac{m}{2}(v_1 + v_2)(v_1 - v_2)$

Exercise 3: Factoring Trinomials

1. $B^2 - 4AC = (-5)^2 - 4(3)(-9)$
$= 133 \sqrt{133} = 11.5$
(not a perfect square)
Not factorable

5. $(-30)^2 - 4(1)(-64) = 1156$
$\sqrt{1156} = 34$
Factorable

9. $x^2 - 10x + 9 = (x - 9)(x - 1)$

13. $x^2 + 7x + 12 = (x + 4)(x + 3)$

17. $x^2 + 6x + 8 = (x + 4)(x + 2)$

21. $b^2 - b - 12 = (b - 4)(b + 3)$

25. $3w^2 + 36w + 96 = 3(w^2 + 12w + 32)$
$= 3(w + 8)(w + 4)$

29. $a^2 + ab - 6b^2 = (a + 3b)(a - 2b)$

33. $a^2 - 20abc - 96b^2c^2 = (a - 24bc)(a + 4bc)$

37. $(a + b)^2 - 7(a + b) - 8 = (a + b + 1)(a + b - 8)$

41. $t^2 - 14t + 24 = (t - 2)(t - 12)$

Exercise 4: Factoring by Grouping

1. $a^3 + 3a^2 + 4a + 12 = a^2(a + 3) + 4(a + 3)$
$= (a^2 + 4)(a + 3)$

5. $x^2 - bx + 3x - 3b = x(x - b) + 3(x - b)$
$= (x - b)(x + 3)$

9. $x^2 + y^2 + 2xy - 4 = (x^2 + 2xy + y^2) - 4$
$= (x + y)^2 - 2^2 = (x + y - 2)(x + y + 2)$

13. $ax + bx + 3a + 3b = x(a + b) + 3(a + b)$
$= (x + 3)(a + b)$

17. $2a + bx^2 + 2b + ax^2 = 2(a + b) + x^2(a + b)$
$= (2 + x^2)(a + b)$

21. $b^2 - bc + ab - ac = b(b - c) + a(b - c)$
$= (a + b)(b - c)$

Exercise 5: General Quadratic Trinomial

1. $4x^2 - 13x + 3 = (4x - 1)(x - 3)$

5. $12b^2 - b - 6 = (3b + 2)(4b - 3)$

9. $5x^2 - 38x + 21 = (x - 7)(5x - 3)$

13. $3x^2 - x - 2 = (3x + 2)(x - 1)$

17. $4a^2 + 4a - 3 = (2a - 1)(2a + 3)$

21. $49x^6 + 14x^3y - 15y^2 = (7x^3 - 3y)(7x^3 + 5y)$

25. $5x^{4n} + 11x^{2n} + 2 = (5x^{2n} + 1)(x^{2n} + 2)$

29. $16t^2 - 82t + 45 = (2t - 9)(8t - 5)$

Exercise 6: The Perfect Square Trinomial

1. $x^2 + 4x + 4 = (x + 2)^2$

5. $2y^2 - 12y + 18 = 2(y^2 - 6y + 9)$
$= 2(y - 3)^2$

9. $9x^2 + 6x + 1 = (3x + 1)^2$

13. $16 + 16a + 4a^2 = 4(4 + 4a + a^2) = 4(2 + a)^2$

17. $x^2 + 2xy + y^2 = (x + y)^2$

21. $x^2 + 10ax + 25a^2 = (x + 5a)^2$

25. $z^6 + 16z^3 + 64 = (z^3 + 8)^2$

29. $a^4 - 2a^2b^2 + b^4 = (a^2 - b^2)^2$
$= [(a - b)(a + b)]^2 = (a - b)^2(a + b)^2$

Exercise 7: Sum or Difference of Two Cubes

1. $62 + x^3 = (4 + x)(16 - 4x + x^2)$

5. $x^3 - 1 = (x - 1)(x^2 + x + 1)$

9. $a^3 + 64 = (a + 4)(a^2 - 4a + 16)$

13. $216 - 8a^3 = 8(3 - a)(9 + 3a + a^2)$

17. $y^9 + 64x^3 = (y^3 + 4x)(y^6 - 4xy^3 + 16x^2)$

21. $8a^{6x} - 125b^{3x}$
$= (2a^{2x} - 5b^x)(4a^{4x} + 10a^{2x}b^x + 25b^{2x})$

25. $\frac{4}{3}\pi r_2^3 - \frac{4}{3}\pi r_1^3 = \frac{4\pi}{3}(r_2^3 - r_1^3)$
$= \frac{4\pi}{3}(r_2 - r_1)(r_2^2 + r_2r_1 + r_1^2)$

Chapter 8 Review Problems

1. $x^2 - 2x - 15 = (x - 5)(x + 3)$

5. $2x^2 + 3x - 2 = (2x - 1)(x + 2)$

9. $\frac{x^2}{y} - \frac{x}{y} = \frac{x}{y}(x - 1)$

13. $x - bx - y + by = x(1 - b) - y(1 - b)$
$= (x - y)(1 - b)$

17. $x^2 - 7x + 12 = (x - 3)(x - 4)$

21. $9a^2 + 12az^2 + 4z^4 = (3a + 2z^2)^2$

25. $(x - y)^2 - z^2 = (x - y - z)(x - y + z)$

29. $9x^4 - x^2 = x^2(9x^2 - 1) = x^2(3x + 1)(3x - 1)$

33. $27a^3 - 8w^3 = (3a - 2w)(9a^2 + 6aw + 4w^2)$

37. $6ab + 2ay + 3bx + xy = 2a(3b + y) + x(3b + y)$
$= (3b + y)(2a + x)$

41. $\frac{V_2^2}{R} - \frac{V_1^2}{R} = \frac{1}{R}(V_2^2 - V_1^2)$
$= \frac{1}{R}(V_2 - V_1)(V_2 + V_1)$

CHAPTER 9: FRACTIONS & FRACTIONAL EQUATIONS

Exercise 1: Simplification of Fractions

1. $\frac{12}{x}$ $x \neq 0$

5. $\frac{7}{x^2-3x+2} = \frac{7}{(x-2)(x-1)}$
$x \neq 2,\ x \neq 1$

9. $\frac{15}{16} = 0.9375$

13. $\frac{4375}{10,000} = \frac{7}{16}$

17.   Let $x = 0.7777\ldots$,
then $100x = 77.777\ldots$
$99x = 77$
$x = \frac{77}{99} = \frac{7}{9}$

21. $\frac{(a-b)(c-d)}{b-a} = -\frac{(a-b)(c-d)}{-(b-a)} = -\frac{(a-b)(c-d)}{(-b+a)}$
$= -\frac{(a-b)(c-d)}{(a-b)} = d - c$

25. $\frac{75}{35} = \frac{15}{7}$

29. $\frac{21m^2p^2}{28mp^4} = \frac{3(7)m^2p^2}{4(7)mp^4} = \frac{3m}{4p^2}$

33. $\frac{x^2+5x}{x^2+4x-5} = \frac{x(x+5)}{(x-1)(x+5)} = \frac{x}{x-1}$

37. $\frac{2m^3n-2m^2n-24mn}{6m^3+6m^2-36m} = \frac{2mn(m^2-m-12)}{6m(m^2+m-6)}$
$= \frac{n(m-4)(m+3)}{3(m+3)(m-2)} = \frac{n(m-4)}{3(m-2)}$

41. $\frac{x^2-z^2}{x^3-z^3} = \frac{(x-z)(x+z)}{(x-z)(x^2+xz+z^2)} = \frac{x+z}{x^2+xz+z^2}$

45. $\frac{x^2-1}{2xy+2y} = \frac{(x+1)(x-1)}{2y(x+1)} = \frac{x-1}{2y}$

49. $\frac{(x+y)^2}{x^2-y^2} = \frac{(x+y)(x+y)}{(x+y)(x-y)} = \frac{x+y}{x-y}$

Exercise 2: Multiplication and Division of Fractions

1. $\frac{1}{3} \times \frac{2}{5} = \frac{2}{15}$

5. $\frac{2}{3} \times 3\frac{1}{5} = \frac{2}{3} \times \frac{16}{5} = \frac{32}{15} = 2\frac{2}{15}$

9. $3 \times \frac{5}{8} \times \frac{4}{5} = \frac{3\cdot5\cdot4}{2\cdot4\cdot5} = \frac{3}{2} = 1\frac{1}{2}$

13. $\frac{5m^2n^2p^4}{3x^2yz^3} \cdot \frac{21xyz^2}{20m^2n^2p^2}$
$= \frac{5\cdot3\cdot7m^2n^2p^4xyz^2}{3\cdot4\cdot5m^2n^2p^2x^2yz^3} = \frac{7p^2}{4xz}$

17. $\frac{a}{x-y} \cdot \frac{b}{x+y} = \frac{ab}{x^2-y^2}$

21. $\frac{x^2-1}{x^2+x-6} \cdot \frac{x^2+2x-8}{x^2-4x-5}$
$= \frac{(x-1)(x+1)(x+4)(x-2)}{(x+3)(x-2)(x-5)(x+1)} = \frac{(x-1)(x+4)}{(x+3)(x-5)}$

25. $\frac{2x^2+x-6}{4x^2-3x-1} \cdot \frac{4x^2+5x+1}{x^2-4x-12}$
$= \frac{(2x-3)(x+2)(4x+1)(x+1)}{(4x+1)(x-1)(x-6)(x+2)}$
$= \frac{(2x-3)(x+1)}{(x-1)(x-6)}$

29. $\frac{9}{16} \div 8 = \frac{9}{16} \cdot \frac{1}{8} = \frac{9}{128}$

33. $2\frac{5}{8} \div \frac{1}{2} = \frac{21}{8} \div \frac{1}{2} = \frac{21}{8} \cdot \frac{2}{1} = \frac{21\cdot2}{2\cdot4} = \frac{21}{4} = 5\frac{1}{4}$

37. $\frac{4a^3x}{6dy^2} \div \frac{2a^2x^2}{8a^2y} = \frac{4a^3x}{6dy^2} \cdot \frac{8a^2y}{2a^2x^2}$
$= \frac{2\cdot2\cdot8a^5xy}{2\cdot3\cdot2a^2dx^2y^2} = \frac{8a^3}{3dxy}$

41. $\frac{ac+ad+bc+bd}{c^2-d^2} \div (a+b) = \frac{(a+b)(c+d)}{(c+d)(c-d)} \cdot \frac{1}{a+b}$
$= \frac{1}{c-d}$

45. $\frac{3x-6}{2x+3} \div \frac{x^2-4}{4x^2+2x-6} = \frac{3(x-2)}{2x+3} \cdot \frac{2(2x^2+x-3)}{(x+2)(x-2)}$
$= \frac{3(x-2)}{2x+3} \cdot \frac{2(2x+3)(x-1)}{(x+2)(x-2)} = \frac{6(x-1)}{x+2}$

49. $\frac{z^2-1}{z^2+z-6} \div \frac{z^2-4z-5}{z^2+2z-8}$
$= \frac{(z+1)(z-1)}{(z+3)(z-2)} \cdot \frac{(z+4)(z-2)}{(z-5)(z+1)}$
$= \frac{(z-1)(z+4)}{(z+3)(z-5)}$

53. $\frac{29}{12} = 2\frac{5}{12}$

57. $\frac{x^2+1}{x} = \frac{x^2}{x} + \frac{1}{x} = x + \frac{1}{x}$

61. $D = \frac{m}{V} = \frac{m}{1} \div \frac{4\pi r^3}{3} = \frac{m}{1} \cdot \frac{3}{4\pi r^3} = \frac{3m}{4\pi r^3}$

65. $M_A = \frac{\pi d^2}{4} \cdot \frac{16d}{3} = \frac{4\cdot4\pi d^3}{4\cdot3} = \frac{4\pi d^3}{3}$

Exercise 3: Addition and Subtraction of Fractions

1. $\frac{3}{5} + \frac{2}{5} = \frac{3+2}{5} = \frac{5}{5} = 1$

5. $\frac{1}{3} - \frac{7}{3} + \frac{11}{3} = \frac{1-7+11}{3} = \frac{5}{3}$

9. $\frac{3}{4} + \frac{7}{16} = \frac{3}{4}\left(\frac{4}{4}\right) + \frac{7}{16} = \frac{12}{16} + \frac{7}{16} = \frac{19}{16}$

13. $2 + \frac{3}{5} = \frac{2}{1} + \frac{3}{5} = \frac{2}{1}\left(\frac{5}{5}\right) + \frac{3}{5} = \frac{10}{5} + \frac{3}{5} = \frac{13}{5}$

17. $\frac{1}{7} + 2 - \frac{3}{7} + \frac{1}{5}$
$= \frac{1}{7}\left(\frac{5}{5}\right) + \frac{2}{1}\left(\frac{35}{35}\right) - \frac{3}{7}\left(\frac{5}{5}\right) + \frac{1}{5}\left(\frac{7}{7}\right)$
$= \frac{5}{35} + \frac{70}{35} - \frac{15}{35} + \frac{7}{35} = \frac{67}{35}$

21. $3\frac{1}{4} = \frac{3}{1}\left(\frac{4}{4}\right) + \frac{1}{4} = \frac{12}{4} + \frac{1}{4} = \frac{13}{4}$

25. $\frac{1}{a} + \frac{5}{a} = \frac{5+1}{a} = \frac{6}{a}$

29. $\frac{5x}{2} - \frac{3x}{2} = \frac{5x-3x}{2} = \frac{2x}{2} = x$

33. $\frac{7}{a+1} + \frac{9}{1+a} - \frac{3}{-a-1}$
$= \frac{7}{a+1} + \frac{9}{a+1} + \frac{3}{-(-a-1)}$
$= \frac{7}{a+1} + \frac{9}{a+1} + \frac{3}{a+1} = \frac{19}{a+1}$

37. $\frac{a}{3} + \frac{2}{x} - \frac{1}{2} = \frac{a}{3}\left(\frac{2x}{2x}\right) + \frac{2}{x}\left(\frac{6}{6}\right) - \frac{1}{2}\left(\frac{3x}{3x}\right)$
$= \frac{2ax}{6x} + \frac{12}{6x} - \frac{3x}{6x} = \frac{2ax-3x+12}{6x}$

41. $\frac{4}{x-1} - \frac{5}{x+1} = \frac{4}{x-1}\left(\frac{x+1}{x+1}\right) - \frac{5}{x+1}\left(\frac{x-1}{x-1}\right)$
$= \frac{4x+4-5x+5}{x^2-1} = \frac{9-x}{x^2-1}$

45. $\frac{1}{2(x-1)} - \frac{1}{2(x+1)} + \frac{1}{x^2}$
$= \frac{1}{2(x-1)}\left(\frac{(x+1)x^2}{(x+1)x^2}\right) - \frac{1}{2(x+1)}\left(\frac{(x-1)x^2}{(x-1)x^2}\right)$
$\quad + \frac{1}{x^2}\left(\frac{2(x-1)(x+1)}{2(x-1)(x+1)}\right)$
$= \frac{x^2(x+1)}{2x^2(x^2-1)} - \frac{x^2(x-1)}{2x^2(x^2-1)} + \frac{2(x^2-1)}{2x^2(x^2-1)}$
$= \frac{x^3+x^2-(x^3-x^2)+2x^2-2}{2x^4-2x^2} = \frac{2x^2-1}{x^4-x^2}$

49. $\frac{x}{x^2-9} + \frac{x}{x+3} = \frac{x}{(x-3)(x+3)} + \frac{x}{x-3}$
$= \frac{x+x(x-3)}{(x-3)(x+3)} = \frac{x+x^2-3x}{(x-3)(x+3)}$
$= \frac{x^2-2x}{(x-3)(x+3)}$

53. $\frac{1}{x+d+1} - \frac{1}{x+1}$
$= \frac{x+1}{(x+1)(x+d+1)} - \frac{x+d+1}{(x+1)(x+d+1)}$
$= \frac{x+1-x-d-1}{(x+1)(x+d+1)} = -\frac{d}{(x+1)(x+d+1)}$

57. $x + \frac{1}{x} = \frac{x}{1}\left(\frac{x}{x}\right) + \frac{1}{x} = \frac{x^2}{x} + \frac{1}{x} = \frac{x^2+1}{x}$

61. $3 - a - \frac{2}{a} = \frac{3}{1}\left(\frac{a}{a}\right) - \frac{a}{1}\left(\frac{a}{a}\right) - \frac{2}{a}$
$= \frac{3a}{a} - \frac{a^2}{a} - \frac{2}{a} = \frac{-a^2+3a-2}{a}$

65. The first crew can grade $\frac{7}{3}$ miles per day
The second crew can grade $\frac{9}{4}$ miles per day
Total $= \frac{7}{3} + \frac{9}{4} = \frac{7\cdot4+9\cdot3}{12} = \frac{28+27}{12}$
$= \frac{55}{12} = 4\frac{7}{12}$ mi

67. Total time = time for cutting + return stroke
$= \frac{1}{15} + \frac{1}{75} = \frac{1}{15}\left(\frac{5}{5}\right) + \frac{1}{75} = \frac{6}{75}$
$= \frac{2}{25}$ min

71. $\frac{(a+b)h}{2} = \frac{\pi d^2}{4} = \frac{2h(a+b)}{4} - \frac{\pi d^2}{4} = \frac{2h(a+b)-\pi d^2}{4}$

Exercise 4: Complex Fractions

1. $\frac{\frac{2}{3}+\frac{3}{4}}{\frac{1}{5}} = \left(\frac{8}{12}+\frac{9}{12}\right)5 = \frac{17}{12}(5) = \frac{85}{12}$

5. $\frac{5-\frac{2}{5}}{6+\frac{1}{3}} = \frac{\frac{25}{5}-\frac{2}{5}}{\frac{18}{3}+\frac{1}{3}} = \frac{23}{5}\left(\frac{3}{19}\right) = \frac{69}{95}$

9. $\frac{1+\frac{x}{y}}{1-\frac{x^2}{y^2}} = \frac{\frac{y}{y}+\frac{x}{y}}{\frac{y^2}{y^2}-\frac{x^2}{y^2}} = \frac{x+y}{y}\left(\frac{y^2}{y^2-x^2}\right)$
$= \frac{(x+y)y}{(y+x)(y-x)} = \frac{y}{y-x}$

13. $\frac{x+\frac{2d}{3ac}}{x+\frac{3d}{2ac}} = \frac{\frac{6acx}{6ac}+\frac{4d}{6ac}}{\frac{6acx}{6ac}+\frac{9d}{6ac}} = \frac{6acx+4d}{6ac}\left(\frac{6ac}{6acx+9d}\right)$
$= \frac{6acx+4d}{6acx+9d} = \frac{2(3acx+2d)}{3(2acx+3d)}$

17. $\frac{1+\frac{1}{x+1}}{1-\frac{1}{x-1}} = \frac{\frac{x+1}{x+1}+\frac{1}{x+1}}{\frac{x-1}{x-1}-\frac{1}{x-1}} = \frac{(x+2)(x-1)}{(x+1)(x-2)}$

28   Chapter 9

21. $\dfrac{1}{a+\dfrac{1}{1+\dfrac{a+1}{3-a}}} = \dfrac{1}{a+\dfrac{1}{\dfrac{3-a+a+1}{3-a}}} = \dfrac{1}{a+\dfrac{3-a}{4}}$

$= \dfrac{1}{\dfrac{4a+3-a}{4}} = \dfrac{4}{4a+3-a} = \dfrac{4}{3a+3}$

$= \dfrac{4}{3(a+1)}$

25. $\dfrac{\dfrac{x^2-y^2-z^2-2yz}{x^2-y^2-z^2+2yz}}{\dfrac{x-y-z}{x+y-z}} = \dfrac{\dfrac{x^2-\left(y^2+2yz+z^2\right)}{x^2-\left(y^2-2yz+z^2\right)}}{\dfrac{x-y-z}{x+y-z}}$

$= \dfrac{\dfrac{x^2-(y+z)^2}{x^2-(y-z)^2}}{\dfrac{x-y-z}{x+y-z}}$

$= \dfrac{[x-(y+z)][x+(y+z)]}{[x-(y-z)][x+(y-z)]} \cdot \dfrac{x+y-z}{x-y-z}$

$= \dfrac{(x-y-z)(x+y+z)}{(x-y+z)(x+y-z)} \cdot \dfrac{x+y-z}{x-y-z} = \dfrac{x+y+z}{x-y+z}$

Exercise 5: Fractional Equations

1. $2x + \dfrac{x}{3} = 28$
$3(2x) + 3\left(\dfrac{x}{3}\right) = 3(28)$
$6x + x = 84$
$7x = 84$
$x = 12$

5. $3x - \dfrac{x}{7} = 40$
$7(3x) - 7\left(\dfrac{x}{7}\right) = 7(40)$
$21x - x = 280$
$20x = 280$
$x = 14$

9. $\dfrac{x}{2} + \dfrac{x}{3} + \dfrac{x}{4} = 26$
$12\left(\dfrac{x}{2}\right) + 12\left(\dfrac{x}{3}\right) + 12\left(\dfrac{x}{4}\right) = 12(26)$
$6x + 4x + 3x = 312$
$13x = 312$
$x = 24$

13. $2x + \dfrac{x}{3} - \dfrac{x}{4} = 50$
$12(2x) + 12\left(\dfrac{x}{3}\right) - 12\left(\dfrac{x}{4}\right) = 12(50)$
$24x + 4x - 3x = 600$
$25x = 60$
$x = 24$

17. $\dfrac{6x-19}{2} = \dfrac{2x-11}{3}$
$6\left(\dfrac{6x-19}{2}\right) = 6\left(\dfrac{2x-11}{3}\right)$
$3(6x - 19) = 2(2x - 11)$
$18x - 57 = 4x - 22$
$14x = 35$
$x = \dfrac{35}{14} = \dfrac{5}{2}$

21. $\dfrac{x-1}{8} - \dfrac{x+1}{18} = 1$
$72\left(\dfrac{x-1}{8}\right) - 72\left(\dfrac{x+1}{18}\right) = 72(1)$
$9(x - 1) - 4(x + 1) = 72$
$9x - 9 - 4x - 4 = 72$
$5x = 85$
$x = 17$

25. $\dfrac{2}{3x} + 6 = 5$
$2 + 18x = 15x$
$3x = -2$
$x = -\dfrac{2}{3}$

29. $\dfrac{x-3}{x+2} = \dfrac{x+4}{x-5}$
$x^2 - 8x + 15 = x^2 + 6x + 8$
$7 = 14x$
$x = \dfrac{1}{2}$

33. $\dfrac{4}{x^2-1} + \dfrac{1}{x-1} + \dfrac{1}{x+1} = 0$
$4 + x + 1 + x - 1 = 0$
$2x = -4$
$x = -2$

37. $\dfrac{9x+20}{36} - \dfrac{x}{4} = \dfrac{4x-12}{5x-4}$
$\dfrac{9x+20-9x}{36} = \dfrac{4x-12}{5x-4}$
$100x - 80 = 144x - 432$
$352 = 44x$
$x = 8$

Exercise 6: Word Problems Leading to Fractional Equations

1. Let $x = $ the number
$4\left(50 - \dfrac{1}{5}x\right) = x - 70$
$200 - \dfrac{4}{5}x = x - 70$
$1000 - 4x = 5x - 350$
$-5x - 4x = -350 - 1000$
$-9x = -1350$
$x = 150$

5. Let $y = $ lower number
$\dfrac{1}{7}(y + 20) = \dfrac{1}{3}y$
$3(y + 20) = 7y$
$3y + 60 = 7y$
$4y = 60$
$y = 15$
$y + 20 = 35$

9. Let h = hrs. for fast train
$h + 2.0$ = hrs. for slow train
$40\,h = 30(h + 2.0)$
$40\,h = 30\,h + 60$
$10\,h = 60$
$h = 6.0\,\text{h}$
Dist. $= (40\,\text{mi/h})(6.0\,\text{h}) = 240\,\text{mi}$

13. Let x = days to do job, together
$\frac{1}{10}x + \frac{1}{14}x = 1$
$14x + 10x = 140$
$24x = 140$
$x = 5\frac{5}{6}$ days

17. $\frac{1}{\frac{1}{8}+\frac{1}{9}} = \frac{1}{\frac{9}{72}+\frac{8}{72}} = \frac{72}{17} = 4.2\,\text{h}$

21. Coal remaining when new boiler begins operation
$$= 10000 - \frac{3}{4}(1500) = 8875 \text{ tons}$$
$$\frac{8875}{\frac{1500}{4}+\frac{2300}{3}} = \frac{8875(12)}{4500+9200} = 7.8 \text{ weeks}$$

25. Let y = time for new panel to collect 35,000 BTU
rate + rate = total rate
$\frac{9000}{7} + \frac{35,000}{y} = \frac{35,000}{5}$
$45,000y + 1,225,000 = 245,000y$
$200,000y = 1,225,000$
$y = 6.1\,\text{h}$

29. Let x = B's capital
$\frac{3}{4}x$ = A's capital
$\frac{3}{4}x - 500 = \frac{1}{2}x$
$\frac{1}{4}x = 500$
$x = \$2000$
$\frac{3}{4}x = \$1500$

Exercise 7: Literal Equations and Formulas

1. $2ax = bc$
$x = \frac{bc}{2a}$

5. $4acx - 3d^2 = a^2d - d^2x$
$4acx + d^2x = a^2d + 3d^2$
$x(4ac + d^2) = a^2d + 3d^2$
$x = \frac{a^2d+3d^2}{4ac+d^2}$

9. $\frac{a}{2}(x - 3w) = z$
$\frac{ax}{2} - \frac{3aw}{2} = z$
$\frac{ax}{2} = z + \frac{3aw}{z}$
$x = \frac{2z}{a} + 3w$

13. $ax - bx = c + dx - m$
$ax - bx - dx = c - m$
$x(a - b - d) = c - m$
$x = \frac{c-m}{a-b-d}$

17. $3(x - b) = 2(bx + c) - c(x - b)$
$3x - 3b = 2bx + 2c - cx + bc$
$3x - 2bx + cx = 2c + bc + 3b$
$x(3 - 2b + c) = 2c + bc + 3b$
$x = \frac{2c+bc+3b}{3-2b+c}$

21. $\frac{p-q}{x} = 3p$
$p - q = 3px$
$x = \frac{p-q}{3p}$

25. $\frac{x}{a} - a = \frac{a}{c} - \frac{x}{c-a}$
$xc(c - a) - a^2c(c - a) = a^2(c - a) - xac$
$xc^2 - xac - a^2c^2 + a^3c = a^2c - a^3 - xac$
$x = \frac{a^2c^2-a^3c+a^2c-a^3}{c^2}$
$x = \frac{a^2(c^2-ac+c-a)}{c^2}$
$x = \frac{a^2(c+1)(c-a)}{c^2}$

29. $\frac{x+a}{x-a} - \frac{x-b}{x+b} = \frac{2(a+b)}{x}$

$x(x+a)(a+b) - x(x-a)(x-b) = 2(a+b)(x-a)(x+b)$

$x^3 + (a+b)x^2 + abx - x^3 + (a+b)x^2 - abx = 2(a+b)x^2 - 2(a^2-b^2)x - 2ab(a+b)$

$2(a+b)x^2 = 2(a+b)x^2 - 2(a^2-b^2)x - 2ab(a+b)$

$2(a^2-b^2)x = -2ab(a+b)$

$2(a-b)x = -2ab$

$x = \frac{ab}{b-a}$

33. Let x = dollars at first

$x - \frac{x}{n} - \frac{x}{m} = a$

$xnm - xm - xn = anm$

$x(nm - m - n) = anm$

$x = \frac{anm}{nm-m-n}$

37. $E = \frac{PL}{ae}$

$Eae = PL$

$a = \frac{PL}{Ee}$

41. $\frac{1}{R} = \frac{1}{R_1} + \frac{1}{R_2}$

$RR_1R_2\left(\frac{1}{R}\right) = RR_1R_2\left(\frac{1}{R_1}\right) + RR_1R_2\left(\frac{1}{R_2}\right)$

$R_1R_2 = RR_2 + RR_1$

$R_1R_2 - RR_2 = RR_1$

$R_2(R_1 - R) = RR_1$

$R_2 = \frac{RR_1}{R_1-R}$

45. $M = R_1L - F(L - x)$

$M = R_1L - FL + Fx$

$R_1L - FL = M - Fx$

$L(R_1 - F) = M - Fx$

$L = \frac{M-Fx}{R_1-F}$

49. $E = mgy + \frac{1}{2}mv^2$

$E = m\left(gy + \frac{1}{2}v^2\right)$

$m = \frac{E}{gy+\frac{1}{2}v^2}$

Chapter 9 Review Problems

1. $\frac{a^2+b^2}{a-b} - a + b = \frac{a^2+b^2}{a-b} - (a - b)$

$= \frac{a^2+b^2}{a-b} - \frac{(a-b)^2}{a-b}$

$= \frac{a^2+b^2-(a^2-2ab+b^2)}{a-b} = \frac{2ab}{a-b}$

5. $(a^2 + 1 + a)\left(1 - \frac{1}{a} + \frac{1}{a^2}\right)$

$= a^2 - a + 1 + 1 - \frac{1}{a} + \frac{1}{a^2} + a - 1 + \frac{1}{a}$

$= a^2 + 1 + \frac{1}{a^2}$

9. $\frac{x^2+8x+15}{x^2+7x+10} - \frac{x-1}{x-2} = \frac{(x+3)(x+5)}{(x+2)(x+5)} - \frac{x-1}{x-2}$

$= \frac{x+3}{x+2} - \frac{x-1}{x-2} = \frac{(x-2)(x+3)-(x-1)(x+2)}{(x+2)(x-2)}$

$= \frac{x^2+x-6-x^2-x+2}{(x+2)(x-2)} = \frac{-4}{x^2-4} = \frac{4}{4-x^2}$

13. $\frac{a^3-b^3}{a^3+b^3} \cdot \frac{a^2-ab+b^2}{a-b}$

$= \frac{(a-b)(a^2+ab+b^2)}{(a+b)(a^2-ab+b^2)} \cdot \frac{a^2-ab+b^2}{a-b}$

$= \frac{a^2+ab+b^2}{a+b}$

17. $\frac{3a^2+6a}{a^2+4a+4} = \frac{3a(a+2)}{(a+2)(a+2)} = \frac{3a}{a+2}$

21. $\frac{(a+b)^2-(c+d)^2}{(a+c)^2-(b+d)^2} = \frac{(a+b+c+d)(a+b-c-d)}{(a+b+c+d)(a+c-b-d)}$

$= \frac{a+b-c-d}{a-b+c-d}$

25. $\frac{5-3x}{4} + \frac{3-5x}{3} = \frac{3}{2} - \frac{5x}{3}$

$15 - 9x + 12 - 20x = 18 - 20x$

$9x = 15 + 12 - 18$

$9x = 9$

$x = 1$

29. $x^2 - (x - p)(x + q) = r$

$x^2 - (x^2 + xq - xp - pq) = r$

$-xq + xp = r - pq$

$x(p - q) = r - pq$

$x = \frac{r-pq}{p-q}$

33. $\frac{x-3}{4} - \frac{x-1}{9} = \frac{x-5}{6}$

$9x - 27 - (4x - 4) = 6x - 30$

$-27 + 4 + 30 = 6x - 9x + 4x$

$x = 7$

37. $\frac{6}{(2x+1)} = \frac{4}{(x-1)}$

$(x-1)(6) = 4(2x+1)$

$6x - 6 = 8x + 4$

$-10 = 2x$

$x = -5$

41. $\frac{x}{15.0} + \frac{x}{11.0} = 1$

$11.0x + 15.0x = 165$

$x = 6.35 \text{ days}$

45. $\frac{3wx^2y^3}{7axyz} \cdot \frac{4a^3xz}{6aw^2y} = \frac{12a^3wx^3y^3z}{42a^2w^2xy^2z}$

$\quad\quad = \frac{2ax^2y}{7w}$

CHAPTER 10: SYSTEMS OF LINEAR EQUATIONS

Exercise 1: Systems of Two Linear Equations

1.

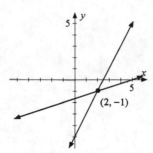

(2, −1)

5.

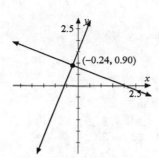

(−0.24, 0.90)

9.
$$4x + 2y = 3$$
$$\underline{-4x + \ \ y = 6}$$
$$3y = 9$$
$$y = 3$$
$$4x + 2(3) = 3$$
$$4x = -3$$
$$x = -\frac{3}{4}$$

13.
$$-6(x + 5y) = -6(11)$$
$$-6x - 30y = -66$$
$$2(3x + 2y) = 2(7)$$
$$6x + \ \ 4y = 14$$
$$\underline{-6x - 30y = -66}$$
$$-26y = -52$$
$$y = 2$$
$$x + 5(2) = 11$$
$$x = 11 - 10 = 1$$

17.
$$-5(7x - 4y) = -5(81)$$
$$-35x + 20y = -405$$
$$7(5x - 3y) = 7(57)$$
$$35x - 21y = 399$$
$$\underline{-35x + 20y = -405}$$
$$-y = -6$$
$$y = 6$$
$$7x - 4(6) = 81$$
$$7x = 81 + 24$$
$$x = \frac{105}{7} = 15$$

21.
$$-2(3x + y) = -2(9)$$
$$-6x - 2y = -18$$
$$\underline{x + 2y = 8}$$
$$-5x = -10$$
$$x = 2$$
$$3(2) + y = 9$$
$$y = 3$$

25.
$$4n = 18 - 3m$$
$$m = 8 - 2n$$
$$4n + 3m = 18$$
$$2n + m = 8$$
$$-6n - 3m = -24$$
$$\underline{4n + 3m = 18}$$
$$-2n = -6$$
$$n = 3$$
$$m = 8 - 2(3)$$
$$m = 2$$

29.
$$3.62x - 4.73y = 11.7$$
$$4.95x - 7.15y = 12.8$$
$$x - 1.3066y = 3.2320$$
$$\underline{x - 1.4444y = 2.5859}$$
$$0.1378y = 0.6461$$
$$y = 4.6887$$
$$x = 3.2320 + 1.3066(4.6887)$$
$$x = 9.36 \qquad y = 4.69$$

Note: The following program for the Gauss-Seidel method is used to solve the set of equations in Problem 7.

35.
```
10'  SEIDEL
20'
30'  This program uses the Gauss-Seidel method
40'  to solve a set of two equations.
50'  Enter your equations on lines 70 and 80.
60'
70   DEF FNA(Y) = (Y + 4) / 3
80   DEF FNB(X) = 11 - 2 * X
90   Y = 0
100  X1 = FNA(Y)
110  Y1 = FNB(X1)
120  IF ABS(Y1 - Y) < .0001 THEN 150
130  Y = Y1
140  GOTO 100
150  LPRINT "The roots are "; X1, Y1
160  END
The roots are 2.99998  5.00004
```

Exercise 2: Other Systems of Equations

1. $\frac{x}{5} + \frac{y}{6} = 18$

$\frac{x}{2} - \frac{y}{4} = 21$

$6x + 5y = 540$

$2x - y = 84$

$6x + 5y = 540$

$\underline{10x - 5y = 420}$

$\overline{16x \qquad = 960}$

$x = \ 60$

$2(60) - y = 84$

$y = 120 - 84 = 36$

5. $\frac{3x}{5} + \frac{2y}{3} = 17$

$\frac{2x}{3} + \frac{3y}{4} = 19$

$9x + 10y = 255$

$8x + 9y = \ 228$

$81x + 90y = 2295$

$\underline{-80x - 90y = -2280}$

$\overline{\quad x \qquad = 15}$

$9(15) + 10y = 255$

$10y = 120$

$y = 12$

9. $\frac{r}{6.20} - \frac{s}{4.30} = \frac{1}{3.10}$

$\frac{r}{4.60} - \frac{s}{2.30} = \frac{1}{3.50}$

$4.30r - 6.20s = 8.60$

$2.30r - 4.60s = 3.02$

$r - 1.44s = \ \ 2.00$

$\underline{-r + \quad 2s = -1.31}$

$\overline{\quad 0.56s = 0.69}$

$s = 1.23$

$r = 2s + 1.31 = 2.46 + 1.31$

$r = 3.77$

* For Problems (10 -15) we let $a = \frac{1}{x}$ and $b = \frac{1}{y}$.

13. $2a + 4b = 14$

$6a - 2b = 14$

$2a + 4b = 14$

$\underline{12a - 4b = 28}$

$\overline{14a \qquad = 42}$

$\frac{1}{a} = \frac{1}{3} = x$

$2(3) + 4b = 14$

$4b = 8$

$\frac{1}{b} = \frac{1}{2} = y$

17. $\frac{1}{5z} + \frac{1}{6w} = 18$

$\frac{1}{4w} - \frac{1}{2z} + 21 = 0$

Let $a = \frac{1}{w}$, $b = \frac{1}{z}$

$\frac{b}{5} + \frac{a}{6} = 18$

$-\frac{b}{2} + \frac{a}{4} = -21$

$5a + 6b = 540$

$a - 2b = -84$

$5a + 6b = 540$

$\underline{3a - 6b = -252}$

$8a \quad = 288$

$a = 36$

$2b = a + 84 = 120$

$b = 60$

$w = \frac{1}{36}$, $z = \frac{1}{60}$

21.

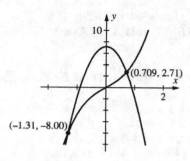

25. $ax + by = r$

$ax + cy = s$

$ax + by = r$

$\underline{-ax - cy = -s}$

$by - cy = r - s$

$y(b - c) = r - s$

$y = \frac{r-s}{b-c}$

$ax + b\left(\frac{r-s}{b-c}\right) = r$

$x = \frac{r(b-c) - b(r-s)}{a(b-c)} = \frac{br - cr - br + bs}{a(b-c)}$

$x = \frac{bs - cr}{a(b-c)}$

Exercise 3: Word Problems with Two Unknowns

1. $x + y = 24$

$\underline{x - y = 8}$

$2x \quad = 32$

$\quad x = 16$

$16 + y = 24$

$\quad y = 8$

5. Let x = tens digit, let y = ones digit

$8(x + y) + 2 = 10x + y$

$8x + 8y + 2 = 10x + y$

$-2x + 7y = -2$

$10x + y - 54 = 10y + x$

$9x - 9y = 54$

$63x - 63y = 378$

$\underline{-18x + 63y = -18}$

$45x \qquad = 360$

$\quad x \qquad = 8$

$9(8) - 9y = 54$

$-9y = 54 - 72$

$y = 2$

number $= 82$

9. Let x = the length of the rectangle and y = the breadth of the rectangle. Then xy = the area of the rectangle, $(x - 4)(y + 3)$ = the area of the rectangle is changed. But, since the new rectangle is a square, its sides are of equal length.

$x - 4 = y + 3$

$(x - 4)(y + 3) = xy$

$x - y = 7$

$3x - 4y = 12$

$3x - 3y = 21$

$y = 9$

$x = 16$

Therefore, the rectangle is 9 by 16.

13. $\frac{280}{A} = \frac{210}{B}$

$A = \frac{280}{210}(B)$

$\frac{245}{B} = \frac{280}{A} + 6\frac{2}{3}$

$735A = 840B + 20AB$

$735\left(\frac{280}{210}B\right) = 840B + 20\left(\frac{280}{210}B\right)B$

$980B - 840B - 26.67B^2 = 0$

$140B - 26.67B^2 = 0$

$B = 5.25$ ft/sec

$A = \frac{280}{210}(B) = 7.00$ ft/sec

17.
$$58.3 = 5.57v_0 + \frac{(5.57)^2}{2}a$$
$$34.8 = 4.28v_0 + \frac{(4.28)^2}{2}a$$
Multiplying the first equation by 4.28
and the second equation by 5.57 we get
$$249.524 = 23.8396v_0 + 66.3933a$$
$$\underline{193.836 = 23.8396v_0 + 51.0167a}$$
$$55.688 = \qquad\qquad 15.3766a$$
$$a = \frac{55.688}{15.3766} = 3.62 \text{ cm/s}^2$$
$$4.28v_0 = 34.8 - \frac{(4.28)^2}{2}\left(\frac{55.688}{15.3766}\right)$$
$$v_0 = 0.381 \text{ cm/s}$$

21. Let x = amt. at 6.2%
and y = amt. at 9.7%

Then $x + y = 4400$
and $0.062x = 0.097y$
or $\qquad 62x = 97y$
So, $\qquad 62x = 97(4400 - x)$
$$159x = 426,800$$
$$x = \$2684$$
and $y = 4400 - 2684 = \$1716$

25. Let m = lbs of first mixture, s = lbs of sand to add
$$0.05m + 255 = 0.12(m + s + 255)$$
$$0.08m + s = 0.15(m + s + 255)$$
$$0.05m + 255 = 0.12m + 0.12s + 30.6$$
$$0.08m + s = 0.15m + 0.15s + 38.25$$
$$0.07m + 0.12s = 224.4$$
$$\underline{-0.07m + 0.85s = 38.25}$$
$$0.97s = 262.65$$
$$s = 271 \text{ lb}$$
$$0.07m + 0.12(271) = 224.4$$
$$m = 2740 \text{ lb}$$

29.
$$T_{1x} - T_{2x} = 0$$
$$T_{1y} + T_{2y} = 572$$
$$\tan 18° = \frac{T_{1x}}{T_{1y}}, \quad T_{1x} = 0.3249T_{1y}$$
$$\tan 35° = \frac{T_{2y}}{T_{2x}}, \quad T_{2y} = 0.7002T_{2x}$$
$$0.3249T_{1y} - T_{2x} = 0$$
$$T_{1y} + 0.7002T_{2x} = 572$$
$$T_{1y} + 0.7002(0.3249T_{1y}) = 572$$
$$T_{1y} = 466.0$$
$$T_{1x} = 0.3249(466.0) = 151.4$$
$$T_{2x} = T_{1x} = 151.4$$
$$T_{2y} = 0.7002(151.4) = 106.0$$
$$T_1 = \sqrt{151.4^2 + 466.0^2} = 49\overline{0} \text{ lb}$$
$$T_2 = \sqrt{151.4^2 + 106.0^2} = 185 \text{ lb}$$

33.
$$3.5x + 4.5y = 117,000$$
$$5.2x + 4.8y = 151,200$$
$$18.2x + 23.4y = 608,400$$
$$\underline{-18.2x - 16.8y = -529,200}$$
$$6.6y = 79,200$$
$$y = 12,000 \text{ gal/h}$$
$$3.5x + 4.5(12,000) = 117,000$$
$$x = 18,000 \text{ gal/h}$$

37.
$$6 - 736I_1 - 386I_1 + 386I_2 = 0$$
$$12 - 375I_2 - 386I_2 + 386I_1 = 0$$
$$1122I_1 - 386I_2 = 6$$
$$-386I_1 + 761I_2 = 12$$
$$216,546I_1 - 74,498I_2 = 1158$$
$$\underline{-216,546I_1 + 426,921I_2 = 6732}$$
$$352,423I_2 = 7890$$
$$I_2 = 0.0224 \text{ A} = 22.4 \text{ mA}$$
$$1122I_1 - 386(0.0224) = 6$$
$$I_1 = 0.0131 \text{ A} = 13.1 \text{ mA}$$

Exercise 4: Systems of Three Linear Equations

1.
$$x + z = 40 \qquad x = 40 - z$$
$$y + z = 45 \qquad y = 45 - z$$
$$x + y = 35$$
$$(40 - z) + (45 - z) = 35$$
$$-2z = -50$$
$$z = 25$$
$$x = 40 - 25 = 15$$
$$y = 45 - 25 = 20$$

5. Adding (2) and (3),
$$2x = 10$$
$$x = 5$$
From (1), $\quad y + z = 13$
From (2), $\quad \underline{-y + z = 1}$
$$2z = 14$$
$$z = 7$$
From (1), $y = 18 - 5 - 7 = 6$

9. Sub. (1) from (3),
$$4y = 16$$
$$y = 4$$
From (1), $x + 2z = 13 \quad$ (4)
From (2), $5x + 6z = 45 \quad$ (5)
Sub. $3 \times$ (4) from (5),
$$2x = 6$$
$$x = 3$$
So $z = \frac{(13 - 3)}{2} = 5$

13. $p + 3q - r = 10$
$5p - 2q + 2r = 6$
$3p + 2q + r = 13$
so $\quad p + 3q - r = 10$
$$\frac{3p + 2q + r = 13}{4p + 5q \quad\quad = 23}$$
and $\quad 5p - 2q + 2r = 6$
$$\frac{-(6p + 4q + 2r = 26)}{-p - 6q \quad\quad = -20}$$
thus $\quad 4p + \quad 5q = 23$
$$\frac{-4p - 24q = -80}{-19q = -57}$$
$q = 3$
$-p = -20 + 6(3)$
$p = 2$
$r = 13 - 3(2) - 2(3)$
$r = 1$

17. Let $a = \frac{1}{x}$, $b = \frac{1}{y}$, $c = \frac{1}{z}$
$a + 2b - c = -3 \quad (1)$
$3a + b + c = 4 \quad (2)$
$a - b + 2c = 6 \quad (3)$
Add (1) and (2),
$4a + 3b = 1 \quad\quad (4)$
Add $2 \times$ (1) to (3),
$3a + 3b = 0 \quad\quad (5)$
Sub. (5) from (4),
$a = 1$
Then $b = -a = -1$
and $c = 4 - 3a - b = 2$
So, $x = 1$, $y = -1$, $z = \frac{1}{2}$

21. $x + 2y + z = 2(a + c) \quad\quad -2x - 4y - 2z = -4(a + c)$
$$\frac{x + y + 2z = 2(b + c)}{-x - 3y = 2(b + c) - 4(a + c)}$$

$2x + y + z = 2(a + b) \quad\quad -2x - y - z = -2(a + b)$
$$\frac{x + 2y + z = 2(a + c)}{-x + y = 2(a + c) - 2(a + b)}$$
$$-x + y = 2(c - b)$$

$-x - 3y = 2(b + c) - 4(a + c)$
$$\frac{x - y = \quad\quad\quad\quad\quad -2(c - b)}{-4y = 2(b + c) - 4(a + c) - 2(c - b)}$$
$$y = (a + c - b)$$
$x - (a + c - b) = -2c + 2b \quad x = a + c - b - 2c + 2b \quad\quad x = a + b - c$
$2x + y + z = 2(a + b) \quad\quad z = 2(a + b) - 2(a + b - c) - (a + c - b) \quad z = -a + b + c$

25. Let $x = 100$'s place, $y = 10$'s place, $z = 1$'s place
Then $x + y + z = 10$
$$x + z = \frac{2y}{3}$$
$$100x + 10y + z - 198 = 100z + 10y + x$$
From which, $x + y + z = 10 \quad\quad (1)$
$3x - 2y + 3z = 0 \quad\quad (2)$
$x \quad\quad - z = 2 \quad\quad (3)$
Adding $2 \times$ (1) to (2), $5x + 5z = 20$
or $\quad\quad\quad\quad\quad\quad\quad x + z = 4 \quad\quad (4)$
Adding (3) and (4), $\quad 2x = 6$
$\quad\quad\quad\quad\quad\quad\quad\quad x = 3$
Substituting back, $\quad z = x - 2 = 1$
$\quad\quad\quad\quad\quad\quad\quad y = 10 - x - z = 6$
So the number is 361.

29.

$$926 + F_{1y} - F_{2y} - F_{3y} = 0 \qquad (1)$$

$$F_{1x} - F_{2x} + F_{3x} = 0 \qquad (2)$$

$$1.50(926) + 3.70\,F_{3y} - 6.30F_{1y} = 0 \qquad (3)$$

Thus, $\quad 926 + F_1\sin 69.3° - F_2\sin 71.1° - F_3\sin 56.6° = 0 \qquad (4)$

$$F_1\cos 69.3° - F_2\cos 71.1° + F_3\cos 56.6° = 0 \qquad (5)$$

$$1.50(926) + 3.70F_3\sin 56.6° - 6.30F_1\sin 69.3° = 0 \qquad (6)$$

Then, $\quad 0.93544F_1 - 0.94609F_2 - 0.83485F_3 = -926 \qquad (7)$

$$0.35347F_1 - 0.32392F_2 + 0.55048F_3 = 0 \qquad (8)$$

$$5.89330F_1 - 3.08894F_3 = 1389 \qquad (9)$$

Solving (9) for F_3 gives

$$F_3 = 1.90787F_1 - 449.669 \qquad (10)$$

Substituting this into (7) and (8) and simplifying, we get

$$F_1 + 1.43925F_2 = 1979.78 \qquad (11)$$

$$\underline{F_1 - 0.23076F_2 = 176.343}$$

$$1.67001F_2 = 1803.437$$

$$F_2 = 10.79.90$$

Substituting back into (11),

$$F_1 = 1979.78 - 1.43925(1079.90) = 425.534$$

Substituting back into (10),

$$F_3 = 1.90787(425.534) - 449.669 = 362.19$$

Thus, $\quad F_1 = 426\,\text{N}, \quad F_2 = 1080\,\text{N}, \quad F_3 = 362\,\text{N}$

Chapter 10 Review Problems

1.
$$4x + 3y = 27$$
$$2x - 5y = -19$$
$$4x + 3y = 27$$
$$\underline{-4x + 10y = 38}$$
$$13y = 65$$
$$y = 5$$
$$2x - 5(5) = -19$$
$$x = 3$$

5.
$$5x + 3y - 2z = 5 \qquad (1)$$
$$3x - 4y + 3z = 13 \qquad (2)$$
$$x + 6y - 4z = -8 \qquad (3)$$

Sub. $5 \times (3)$ from (1), $\quad -27y + 18z = 45$

or, $\qquad\qquad\qquad -3y + 2z = 5 \quad (4)$

Sub. $3 \times (3)$ from (2), $-22y + 15z = 37 \quad (5)$

Mult. (4) by -15, $\qquad 45y - 30z = -75$

Mult. (5) by 2, $\qquad \underline{-44y + 30z = 74}$

Adding, $\qquad\qquad\quad y = -1$

Substituting back,

$$z = \frac{(5 + 3y)}{2} = 1$$
$$x = 4z - 6y - 8 = 2$$

9.
$$2x - 3y = -14$$
$$2x = 3y - 14$$
$$x = \tfrac{3}{2}y - 7$$
$$3\left(\tfrac{3}{2}y - 7\right) + 2y = 44$$
$$\tfrac{9}{2}y + 2y = 44 + 21$$
$$\tfrac{13}{2}y = 65$$
$$y = 10$$
$$x = \tfrac{3}{2}(10) - 7 = 8$$

13.
$$x + y = a \qquad (1)$$
$$x + z = b \qquad (2)$$
$$y + z = c \qquad (3)$$

Sub. (2) from (1),

$$y - z = a - b \qquad (4)$$

Add (3) and (4),

$$2y = a - b + c$$

$$y = \frac{a - b + c}{2}$$

Then $\quad x = a - y = \frac{2a - a + b - c}{2}$

$$= \frac{a + b - c}{2}$$

and, $\quad z = b - x = \frac{2b - a - b + c}{2}$

$$= \frac{b - a + c}{2}$$

17.　Add $3 \times (3)$ to (1),
$$5x + 5y = 70 \qquad (4)$$
Sub. $2 \times (3)$ from (2),
$$x - 5y = -34 \qquad (5)$$
Add (4) and (5),
$$6x = 36$$
$$x = 6$$
Then $y = \dfrac{(x+34)}{5} = 8$
and $z = x + 3y - 20 = 10$

21.
$$2x - y = 9 \qquad\qquad (1)$$
$$5x - 3y = 14 \qquad\qquad (2)$$

$(1) \times 3 \qquad 6x - 3y = 27$
Subtract (2) $\underline{\quad 5x - 3y = 14 \quad}$
$$ x = 13$$
$$y = 2(13) - 9 = 17$$

25.　Let $x =$ amt. A alone can do in 1 day
　　　　$y =$ amt. B alone can do in 1 day
　　　　$z =$ amt. C alone can do in 1 day
Then　$12(x + y) = 1$ job
　　　　$16(y + z) = 1$ job
　　　　$x = 1.5z$
or　　　$12x + 12y = 1$
　　　　$16y + 16z = 1$
　　　　$x = 1.5z$
also　　$y = \dfrac{(1 - 12x)}{12}$
　　　　$z = \dfrac{(1 - 16y)}{16}$
　　　　$z = \dfrac{2x}{3}$
By substitution,
$$x = 1.5z = \left(\frac{3}{2}\right)\frac{(1 - 16y)}{16}$$
$$32x = 3 - 48y = 3 - 48\frac{(1 - 12x)}{12} = 48x - 1$$
$$x = \tfrac{1}{16} \text{ job/day}$$
$$y = \frac{(1 - 12x)}{12} = \frac{(1 - 12/16)}{12} = \tfrac{1}{48} \text{ job/day}$$
$$z = \frac{2x}{3} = \frac{2(1/16)}{3} = \tfrac{1}{24} \text{ job/day}$$
Together they do $\tfrac{1}{16} + \tfrac{1}{24} + \tfrac{1}{48} = \tfrac{1}{8}$ job per day,
so it will take them 8 days for one job.

CHAPTER 11: DETERMINANTS

Exercise 1: Second-Order Determinants

1. $(3)(-4) - 2(1) = -12 - 2 = -14$

5. $-4(-2) - 5(7) = 8 - 35 = -27$

9. $\left(-\frac{2}{3}\right)\left(\frac{4}{5}\right) - \left(\frac{2}{5}\right)\left(-\frac{1}{3}\right)$
$= -\frac{8}{15} + \frac{2}{15} = -\frac{6}{15} = -\frac{2}{5}$

13. $\sin\theta\,(\sin\theta) - 3(\tan\theta) = \sin^2\theta - 3\tan\theta$

17. $\triangle = \begin{vmatrix} 2 & 1 \\ 3 & -1 \end{vmatrix} = 2(-1) - 3(1) = -5$

$x = \dfrac{\begin{vmatrix} 11 & 1 \\ 4 & -1 \end{vmatrix}}{\triangle} = \dfrac{11(-1)-4(1)}{-5} = \dfrac{-15}{-5} = 3$

$y = \dfrac{\begin{vmatrix} 2 & 11 \\ 3 & 4 \end{vmatrix}}{\triangle} = \dfrac{2(4)-3(11)}{-5} = \dfrac{-25}{-5} = 5$

21. $\triangle = \begin{vmatrix} 1 & 5 \\ 3 & 2 \end{vmatrix} = 1(2) - 3(5) = -13$

$x = \dfrac{\begin{vmatrix} 11 & 5 \\ 7 & 2 \end{vmatrix}}{\triangle} = \dfrac{11(2)-7(5)}{-13}$
$= \dfrac{-13}{-13} = 1$

$y = \dfrac{\begin{vmatrix} 1 & 11 \\ 3 & 7 \end{vmatrix}}{\triangle} = \dfrac{1(7)-3(11)}{-13}$
$= \dfrac{-26}{-13} = 2$

25. $\triangle = \begin{vmatrix} 7 & -4 \\ 5 & -3 \end{vmatrix} = 7(-3) - 5(-4) = -1$

$x = \dfrac{\begin{vmatrix} 81 & -4 \\ 57 & -3 \end{vmatrix}}{\triangle} = \dfrac{81(-3)-57(-4)}{-1}$
$= \dfrac{-15}{-1} = 15$

$y = \dfrac{\begin{vmatrix} 7 & 81 \\ 5 & 57 \end{vmatrix}}{\triangle} = \dfrac{7(57)-5(81)}{-1}$
$= \dfrac{-6}{-1} = 6$

29. $y = 9 - 3x \Rightarrow 3x + y = 9$
$x = 8 - 2y \Rightarrow x + 2y = 8$

$\triangle = \begin{vmatrix} 3 & 1 \\ 1 & 2 \end{vmatrix} = 3(2) - 1(1) = 5$

$x = \dfrac{\begin{vmatrix} 9 & 1 \\ 8 & 2 \end{vmatrix}}{\triangle} = \dfrac{9(2)-8(1)}{5}$
$= \dfrac{10}{5} = 2$

$y = \dfrac{\begin{vmatrix} 3 & 9 \\ 1 & 8 \end{vmatrix}}{\triangle} = \dfrac{3(8)-1(9)}{5}$
$= \dfrac{15}{5} = 3$

33. $4n = 18 - 3m \Rightarrow 3m + 4n = 18$
$m = 8 - 2n \Rightarrow m + 2n = 8$

$\triangle = \begin{vmatrix} 3 & 4 \\ 1 & 2 \end{vmatrix} = 3(2) - 1(4) = 2$

$m = \dfrac{\begin{vmatrix} 18 & 4 \\ 8 & 2 \end{vmatrix}}{\triangle} = \dfrac{18(2)-8(4)}{2}$
$= \dfrac{4}{2} = 2$

$n = \dfrac{\begin{vmatrix} 3 & 18 \\ 1 & 8 \end{vmatrix}}{\triangle} = \dfrac{3(8)-1(18)}{2}$
$= \dfrac{6}{2} = 3$

37. $x = \dfrac{1\left(-\frac{1}{3}\right)-\frac{1}{2}(-1)}{\frac{1}{3}\left(-\frac{1}{3}\right)-\frac{1}{2}\left(\frac{1}{2}\right)} = \dfrac{-\frac{1}{3}+\frac{1}{2}}{-\frac{1}{9}-\frac{1}{4}} = \dfrac{-\frac{2}{6}+\frac{3}{6}}{-\frac{4}{36}-\frac{9}{36}}$
$= \dfrac{1}{6}\cdot-\dfrac{36}{13} = -\dfrac{6}{13}$

$y = \dfrac{\frac{1}{3}(-1)-1\left(\frac{1}{2}\right)}{-\frac{13}{36}} = \dfrac{-\frac{1}{3}-\frac{1}{2}}{-\frac{13}{36}} = \dfrac{-\frac{2}{6}-\frac{3}{6}}{-\frac{13}{36}}$
$= -\dfrac{5}{6}\cdot-\dfrac{36}{13} = \dfrac{30}{13}$

41. $\dfrac{1}{x} = \dfrac{3(-4)-8(4)}{3(-4)-8(15)} = \dfrac{-12-32}{-12-120}$
$= \dfrac{-44}{-132} = \dfrac{1}{3} \quad x = 3$

$\dfrac{1}{y} = \dfrac{3(4)-3(15)}{-132} = \dfrac{12-45}{-132}$
$= \dfrac{-33}{-132} = \dfrac{1}{4} \quad y = 4$

45. $0.310(325) + x = 0.450(325 + x + y)$
$100.75 + x = 146.25 + 0.450x + 0.450y$
$0.550x - 0.450y = 45.5$
$0.0350(325) + y = 0.0480(325 + x + y)$
$11.375 + y = 15.6 + 0.0480x + 0.0480y - 0.0480x + 0.952y = 4.225$

$$\triangle = \begin{vmatrix} 0.550 & -0.450 \\ -0.0480 & 0.952 \end{vmatrix} = 0.550(0.952) - (-0.0480)(-0.450) = 0.502$$

$$x = \frac{\begin{vmatrix} 45.5 & -0.450 \\ 4.225 & 0.952 \end{vmatrix}}{\triangle} = \frac{45.5(0.952) - 4.225(-0.450)}{0.502} = 90.1 \text{ kg}$$

$$y = \frac{\begin{vmatrix} 0.550 & 45.5 \\ -0.0480 & 4.225 \end{vmatrix}}{\triangle} = \frac{0.550(4.225) - 45.5(-0.0480)}{0.502} = 8.98 \text{ kg}$$

Exercise 2: Third-Order Determinants

1. $$\begin{vmatrix} 1 & 0 & 2 \\ 3 & 1 & 0 \\ 1 & 2 & 1 \end{vmatrix} = 1\begin{vmatrix} 1 & 0 \\ 2 & 1 \end{vmatrix} - 0 + 2\begin{vmatrix} 3 & 1 \\ 1 & 2 \end{vmatrix} = 1[1(1) - 2(0)] + 2[3(2) - 1(1)] = 1 + 10 = 11$$

5. $$\begin{vmatrix} 5 & 1 & 2 \\ -3 & 2 & -1 \\ 4 & -3 & 5 \end{vmatrix} = 5\begin{vmatrix} 2 & -1 \\ -3 & 5 \end{vmatrix} - 1\begin{vmatrix} -3 & -1 \\ 4 & 5 \end{vmatrix} + 2\begin{vmatrix} -3 & 2 \\ 4 & -3 \end{vmatrix}$$
$$= 5[2(5) - (-3)(-1)] - 1[-3(5) - 4(-1)] + 2[-3(-3) - 4(2)] = 35 + 11 + 2 = 48$$

9. $$\triangle = \begin{vmatrix} 1 & 1 & 1 \\ 1 & -1 & 1 \\ 1 & 1 & -1 \end{vmatrix} = \begin{vmatrix} -1 & 1 \\ 1 & -1 \end{vmatrix} - \begin{vmatrix} 1 & 1 \\ 1 & -1 \end{vmatrix} + \begin{vmatrix} 1 & 1 \\ -1 & 1 \end{vmatrix} = 0 + 2 + 2 = 4$$

$$x \cdot \triangle = \begin{vmatrix} 18 & 1 & 1 \\ 6 & -1 & 1 \\ 4 & 1 & -1 \end{vmatrix} = \begin{vmatrix} 6 & -1 \\ 4 & 1 \end{vmatrix} - \begin{vmatrix} 18 & 1 \\ 4 & 1 \end{vmatrix} - \begin{vmatrix} 18 & 1 \\ 6 & -1 \end{vmatrix} = 10 - 14 + 24 = 20$$

$$x = \frac{20}{4} = 5$$

$$y \cdot \triangle = \begin{vmatrix} 1 & 18 & 1 \\ 1 & 6 & 1 \\ 1 & 4 & -1 \end{vmatrix} = \begin{vmatrix} 6 & 1 \\ 4 & -1 \end{vmatrix} - \begin{vmatrix} 18 & 1 \\ 4 & -1 \end{vmatrix} + \begin{vmatrix} 18 & 1 \\ 6 & 1 \end{vmatrix} = -10 + 22 + 12 = 24$$

$$y = \frac{24}{4} = 6$$

$$z = x + y - 4 = 7$$

13. $$\triangle = \begin{vmatrix} 1 & 2 & 3 \\ 2 & 1 & 2 \\ 3 & 4 & -3 \end{vmatrix} = \begin{vmatrix} 1 & 2 \\ 4 & -3 \end{vmatrix} - 2\begin{vmatrix} 2 & 3 \\ 4 & -3 \end{vmatrix} + 3\begin{vmatrix} 2 & 3 \\ 1 & 2 \end{vmatrix} = -11 + 36 + 3 = 28$$

$$x \cdot \triangle = \begin{vmatrix} 14 & 2 & 3 \\ 10 & 1 & 2 \\ 2 & 4 & -3 \end{vmatrix} = -2\begin{vmatrix} 10 & 2 \\ 2 & -3 \end{vmatrix} + \begin{vmatrix} 14 & 3 \\ 2 & -3 \end{vmatrix} - 4\begin{vmatrix} 14 & 3 \\ 10 & 2 \end{vmatrix} = 68 - 48 + 8 = 28$$

$$x = \frac{28}{28} = 1$$

$$y \cdot \triangle = \begin{vmatrix} 1 & 14 & 3 \\ 2 & 10 & 2 \\ 3 & 2 & -3 \end{vmatrix} = \begin{vmatrix} 10 & 2 \\ 2 & -3 \end{vmatrix} - 2\begin{vmatrix} 14 & 3 \\ 2 & -3 \end{vmatrix} + 3\begin{vmatrix} 14 & 3 \\ 10 & 2 \end{vmatrix} = -34 + 96 - 6 = 56$$

$$y = \frac{56}{28} = 2$$

$$z = \frac{(14 - x - 2y)}{3} = 3$$

17.

$$\Delta = \begin{vmatrix} 2 & -4 & 3 \\ 3 & 1 & -2 \\ 1 & 3 & -1 \end{vmatrix} = 3\begin{vmatrix} 3 & 1 \\ 1 & 3 \end{vmatrix} + 2\begin{vmatrix} 2 & -4 \\ 1 & 3 \end{vmatrix} - \begin{vmatrix} 2 & -4 \\ 3 & 1 \end{vmatrix} = 24 + 20 - 14 = 30$$

$$x \cdot \Delta = \begin{vmatrix} 10 & -4 & 3 \\ 6 & 1 & -2 \\ 20 & 3 & -1 \end{vmatrix} = 3\begin{vmatrix} 6 & 1 \\ 20 & 3 \end{vmatrix} + 2\begin{vmatrix} 10 & -4 \\ 20 & 3 \end{vmatrix} - \begin{vmatrix} 10 & -4 \\ 6 & 1 \end{vmatrix} = -6 + 220 - 34 = 180$$

$$x = \frac{180}{30} = 6$$

$$y \cdot \Delta = \begin{vmatrix} 2 & 10 & 3 \\ 3 & 6 & -2 \\ 1 & 20 & -1 \end{vmatrix} = 3\begin{vmatrix} 3 & 6 \\ 1 & 20 \end{vmatrix} + 2\begin{vmatrix} 2 & 10 \\ 1 & 20 \end{vmatrix} - \begin{vmatrix} 2 & 10 \\ 3 & 6 \end{vmatrix} = 162 + 60 + 18 = 240$$

$$y = \frac{240}{30} = 8$$
$$z = x + 3y - 20 = 10$$

21. Multiplying each equation by 3 gives

$$3x + y = 15$$
$$3x + z = 18$$
$$3y + z = 27$$

$$\Delta = \begin{vmatrix} 3 & 1 & 0 \\ 3 & 0 & 1 \\ 0 & 3 & 1 \end{vmatrix} = -\begin{vmatrix} 3 & 1 \\ 0 & 3 \end{vmatrix} + \begin{vmatrix} 3 & 1 \\ 3 & 0 \end{vmatrix} = -9 - 3 = -12$$

$$x \cdot \Delta = \begin{vmatrix} 15 & 1 & 0 \\ 18 & 0 & 1 \\ 27 & 3 & 1 \end{vmatrix} = -\begin{vmatrix} 15 & 1 \\ 27 & 3 \end{vmatrix} + \begin{vmatrix} 15 & 1 \\ 18 & 0 \end{vmatrix} = -18 - 18 = -36$$

$$x = \frac{-36}{-12} = 3$$

$$y \cdot \Delta = \begin{vmatrix} 3 & 15 & 0 \\ 3 & 18 & 1 \\ 0 & 27 & 1 \end{vmatrix} = -\begin{vmatrix} 3 & 15 \\ 0 & 27 \end{vmatrix} + \begin{vmatrix} 3 & 15 \\ 3 & 18 \end{vmatrix} = -81 + 9 = -72$$

$$y = \frac{-72}{-12} = 6$$
$$z = 18 - 3x = 9$$

25.

$$283I_1 - 274I_2 + 163I_3 = 352$$
$$428I_1 + 163I_2 + 373I_3 = 169$$
$$338I_1 - 112I_2 - 227I_3 = 825$$

$$\Delta = \begin{vmatrix} 283 & -274 & 163 \\ 428 & 163 & 373 \\ 338 & -112 & -227 \end{vmatrix} = 283\begin{vmatrix} 163 & 373 \\ -112 & -227 \end{vmatrix} - 428\begin{vmatrix} -274 & 163 \\ -112 & -227 \end{vmatrix} + 338\begin{vmatrix} -274 & 163 \\ 163 & 373 \end{vmatrix}$$

$$= 283(4775) - 428(80,454) + 338(-128,771) = -76,607,585$$

$$I_1 \cdot \Delta = \begin{vmatrix} 352 & -274 & 163 \\ 169 & 163 & 373 \\ 825 & -112 & -227 \end{vmatrix} = 352\begin{vmatrix} 163 & 373 \\ -112 & -227 \end{vmatrix} - 169\begin{vmatrix} -274 & 163 \\ -112 & -227 \end{vmatrix} + 825\begin{vmatrix} -274 & 163 \\ 163 & 373 \end{vmatrix}$$

$$= 352(4775) - 169(80,454) + 825(-128,771) = -118,152,001$$

$$I_1 = \frac{-118,152,001}{-76,607,585} = 1.54$$

$$I_2 \cdot \Delta = \begin{vmatrix} 283 & 352 & 163 \\ 428 & 169 & 373 \\ 338 & 825 & -227 \end{vmatrix} = 283\begin{vmatrix} 169 & 373 \\ 825 & -227 \end{vmatrix} - 428\begin{vmatrix} 352 & 163 \\ 825 & -227 \end{vmatrix} + 338\begin{vmatrix} 352 & 163 \\ 169 & 373 \end{vmatrix}$$

$$= 283(-346,088) - 428(-214,379) + 338(103,749) = 28,878,470$$

$$I_2 = \frac{28,878,470}{-76,607,585} = -0.377$$

$$283(1.54) - 274(-0.377) + 163I_3 = 352$$
$$I_3 = -1.15$$

Exercise 3: Higher-Order Determinants

1. Two row interchanges made.

$$\begin{vmatrix} 4.00 & 3.00 & 1.00 & 0.00 \\ 0.00 & 1.00 & -1.00 & 2.00 \\ 0.00 & 2.00 & -3.00 & 5.00 \\ -1.00 & 2.00 & -3.00 & 5.00 \end{vmatrix} = 4\begin{vmatrix} 1.00 & -1.00 & 2.00 \\ 2.00 & -3.00 & 5.00 \\ 2.75 & -2.75 & 5.00 \end{vmatrix} = 4\begin{vmatrix} -1.00 & 1.00 \\ 0.00 & -0.50 \end{vmatrix} = -4 \ |-0.50| = 2$$

5.

$$\begin{vmatrix} 3.00 & 1.00 & 0.00 & 2.00 & 4.00 \\ 1.00 & 2.00 & 4.00 & 0.00 & 1.00 \\ 2.00 & 3.00 & 1.00 & 4.00 & 2.00 \\ 1.00 & 2.00 & 0.00 & 2.00 & 1.00 \\ 3.00 & 4.00 & 1.00 & 3.00 & 1.00 \end{vmatrix} = 3\begin{vmatrix} 1.67 & 4.00 & -0.67 & -0.33 \\ 2.33 & 1.00 & 2.67 & -0.67 \\ 1.67 & 0.00 & 1.33 & -0.33 \\ 3.00 & 1.00 & 1.00 & -3.00 \end{vmatrix} = 5\begin{vmatrix} -4.60 & 3.60 & -0.20 \\ -4.00 & 2.00 & 0.00 \\ -6.20 & 2.20 & -2.40 \end{vmatrix}$$

$$= -23\begin{vmatrix} -1.13 & 0.17 \\ -2.65 & -2.13 \end{vmatrix} = 26|-2.54| = -66$$

9.

$$\triangle = \begin{vmatrix} 3.00 & -2.00 & -1.00 & 1.00 \\ -1.00 & -1.00 & 3.00 & 2.00 \\ 1.00 & 3.00 & -2.00 & 1.00 \\ 2.00 & -1.00 & -1.00 & -3.00 \end{vmatrix} = 3\begin{vmatrix} -1.67 & 2.67 & 2.33 \\ 3.67 & -1.67 & 0.67 \\ 0.33 & -0.33 & -3.67 \end{vmatrix} = -5\begin{vmatrix} 4.20 & 5.80 \\ 0.20 & -3.20 \end{vmatrix}$$

$$= -21 \ |-3.48| = 73$$

$$x \cdot \triangle = \begin{vmatrix} -3.00 & -2.00 & -1.00 & 1.00 \\ 23.00 & -1.00 & 3.00 & 2.00 \\ -12.00 & 3.00 & -2.00 & 1.00 \\ -22.00 & -1.00 & -1.00 & -3.00 \end{vmatrix} = -3\begin{vmatrix} -16.33 & -4.67 & 9.67 \\ 11.00 & 2.00 & -3.00 \\ 13.67 & 6.33 & -10.33 \end{vmatrix} = 49\begin{vmatrix} -1.14 & 3.51 \\ 2.43 & -2.24 \end{vmatrix}$$

$$= -56 \ |5.21| = -292$$

$$y \cdot \triangle = \begin{vmatrix} 3.00 & -3.00 & -1.00 & 1.00 \\ -1.00 & 23.00 & 3.00 & 2.00 \\ 1.00 & -12.00 & -2.00 & 1.00 \\ 2.00 & -22.00 & -1.00 & -3.00 \end{vmatrix} = 3\begin{vmatrix} 22.00 & 2.67 & 2.33 \\ -11.00 & -1.67 & 0.67 \\ -20.00 & -0.33 & -3.67 \end{vmatrix} = 66\begin{vmatrix} -0.33 & 1.83 \\ 2.09 & -1.55 \end{vmatrix}$$

$$= -22|9.95| = -219$$

$$z \cdot \triangle = \begin{vmatrix} 3.00 & -2.00 & -3.00 & 1.00 \\ -1.00 & -1.00 & 23.00 & 2.00 \\ 1.00 & 3.00 & -12.00 & 1.00 \\ 2.00 & -1.00 & -22.00 & -3.00 \end{vmatrix} = 3\begin{vmatrix} -1.67 & 22.00 & 2.33 \\ 3.67 & -11.00 & 0.67 \\ 0.33 & -20.00 & -3.67 \end{vmatrix} = -5\begin{vmatrix} 37.40 & 5.80 \\ -15.60 & -3.20 \end{vmatrix}$$

$$= -187|-0.78| = 146$$

$$w \cdot \triangle = \begin{vmatrix} 3.00 & -2.00 & -1.00 & -3.00 \\ -1.00 & -1.00 & 3.00 & 23.00 \\ 1.00 & 3.00 & -2.00 & -12.00 \\ 2.00 & -1.00 & -1.00 & -22.00 \end{vmatrix} = 3\begin{vmatrix} -1.67 & 2.67 & 22.00 \\ 3.67 & -1.67 & -11.00 \\ 0.33 & -0.33 & -20.00 \end{vmatrix} = -5\begin{vmatrix} 4.20 & 37.40 \\ 0.20 & -15.60 \end{vmatrix}$$

$$= -21|-17.38| = 365$$

$$x = -\frac{272}{73} = -4 \qquad y = -\frac{219}{73} = -3$$
$$z = \frac{146}{73} = 2 \qquad w = \frac{365}{73} = 5$$

13.
$$\triangle = \begin{vmatrix} 0 & 0 & 1 & 1 & 0 \\ 0 & 0 & 0 & 1 & 1 \\ 0 & 1 & 0 & 0 & 1 \\ 1 & 1 & 0 & 0 & 0 \\ 1 & 0 & 1 & 0 & 0 \end{vmatrix} = \begin{vmatrix} 0 & 0 & 1 & 1 \\ 0 & 1 & 0 & 1 \\ 1 & 1 & 0 & 0 \\ 1 & 0 & 0 & 0 \end{vmatrix} + \begin{vmatrix} 0 & 0 & 1 & 0 \\ 0 & 0 & 1 & 1 \\ 0 & 1 & 0 & 1 \\ 1 & 1 & 0 & 0 \end{vmatrix}$$

$$= \begin{vmatrix} 0 & 1 & 1 \\ 1 & 1 & 0 \\ 1 & 0 & 0 \end{vmatrix} - \begin{vmatrix} 0 & 1 & 0 \\ 0 & 1 & 1 \\ 1 & 0 & 1 \end{vmatrix}$$

$$= \begin{vmatrix} 1 & 1 \\ 1 & 0 \end{vmatrix} - \begin{vmatrix} 1 & 0 \\ 1 & 1 \end{vmatrix} = -2$$

$$u\triangle = \begin{vmatrix} 9 & 0 & 1 & 1 & 0 \\ 11 & 0 & 0 & 1 & 1 \\ 13 & 1 & 0 & 0 & 1 \\ 15 & 1 & 0 & 0 & 0 \\ 12 & 0 & 1 & 0 & 0 \end{vmatrix} = 9\begin{vmatrix} 0 & 0 & 1 & 1 \\ 1 & 0 & 0 & 1 \\ 1 & 0 & 0 & 0 \\ 0 & 1 & 0 & 0 \end{vmatrix} + \begin{vmatrix} 11 & 0 & 1 & 1 \\ 13 & 1 & 0 & 1 \\ 15 & 1 & 0 & 0 \\ 12 & 0 & 0 & 0 \end{vmatrix} - \begin{vmatrix} 11 & 0 & 0 & 1 \\ 13 & 1 & 0 & 1 \\ 15 & 1 & 0 & 0 \\ 12 & 0 & 1 & 0 \end{vmatrix}$$

$$= 9\begin{vmatrix} 0 & 1 & 1 \\ 0 & 0 & 1 \\ 1 & 0 & 0 \end{vmatrix} - 12\begin{vmatrix} 0 & 1 & 1 \\ 1 & 0 & 1 \\ 1 & 0 & 0 \end{vmatrix} + \begin{vmatrix} 11 & 0 & 1 \\ 13 & 1 & 1 \\ 15 & 1 & 0 \end{vmatrix}$$

$$= 9\begin{vmatrix} 1 & 1 \\ 0 & 1 \end{vmatrix} - 12\begin{vmatrix} 1 & 1 \\ 0 & 1 \end{vmatrix} + 11\begin{vmatrix} 1 & 1 \\ 1 & 0 \end{vmatrix} + \begin{vmatrix} 13 & 1 \\ 15 & 1 \end{vmatrix} = -16$$

$u = \frac{-16}{-2} = 8$

$x = 12 - u = 4$

$y = 9 - x = 5$

$z = 11 - y = 6$

$w = 13 - z = 7$

21. Use a calculator or computer to solve the system of equations.
Answer: $I_1 = -4.31$, $I_2 = 7.57$, $I_3 = 6.27$, $I_4 = 3.71$

Chapter 11 Review Problems

1.
$$\begin{vmatrix} 6 & \frac{1}{2} & -2 \\ 3 & \frac{1}{4} & 4 \\ 2 & -\frac{1}{2} & 3 \end{vmatrix} = -\frac{1}{2}\begin{vmatrix} 3 & 4 \\ 2 & 3 \end{vmatrix} + \frac{1}{4}\begin{vmatrix} 6 & -2 \\ 2 & 3 \end{vmatrix} + \frac{1}{2}\begin{vmatrix} 6 & -2 \\ 3 & 4 \end{vmatrix}$$
$$= -\frac{1}{2}(9-8) + \frac{1}{4}(18+4) + \frac{1}{2}(24+6) = 20$$

5. Expanding 1st row:
$$\begin{vmatrix} 25 & 23 & 19 \\ 14 & 11 & 9 \\ 21 & 17 & 14 \end{vmatrix} = 25\begin{vmatrix} 11 & 9 \\ 17 & 14 \end{vmatrix} - 23\begin{vmatrix} 14 & 9 \\ 21 & 14 \end{vmatrix} + 19\begin{vmatrix} 14 & 11 \\ 21 & 17 \end{vmatrix}$$
$$= 25(154 - 153) - 23(196 - 189) + 19(238 - 231)$$
$$= 25(1) - 23(7) + 19(7)$$
$$= 25 - 161 + 133$$
$$= -3$$

9. Expanding 1st row:

$$\begin{vmatrix} 1 & 2 & 3 \\ 3 & 1 & 2 \\ 2 & 3 & 1 \end{vmatrix} = \begin{vmatrix} 1 & 2 \\ 3 & 1 \end{vmatrix} - 2\begin{vmatrix} 3 & 2 \\ 2 & 1 \end{vmatrix} + 3\begin{vmatrix} 3 & 1 \\ 2 & 3 \end{vmatrix}$$

$$= -5 - 2(-1) + 3(7) = 18$$

13. Expanding 1st row:

$$\begin{vmatrix} 3 & 1 & 5 & 2 \\ 4 & 10 & 14 & 6 \\ 8 & 9 & 1 & 4 \\ 6 & 15 & 21 & 9 \end{vmatrix} = 3\begin{vmatrix} 10 & 14 & 6 \\ 9 & 1 & 4 \\ 15 & 21 & 9 \end{vmatrix} - \begin{vmatrix} 4 & 14 & 6 \\ 8 & 1 & 4 \\ 6 & 21 & 9 \end{vmatrix} + 5\begin{vmatrix} 4 & 10 & 6 \\ 8 & 9 & 4 \\ 6 & 15 & 9 \end{vmatrix} - 2\begin{vmatrix} 4 & 10 & 14 \\ 8 & 9 & 1 \\ 6 & 15 & 21 \end{vmatrix}$$

Expanding 1st row:

$$= 3\left[10\begin{vmatrix} 1 & 4 \\ 21 & 9 \end{vmatrix} - 14\begin{vmatrix} 9 & 4 \\ 15 & 9 \end{vmatrix} + 6\begin{vmatrix} 9 & 1 \\ 15 & 21 \end{vmatrix} \right]$$

Expanding 1st row:

$$- \left[4\begin{vmatrix} 1 & 4 \\ 21 & 9 \end{vmatrix} - 14\begin{vmatrix} 8 & 4 \\ 6 & 9 \end{vmatrix} + 6\begin{vmatrix} 8 & 1 \\ 6 & 21 \end{vmatrix} \right]$$

Expanding 1st row:

$$+ 5\left[4\begin{vmatrix} 9 & 4 \\ 15 & 9 \end{vmatrix} - 10\begin{vmatrix} 8 & 4 \\ 6 & 9 \end{vmatrix} + 6\begin{vmatrix} 8 & 9 \\ 6 & 15 \end{vmatrix} \right]$$

Expanding 1st row:

$$- 2\left[4\begin{vmatrix} 9 & 1 \\ 15 & 21 \end{vmatrix} - 10\begin{vmatrix} 8 & 1 \\ 6 & 21 \end{vmatrix} + 14\begin{vmatrix} 8 & 9 \\ 6 & 15 \end{vmatrix} \right]$$

$$= 3(-750 - 294 + 1044) - (-300 - 672 + 972) + 5(84 - 480 + 396) - 2(696 - 1620 + 924)$$
$$= 3(0) - 0 + 5(0) - 2(0) = 0$$

17. Expanding 2nd column:

$$\begin{vmatrix} 6 & 4 & 7 \\ 9 & 0 & 8 \\ 5 & 3 & 2 \end{vmatrix} = -4\begin{vmatrix} 9 & 8 \\ 5 & 2 \end{vmatrix} - 3\begin{vmatrix} 6 & 7 \\ 9 & 8 \end{vmatrix} = -4(-22) - 3(-15) = 133$$

21. $4x + 3y = 27$
$2x - 5y = -19$

$$\triangle = \begin{vmatrix} 4 & 3 \\ 2 & -5 \end{vmatrix} = -20 - 6 = -26$$

$$x \cdot \triangle = \begin{vmatrix} 27 & 3 \\ -19 & -5 \end{vmatrix} = -135 + 57 = -78$$

So $x = \frac{-78}{-26} = 3$

$$y \cdot \triangle = \begin{vmatrix} 4 & 27 \\ 2 & -19 \end{vmatrix} = -76 - 54 = -130$$

So $y = \frac{-130}{-26} = 5$

25. $x = \frac{7(2) - (-3)(27)}{2(2) - (-3)(5)} = \frac{14 + 81}{4 + 15} = \frac{95}{19} = 5$

$y = \frac{2(27) - 7(5)}{19} = \frac{54 - 35}{19} = \frac{19}{19} = 1$

33. $x = \frac{25(3) - 4(21)}{3(3) - 4(4)} = \frac{75 - 84}{9 - 16} = \frac{-9}{-7} = \frac{9}{7}$

$y = \frac{3(21) - 25(4)}{-7} = \frac{63 - 100}{-7} = \frac{-37}{-7} = \frac{37}{7}$

29. $x = \frac{3(5) - (-5)(11)}{4(5) - (-5)(3)} = \frac{15 + 55}{20 + 15} = \frac{70}{35} = 2$

$y = \frac{4(11) - 3(3)}{35} = \frac{44 - 9}{35} = \frac{35}{35} = 1$

37. $72.3x + 54.2y + 83.3z = 52.5$
$52.2x - 26.6y + 83.7z = 75.2$
$33.4x + 61.6y + 30.2z = 58.5$

$$\triangle = \begin{vmatrix} 72.3 & 54.2 & 83.3 \\ 52.2 & -26.6 & 83.7 \\ 33.4 & 61.6 & 30.2 \end{vmatrix}$$

$$= 72.3 \begin{vmatrix} -26.6 & 83.7 \\ 61.6 & 30.2 \end{vmatrix} - 52.2 \begin{vmatrix} 54.2 & 83.3 \\ 61.6 & 30.2 \end{vmatrix}$$

$$+ 33.4 \begin{vmatrix} 54.2 & 83.3 \\ -26.6 & 83.7 \end{vmatrix}$$

$$= 72.3(-5959.24) - 52.2(-3494.44)$$
$$+ 33.4(6752.32)$$
$$= -22{,}915.8$$

$$x \cdot \triangle = \begin{vmatrix} 52.2 & 54.2 & 83.3 \\ 75.2 & -26.6 & 83.7 \\ 58.5 & 61.6 & 30.2 \end{vmatrix}$$

$$= 52.5 \begin{vmatrix} -26.6 & 83.7 \\ 61.6 & 30.2 \end{vmatrix} - 75.2 \begin{vmatrix} 54.2 & 83.3 \\ 61.6 & 30.2 \end{vmatrix}$$

$$+ 58.5 \begin{vmatrix} 54.2 & 83.3 \\ -26.6 & 83.7 \end{vmatrix}$$

$$= 52.5(-5959.24) - 75.2(-3494.44)$$
$$+ 58.5(6752.32)$$
$$= 344{,}933$$

$$x = \frac{344{,}933}{-22{,}915.8} = -15.0522$$

$$y \cdot \triangle = \begin{vmatrix} 72.3 & 52.5 & 83.3 \\ 52.2 & 75.2 & 83.7 \\ 33.4 & 58.5 & 30.2 \end{vmatrix}$$

$$= 72.3 \begin{vmatrix} 75.2 & 83.7 \\ 58.5 & 30.2 \end{vmatrix} - 52.2 \begin{vmatrix} 52.5 & 83.3 \\ 58.5 & 30.2 \end{vmatrix}$$

$$+ 33.4 \begin{vmatrix} 52.5 & 83.3 \\ 75.2 & 83.7 \end{vmatrix}$$

$$= 72.3(-2625.41) - 52.2(-3287.55)$$
$$+ 33.4(-1869.91)$$
$$= -80{,}662.0$$

$$y = \frac{-80{,}662.0}{-22{,}915.8} = 3.51993$$

Substituting into an original equation for z,
$72.3(-15.0522) + 54.2(3.51993) + 83.3z = 52.5$
$z = 11.4045$
$x = -15.1, \ y = 3.52, \ z = 11.4$

CHAPTER 12: MATRICES

Exercise 1: Definitions

1. **A, B, D, E, F, I**

5. **I**

9. **F**

13. 4×3

17. $\begin{pmatrix} 0 & 0 \\ 0 & 0 \\ 0 & 0 \\ 0 & 0 \end{pmatrix}$

Exercise 2: Operations with Matrices

1. $(3 \quad 8 \quad 2 \quad 1) + (9 \quad 5 \quad 3 \quad 7) = (12 \quad 13 \quad 5 \quad 8)$

5. $\begin{pmatrix} 1 & 5 & -2 & 1 \\ 0 & -2 & 3 & 5 \\ -2 & 9 & 3 & -2 \\ 5 & -1 & 2 & 6 \end{pmatrix} - \begin{pmatrix} -6 & 4 & 2 & 8 \\ 9 & 2 & -6 & 3 \\ -5 & 2 & 6 & 7 \\ 6 & 3 & 7 & 0 \end{pmatrix} = \begin{pmatrix} 7 & 1 & -4 & -7 \\ -9 & -4 & 9 & 2 \\ 3 & 7 & -3 & -9 \\ -1 & -4 & -5 & 6 \end{pmatrix}$

9. $-3 \times \begin{pmatrix} 4 & 2 & -3 & 0 \\ -1 & 6 & 3 & -4 \end{pmatrix} = \begin{pmatrix} -12 & -6 & 9 & 0 \\ 3 & -18 & -9 & 12 \end{pmatrix}$

13. $3\mathbf{A} + \mathbf{B} = \begin{pmatrix} 6 & 15 & -3 \\ -9 & 6 & 18 \\ 27 & -12 & 15 \end{pmatrix} + \begin{pmatrix} 0 & 1 & 6 \\ -7 & 2 & 1 \\ 4 & -2 & 0 \end{pmatrix} = \begin{pmatrix} 6 & 16 & 3 \\ -16 & 8 & 19 \\ 31 & -14 & 15 \end{pmatrix}$

17. $(5 \quad 3 \quad 7)\begin{pmatrix} 2 \\ 6 \\ 1 \end{pmatrix} = 35$

21. $(3 \quad -1)\begin{pmatrix} 4 & 7 & -2 & 4 & 0 \\ 2 & -6 & 8 & -3 & 7 \end{pmatrix} = (10 \quad 27 \quad -14 \quad 15 \quad -7)$

25. $(2 \quad 4 \quad 3 \quad -1 \quad 0)\begin{pmatrix} 4 & 3 & 5 & -1 & 0 \\ 1 & 3 & 5 & -2 & 6 \\ 2 & -4 & 3 & 6 & 5 \\ 3 & 4 & -1 & 6 & 5 \\ 2 & 3 & 1 & 4 & -6 \end{pmatrix} = (15 \quad 2 \quad 40 \quad 2 \quad 34)$

29. $\begin{pmatrix} 5 & 1 \\ -2 & 0 \\ 1 & 3 \end{pmatrix}\begin{pmatrix} -4 & 1 \\ 0 & 5 \end{pmatrix} = \begin{pmatrix} -20 & 10 \\ 8 & -2 \\ -4 & 16 \end{pmatrix}$

33. $\begin{pmatrix} 3 & 1 \\ -2 & 0 \\ 1 & -3 \\ 5 & 0 \end{pmatrix} \begin{pmatrix} -1 & 0 \\ 3 & 2 \end{pmatrix} = \begin{pmatrix} 0 & 2 \\ 2 & 0 \\ -10 & -6 \\ -5 & 0 \end{pmatrix}$

37. $\mathbf{AB} = \begin{pmatrix} 2 & -1 & 0 \\ 1 & 0 & 2 \\ -2 & 1 & 0 \end{pmatrix} \begin{pmatrix} 2 & 1 & 0 \\ 3 & 0 & 1 \\ -2 & 4 & 1 \end{pmatrix} = \begin{pmatrix} 1 & 2 & -1 \\ -2 & 9 & 2 \\ -1 & -2 & 1 \end{pmatrix}$

$\mathbf{BA} = \begin{pmatrix} 2 & 1 & 0 \\ 3 & 0 & 1 \\ -2 & 4 & 1 \end{pmatrix} \begin{pmatrix} 2 & -1 & 0 \\ 1 & 0 & 2 \\ -2 & 1 & 0 \end{pmatrix} = \begin{pmatrix} 5 & -2 & 2 \\ 4 & -2 & 0 \\ -2 & 3 & 8 \end{pmatrix}$

41. $\begin{pmatrix} 2 & 0 & 1 \\ 1 & 5 & 2 \end{pmatrix} \begin{pmatrix} 3 \\ 0 \\ -1 \end{pmatrix} = \begin{pmatrix} 5 \\ 1 \end{pmatrix}$

45. $(81 \quad 72 \quad 93 \quad 69) \begin{pmatrix} 2 \\ 3 \\ 7 \\ 4 \end{pmatrix} = \big(81(2) + 72(3) + 93(7) + 4(69)\big) = (1305)$ or \$1,305

49. $\mathbf{P} = \begin{pmatrix} 9.93 \\ 10.45 \\ 11.20 \end{pmatrix}, \mathbf{M} = \begin{pmatrix} 32 & 26 & 19 \\ 41 & 33 & 43 \end{pmatrix}, \mathbf{T} = \begin{pmatrix} 35 & 18 & 22 \\ 33 & 26 & 32 \end{pmatrix}$

(a) $\mathbf{M} + \mathbf{T} = \begin{pmatrix} 32 & 26 & 19 \\ 41 & 33 & 43 \end{pmatrix} + \begin{pmatrix} 35 & 18 & 22 \\ 33 & 26 & 32 \end{pmatrix} = \begin{pmatrix} 67 & 44 & 41 \\ 74 & 59 & 75 \end{pmatrix}$

(b) $\mathbf{T} - \mathbf{M} = \begin{pmatrix} 35 & 18 & 22 \\ 33 & 26 & 32 \end{pmatrix} - \begin{pmatrix} 32 & 26 & 19 \\ 41 & 33 & 43 \end{pmatrix} = \begin{pmatrix} 3 & -8 & 3 \\ -8 & -7 & -11 \end{pmatrix}$

(c) $\mathbf{MP} = \begin{pmatrix} 32 & 26 & 19 \\ 41 & 33 & 43 \end{pmatrix} \begin{pmatrix} 9.93 \\ 10.45 \\ 11.20 \end{pmatrix} = \begin{pmatrix} 802.26 \\ 1233.58 \end{pmatrix}$

$\mathbf{TP} = \begin{pmatrix} 35 & 18 & 22 \\ 33 & 26 & 32 \end{pmatrix} \begin{pmatrix} 9.93 \\ 10.45 \\ 11.20 \end{pmatrix} = \begin{pmatrix} 782.05 \\ 957.79 \end{pmatrix}$

Exercise 3: The Inverse of a Matrix

The following matrices were inverted by computer.

1. $\left(\begin{array}{cc|cc} 4.000 & 8.000 & 1.000 & 0.000 \\ -5.000 & 0.000 & 0.000 & 1.000 \end{array} \right)$

$\left(\begin{array}{cc|cc} 1.000 & 2.000 & 0.250 & 0.000 \\ -5.000 & 0.000 & 0.000 & 1.000 \end{array} \right)$

$\left(\begin{array}{cc|cc} 1.000 & 2.000 & 0.250 & 0.000 \\ 0.000 & 10.000 & 1.250 & 1.000 \end{array} \right)$

$\left(\begin{array}{cc|cc} 1.000 & 2.000 & 0.250 & 0.000 \\ 0.000 & 1.000 & 0.125 & 0.100 \end{array} \right)$

$\left(\begin{array}{cc|cc} 1.000 & 0.000 & 0.000 & -0.200 \\ 0.000 & 1.000 & 0.125 & 0.100 \end{array} \right)$

5.
$$\left(\begin{array}{rrr|rrr} -1.000 & 3.000 & 7.000 & 1.000 & 0.000 & 0.000 \\ 4.000 & -2.000 & 0.000 & 0.000 & 1.000 & 0.000 \\ 7.000 & 2.000 & -9.000 & 0.000 & 0.000 & 1.000 \end{array}\right)$$

$$\left(\begin{array}{rrr|rrr} 1.000 & -3.000 & -7.000 & -1.000 & 0.000 & 0.000 \\ 4.000 & -2.000 & 0.000 & 0.000 & 1.000 & 0.000 \\ 7.000 & 2.000 & -9.000 & 0.000 & 0.000 & 1.000 \end{array}\right)$$

$$\left(\begin{array}{rrr|rrr} 1.000 & -3.000 & -7.000 & -1.000 & 0.000 & 0.000 \\ 0.000 & 10.000 & 28.000 & 4.000 & 1.000 & 0.000 \\ 0.000 & 23.000 & 40.000 & 7.000 & 0.000 & 1.000 \end{array}\right)$$

$$\left(\begin{array}{rrr|rrr} 1.000 & -3.000 & -7.000 & -1.000 & 0.000 & 0.000 \\ 0.000 & 1.000 & 2.800 & 0.400 & 0.100 & 0.000 \\ 0.000 & 23.000 & 40.000 & 7.000 & 0.000 & 1.000 \end{array}\right)$$

$$\left(\begin{array}{rrr|rrr} 1.000 & -3.000 & -7.000 & -1.000 & 0.000 & 0.000 \\ 0.000 & 1.000 & 2.800 & 0.400 & 0.100 & 0.000 \\ 0.000 & 0.000 & -24.400 & -2.200 & -2.300 & 1.000 \end{array}\right)$$

$$\left(\begin{array}{rrr|rrr} 1.000 & -3.000 & -7.000 & -1.000 & 0.000 & 0.000 \\ 0.000 & 1.000 & 2.800 & 0.400 & 0.100 & 0.000 \\ 0.000 & 0.000 & 1.000 & 0.090 & 0.094 & -0.041 \end{array}\right)$$

$$\left(\begin{array}{rrr|rrr} 1.000 & -3.000 & 0.000 & -0.369 & 0.660 & -0.287 \\ 0.000 & 1.000 & 0.000 & 0.148 & -0.164 & 0.115 \\ 0.000 & 0.000 & 1.000 & 0.090 & 0.094 & -0.041 \end{array}\right)$$

$$\left(\begin{array}{rrr|rrr} 1.000 & 0.000 & 0.000 & 0.074 & 0.168 & 0.057 \\ 0.000 & 1.000 & 0.000 & 0.148 & -0.164 & 0.115 \\ 0.000 & 0.000 & 1.000 & 0.090 & 0.094 & -0.041 \end{array}\right)$$

Chapter 12 Review Problems

1. $(4 \quad 9 \quad -2 \quad 6) + (2 \quad 4 \quad 9 \quad -1) - (4 \quad 1 \quad -3 \quad 7) = (2 \quad 12 \quad 10 \quad -2)$

5.
$$\left(\begin{array}{rr|rr} 4.000 & -4.000 & 1.000 & 0.000 \\ 3.000 & 1.000 & 0.000 & 1.000 \end{array}\right)$$

$$\left(\begin{array}{rr|rr} 1.000 & -1.000 & 0.250 & 0.000 \\ 3.000 & 1.000 & 0.000 & 1.000 \end{array}\right)$$

$$\left(\begin{array}{rr|rr} 1.000 & -1.000 & 0.250 & 0.000 \\ 0.000 & 4.000 & -0.750 & 1.000 \end{array}\right)$$

$$\left(\begin{array}{rr|rr} 1.000 & -1.000 & 0.250 & 0.000 \\ 0.000 & 1.000 & -0.188 & 0.250 \end{array}\right)$$

$$\left(\begin{array}{rr|rr} 1.000 & 0.000 & 0.063 & 0.250 \\ 0.000 & 1.000 & -0.188 & 0.250 \end{array}\right)$$

9. To solve by matrix inversion, append the unit matrix to $\mathbf{A}$ to find $\mathbf{A}^{-1}$

$$\left(\begin{array}{cccc|cccc} 1 & 1 & 1 & 1 & 1 & 0 & 0 & 0 \\ 1 & 2 & 3 & 4 & 0 & 1 & 0 & 0 \\ 1 & 3 & 6 & 10 & 0 & 0 & 1 & 0 \\ 1 & 4 & 10 & 20 & 0 & 0 & 0 & 1 \end{array}\right)$$

$$\left(\begin{array}{cccc|cccc} 1 & 1 & 1 & 1 & 1 & 0 & 0 & 0 \\ 0 & 1 & 2 & 3 & -1 & 1 & 0 & 0 \\ 0 & 2 & 5 & 9 & -1 & 0 & 1 & 0 \\ 0 & 3 & 9 & 19 & -1 & 0 & 0 & 1 \end{array}\right)$$

$$\left(\begin{array}{cccc|cccc} 1 & 0 & -1 & -2 & 2 & -1 & 0 & 0 \\ 0 & 1 & 2 & 3 & -1 & 1 & 0 & 0 \\ 0 & 2 & 5 & 9 & -1 & 0 & 1 & 0 \\ 0 & 3 & 9 & 19 & -1 & 0 & 0 & 1 \end{array}\right)$$

$$\left(\begin{array}{cccc|cccc} 1 & 0 & -1 & -2 & 2 & -1 & 0 & 0 \\ 0 & 1 & 2 & 3 & -1 & 1 & 0 & 0 \\ 0 & 0 & 1 & 3 & 1 & -2 & 1 & 0 \\ 0 & 3 & 9 & 19 & -1 & 0 & 0 & 1 \end{array}\right)$$

$$\left(\begin{array}{cccc|cccc} 1 & 0 & -1 & -2 & 2 & -1 & 0 & 0 \\ 0 & 1 & 2 & 3 & -1 & 1 & 0 & 0 \\ 0 & 0 & 1 & 3 & 1 & -2 & 1 & 0 \\ 0 & 0 & 3 & 10 & 2 & -3 & 0 & 1 \end{array}\right)$$

$$\left(\begin{array}{cccc|cccc} 1 & 0 & 0 & 1 & 3 & -3 & 1 & 0 \\ 0 & 1 & 2 & 3 & -1 & 1 & 0 & 0 \\ 0 & 0 & 1 & 3 & 1 & -2 & 1 & 0 \\ 0 & 0 & 3 & 10 & 2 & -3 & 0 & 1 \end{array}\right)$$

$$\left(\begin{array}{cccc|cccc} 1 & 0 & 0 & 1 & 3 & -3 & 1 & 0 \\ 0 & 1 & 0 & -3 & -3 & 5 & -2 & 0 \\ 0 & 0 & 1 & 3 & 1 & -2 & 1 & 0 \\ 0 & 0 & 3 & 10 & 2 & -3 & 0 & 1 \end{array}\right)$$

$$\left(\begin{array}{cccc|cccc} 1 & 0 & 0 & 1 & 3 & -3 & 1 & 0 \\ 0 & 1 & 0 & -3 & -3 & 5 & -2 & 0 \\ 0 & 0 & 1 & 3 & 1 & -2 & 1 & 0 \\ 0 & 0 & 0 & 1 & -1 & 3 & -3 & 1 \end{array}\right)$$

$$\left(\begin{array}{cccc|cccc} 1 & 0 & 0 & 0 & 4 & -6 & 4 & -1 \\ 0 & 1 & 0 & -3 & -3 & 5 & -2 & 0 \\ 0 & 0 & 1 & 3 & 1 & -2 & 1 & 0 \\ 0 & 0 & 0 & 1 & -1 & 3 & -3 & 1 \end{array}\right)$$

$$\left(\begin{array}{cccc|cccc} 1 & 0 & 0 & 0 & 4 & -6 & 4 & -1 \\ 0 & 1 & 0 & 0 & -6 & 14 & -11 & 3 \\ 0 & 0 & 1 & 3 & 1 & -2 & 1 & 0 \\ 0 & 0 & 0 & 1 & -1 & 3 & -3 & 1 \end{array}\right)$$

$$\left(\begin{array}{cccc|cccc} 1 & 0 & 0 & 0 & 4 & -6 & 4 & -1 \\ 0 & 1 & 0 & 0 & -6 & 14 & -11 & 3 \\ 0 & 0 & 1 & 0 & 4 & -11 & 10 & -3 \\ 0 & 0 & 0 & 1 & -1 & 3 & -3 & 1 \end{array}\right)$$

Since $\mathbf{A}^{-1}\mathbf{B} = \mathbf{X}$,

$$\left(\begin{array}{cccc} 4 & -6 & 4 & -1 \\ -6 & 14 & -11 & 3 \\ 4 & -11 & 10 & -3 \\ -1 & 3 & -3 & 1 \end{array}\right)\left(\begin{array}{c} -4 \\ 0 \\ 9 \\ 24 \end{array}\right) = \left(\begin{array}{c} -4 \\ -3 \\ 2 \\ 1 \end{array}\right)$$

Thus $x = -4$, $y = -3$, $z = 2$ and $w = 1$

13. The coefficient matrix is $\mathbf{A} = \begin{pmatrix} 4 & 3 \\ 2 & -5 \end{pmatrix}$ from which we compute $\mathbf{A}^{-1}$ by interchanging the 4 and -5, reversing the signs of the 2 and 3, dividing by the determinant of $\mathbf{A}$, which equals, $4(-5) - 2 \cdot 3 = -26$.

$$\mathbf{A}^{-1} = \frac{\begin{pmatrix} -5 & -3 \\ -2 & 4 \end{pmatrix}}{-26} = \begin{pmatrix} \frac{5}{26} & \frac{3}{26} \\ \frac{2}{26} & -\frac{4}{26} \end{pmatrix}$$

Using $\mathbf{X} = \mathbf{A}^{-1}\mathbf{B}$, we obtain,

$$\begin{pmatrix} x \\ y \end{pmatrix} = \begin{pmatrix} \frac{5}{26} & \frac{3}{26} \\ \frac{2}{26} & -\frac{4}{26} \end{pmatrix} \begin{pmatrix} 27 \\ -19 \end{pmatrix} = \begin{pmatrix} \frac{135}{26} - \frac{57}{26} \\ \frac{54}{26} + \frac{76}{26} \end{pmatrix} = \begin{pmatrix} \frac{78}{26} \\ \frac{130}{26} \end{pmatrix} = \begin{pmatrix} 3 \\ 5 \end{pmatrix}$$

Thus, $x = 3$, and $y = 5$.

17. $(181 \quad 62 \quad 33 \quad 49) \begin{pmatrix} 232 & 273 \\ 373 & 836 \\ 737 & 361 \\ 244 & 227 \end{pmatrix} = (101,395 \quad 124,281) = \$101,395, \$124,281$

CHAPTER 13: EXPONENTS & RADICALS

Exercise 1: Integral Exponents

1. $3x^{-1} = \dfrac{3}{x}$

5. $a(2b)^{-2} = \dfrac{a}{(2b)^2} = \dfrac{a}{4b^2}$

9. $p^3 q^{-1} = \dfrac{p^3}{q}$

13. $(4a^3 b^2 c^6)^{-2} = \dfrac{1}{(4a^3 b^2 c^6)^2} = \dfrac{1}{16a^6 b^4 c^{12}}$

17. $(m^{-2} - 6n)^{-2} = \dfrac{1}{(m^{-2}-6n)^{-2}}$

$\quad = \dfrac{1}{m^{-4} - 12m^{-2}n + 36n^2}$

$\quad = \dfrac{1}{\dfrac{1}{m^4} + \dfrac{12n}{m^2} + 36n^2}$

$\quad = \dfrac{1}{\dfrac{1 - 12m^2 n + 36m^4 n^2}{m^4}}$

$\quad = \dfrac{m^4}{(1 - 6m^2 n)^2}$

21. $(3m)^{-3} - 2n^{-2} = \dfrac{1}{(3m)^3} - \dfrac{2}{n^2}$

25. $\dfrac{16x^6 y^0}{8x^4 y} \div 4xy^6 = \dfrac{2x^2}{y} \cdot \dfrac{1}{4xy^6} = \dfrac{x}{2y^7}$

29. $\left(\dfrac{x^{m+n}}{x^n}\right)^m = (x^{m+n-n})^m = (x^m)^m = x^{m^2}$

33. $\left(\dfrac{2x}{3y}\right)^{-2} = \left(\dfrac{3y}{2x}\right)^2 = \dfrac{9y^2}{4x^2}$

37. $\left(\dfrac{3a^4 b^3}{5x^2 y}\right)^2 = \dfrac{9a^8 b^6}{25x^4 y^2}$

41. $\dfrac{3q^{-2} x^{-4}}{2p^{-2} z^3} = \dfrac{3p^2}{2q^2 x^4 z^3}$

45. $\left(\dfrac{5m^{-3} x^4}{3n^{-2} y^3}\right)^2 = \left(\dfrac{5x^4 n^2}{3y^3 m^3}\right)^2 = \dfrac{25n^4 x^8}{9m^6 y^6}$

49. $\dfrac{(x^2 - xy)^7}{(x-y)^5} = \dfrac{x^7 (x-y)^7}{(x-y)^5} = x^7 (x-y)^2$

53. $\dfrac{(x+1)^3}{(y+1)^3} \cdot \dfrac{(y+1)(y-1)}{(x+1)(x^2 - x + 1)} = \dfrac{(x+1)^2 (y-1)}{(y+1)^2 (x^2 - x + 1)}$

57. $\dfrac{128a^4 b^3 - 48a^5 b^2 - 40a^6 b + 15a^7}{a(3a - 8b)}$

$\quad = \dfrac{128a^3 b^3 - 48a^4 b^2 - 40a^5 b + 15a^6}{3a - 8b}$

$\quad = \dfrac{a^3 (128b^3 - 48ab^2 - 40a^2 b + 15a^3)}{3a - 8b}$

$\quad = \dfrac{a^3 (3a - 8b)(5a^2 - 16b^2)}{3a - 8b}$

$\quad = a^3 (5a^2 - 16b^2)$

61. $\dfrac{3x^4 y^3 + 4x^5 y^2 + 3x^6 y + 4x^7}{4x^2 + 3xy}$

$\quad = \dfrac{x^4 (3y^3 + 4xy^2 + 3x^2 y + 4x^3)}{x(4x + 3y)}$

$\quad = \dfrac{x^3 [y^2 (3y + 4x) + x^2 (3y + 4x)]}{4x + 3y}$

$\quad = x^3 (x^2 + y^2)$

65. $R = \dfrac{P}{I^2}$

$\quad R = PI^{-2}$

Exercise 2: Simplification of Radicals

1. $a^{1/4} = \sqrt[4]{a}$

5. $(m - n)^{1/2} = \sqrt{m - n}$

9. $\sqrt{b} = b^{1/2}$

13. $\sqrt[n]{a + b} = (a + b)^{1/n}$

17. $\sqrt{18} = \sqrt{9} \cdot \sqrt{2} = 3\sqrt{2}$

21. $\sqrt[3]{-56} = \sqrt[3]{-8} \cdot \sqrt[3]{7} = -2\sqrt[3]{7}$

25. $\sqrt{36x^2 y} = \sqrt{6^2 x^2 y} = 6x\sqrt{y}$

29. $3\sqrt[5]{32xy^{11}} = 3\sqrt[5]{2^5 xy^5 y^5 y} = 6y^2 \sqrt[5]{xy}$

33. $\sqrt{9m^3 + 18n} = \sqrt{9(m^3 + 2n)} = 3\sqrt{m^3 + 2n}$

37. $(a + b)\sqrt{a^3 - 2a^2 b + ab^2}$

$\quad = (a + b)\sqrt{a(a^2 - 2ab + b^2)}$

$\quad = (a + b)\sqrt{a(a - b)^2} = (a - b)(a + b)\sqrt{a}$

41. $\sqrt[3]{\dfrac{1}{4}} = \dfrac{\sqrt[3]{1}}{\sqrt[3]{4}} \cdot \dfrac{\sqrt[3]{2}}{\sqrt[3]{2}} = \dfrac{\sqrt[3]{2}}{2}$

45. $\sqrt{\frac{1}{2x}} = \frac{\sqrt{1}}{\sqrt{2x}} \cdot \frac{\sqrt{2x}}{\sqrt{2x}} = \frac{\sqrt{2x}}{2x}$

49. $\sqrt[3]{\frac{1}{x^2}} = \frac{\sqrt[3]{1}}{\sqrt[3]{x^2}} \cdot \frac{\sqrt[3]{x}}{\sqrt[3]{x}} = \frac{\sqrt[3]{x}}{x}$

53. $(x^2 - y^2)\sqrt{\frac{x}{x+y}} = (x^2-y^2)\frac{\sqrt{x}}{\sqrt{x+y}} \cdot \frac{\sqrt{x+y}}{\sqrt{x+y}}$
$$= \frac{(x^2-y^2)\sqrt{x}\sqrt{x+y}}{x+y} = (x-y)\sqrt{x^2+xy}$$

57. $\sqrt{\frac{8a^2-48a+72}{3a}} = \sqrt{8}\sqrt{\frac{a^2-6a+9}{3a}} = \sqrt{8}\sqrt{\frac{(a-3)^2}{3a}}$
$$= \sqrt{8}(a-3)\sqrt{\frac{1}{3a}} = \frac{\sqrt{8}(a-3)}{\sqrt{3a}}$$
$$= \frac{\sqrt{3a}\sqrt{8}(a-3)}{3a} = \frac{\sqrt{24a}(a-3)}{3a} = \frac{2(a-3)\sqrt{6a}}{3a}$$
$$= \frac{(2a-6)\sqrt{6a}}{3a}$$

61. $c = \sqrt{a^2+b^2} = \sqrt{a^2+(3a)^2}$
$$= \sqrt{a^2+9a^2} = \sqrt{10a^2} = a\sqrt{10}$$

Exercise 3: Operations with Radicals

1. $2\sqrt{24} - \sqrt{54} = 2\cdot 2\sqrt{6} - 3\sqrt{6} = \sqrt{6}$

5. $2\sqrt{50} + \sqrt{72} + 3\sqrt{18}$
$$= 2\cdot 5\sqrt{2} + 6\sqrt{2} + 3\cdot 3\sqrt{2} = 25\sqrt{2}$$

9. $\sqrt[3]{625} - 2\sqrt[3]{135} - \sqrt[3]{320}$
$$= 5\sqrt[3]{5} - 2\cdot 3\sqrt[3]{5} - 4\sqrt[3]{5} = -5\sqrt[3]{5}$$

13. $\sqrt{128x^2y} - \sqrt{98x^2y} + \sqrt{162x^2y}$
$$= \sqrt{64\cdot 2x^2y} - \sqrt{49\cdot 2x^2y} + \sqrt{81\cdot 2x^2y}$$
$$= 8x\sqrt{2y} - 7x\sqrt{2y} + 9x\sqrt{2y} = 10x\sqrt{2y}$$

17. $\sqrt{a^2x} + \sqrt{b^2x} = a\sqrt{x} + b\sqrt{x} = (a+b)\sqrt{x}$

21. $4\sqrt{3a^2x} - 2a\sqrt{48x} = 4a\sqrt{3x} - 2a\cdot 4\sqrt{3x}$
$$= -4a\sqrt{3x}$$

25. $\sqrt[5]{a^{13}b^{11}c^{12}} - 2\sqrt[5]{a^8bc^2} + \sqrt[5]{a^3b^6c^7}$
$$= a^2b^2c^2\sqrt[5]{a^3bc^2} - 2a\sqrt[5]{a^3bc^2} + bc\sqrt[5]{a^3bc^2}$$
$$= (a^2b^2c^2 - 2a + bc)\sqrt[5]{a^3bc^2}$$

29. $3\sqrt{\frac{3x}{4y^2}} + 2\sqrt{\frac{x}{27y^2}} - \sqrt{\frac{x}{3y^2}}$
$$= \frac{3}{2y}\sqrt{3x} + \frac{2}{3y}\sqrt{\frac{x}{3}} - \frac{1}{y}\sqrt{\frac{x}{3}}$$
$$= \frac{3}{2y}\sqrt{3x} + \frac{2}{3y}\left(\frac{1}{3}\right)\sqrt{3x} - \frac{1}{y}\left(\frac{1}{3}\right)\sqrt{3x}$$
$$= \left(\frac{3}{2y} + \frac{2}{9y} - \frac{1}{3y}\right)\sqrt{3x}$$
$$= \frac{25}{18y}\sqrt{3x}$$

33. $\left(\sqrt{8}\right)\left(\sqrt{160}\right) = \left(2\sqrt{2}\right)\left(4\sqrt{10}\right)$
$$= 8\sqrt{20} = 16\sqrt{5}$$

37. $\left(3\sqrt{2}\right)\left(2\sqrt[3]{3}\right) = 6\cdot 2^{1/2}\cdot 3^{1/3}$
$$= 6\cdot 2^{3/6}\cdot 3^{2/6} = 6\sqrt[6]{2^3\cdot 3^2} = 6\sqrt[6]{72}$$

41. $\left(2x\sqrt{3a}\right)\left(3\sqrt{y}\right) = 6x\sqrt{3ay}$

45. $\left(\sqrt{a}\right)\left(\sqrt[4]{b}\right) = a^{1/2}b^{1/4} = a^{2/4}b^{1/4} = \sqrt[4]{a^2b}$

49. $\left(\sqrt{5}-\sqrt{3}\right)\left(2\sqrt{3}\right)$
$$= 2\sqrt{15} - 2\cdot 3 = 2\sqrt{15} - 6$$

53. $\left(a+\sqrt{b}\right)\left(a-\sqrt{b}\right) = a^2 - b$

57. $\left(3\sqrt{y}\right)^2 = 9y$

61. $\left(3-5\sqrt{a}\right)^2 = 9 - 15\sqrt{a} - 15\sqrt{a} + 25a$
$$= 9 - 30\sqrt{a} + 25a$$

65. $\left(5\sqrt[3]{2ax}\right)^3 = 125(2ax) = 250ax$

69. $\frac{5\sqrt[3]{12}}{10\sqrt{8}} \cdot \frac{\sqrt{2}}{\sqrt{2}} = \frac{5\cdot 12^{1/3}\cdot 2^{1/2}}{40} = \frac{12^{2/6}\cdot 2^{3/6}}{8}$
$$= \frac{\sqrt[6]{12^2\cdot 2^3}}{8} = \frac{\sqrt[6]{1152}}{8}$$
$$= \frac{\sqrt[6]{64\cdot 18}}{8} = \frac{2\sqrt[6]{18}}{8} = \frac{\sqrt[6]{18}}{4}$$

73. $\frac{4\sqrt[3]{4}+2\sqrt[3]{3}+3\sqrt[3]{6}}{\sqrt[3]{6}} \cdot \frac{\sqrt[3]{36}}{\sqrt[3]{36}} = \frac{4\sqrt[3]{144}+2\sqrt[3]{108}+18}{6}$
$$= \frac{4\cdot 2\sqrt[3]{18}+2\cdot 3\sqrt[3]{4}+18}{6}$$
$$= \frac{4\sqrt[3]{18}}{3} + \sqrt[3]{4} + 3$$

77. $\frac{8\sqrt[3]{ab}}{4\sqrt{ac}} = \frac{2(ab)^{1/3}}{(ac)^{1/2}} = 2\sqrt[6]{\frac{(ab)^2}{(ac)^3}} = 2\sqrt[6]{\frac{b^2}{ac^3} \cdot \frac{a^5c^3}{a^5c^3}}$
$$= \frac{2\sqrt[6]{a^5b^2c^3}}{ac}$$

81. $\dfrac{3+\sqrt{2}}{2-\sqrt{2}} \cdot \dfrac{2+\sqrt{2}}{2+\sqrt{2}} = \dfrac{6+3\sqrt{2}+2\sqrt{2}+2}{4-2} = \dfrac{8+5\sqrt{2}}{2}$

85. $\dfrac{12}{\sqrt[3]{4x^2}} \cdot \dfrac{\sqrt[3]{2x}}{\sqrt[3]{2x}} = \dfrac{12\sqrt[3]{2x}}{2x} = \dfrac{6\sqrt[3]{2x}}{x}$

89. $\dfrac{a}{a+\sqrt{b}} \cdot \dfrac{a-\sqrt{b}}{a-\sqrt{b}} = \dfrac{a^2-a\sqrt{b}}{a^2-b}$

93. $\dfrac{5\sqrt{x}+\sqrt{2y}}{2\sqrt{3x}+2\sqrt{y}} = \dfrac{5\sqrt{x}+\sqrt{2y}}{2\left(\sqrt{3x}+\sqrt{y}\right)} \cdot \dfrac{\sqrt{3x}-\sqrt{y}}{\sqrt{3x}-\sqrt{y}}$

$\qquad = \dfrac{5\sqrt{3x}-5\sqrt{xy}+\sqrt{6xy}-\sqrt{2y}}{6x-2y}$

Exercise 4: Radical Equations

1. $\sqrt{x}=6$
$x = 36$

5. $\sqrt{2.95x-1.84}=6.23$
$2.95x - 1.84 = 38.81$
$2.95x = 40.65$
$x = 13.8$

9. $\sqrt{3x+1}=5$
$3x+1=25$
$3x=24$
$x=8$

13. $\sqrt{x^2-7.25}=8.75-x$
$x^2-7.25=76.56-17.5x+x^2$
$17.5x=83.8$
$x=4.79$

17. $\sqrt[5]{x-7}=1$
$x-7=1$
$x=8$

21. $\sqrt{x-2}-\sqrt{x}=\dfrac{1}{\sqrt{x-2}}$
$x-2-\sqrt{x^2-2x}=1$
$\sqrt{x^2-2x}=x-3$
$x^2-2x=x^2-6x+9$
$4x=9$
$x=\dfrac{9}{4}=2\dfrac{1}{4}$
Doesn't check.
No solution.

25. $x+25.3+\sqrt{x^2+(25.3)^2}=68.4$
$\sqrt{x^2+(25.3)^2}=43.1-x$
$x^2+640.1=1857.6-86.2x+x^2$
$86.2x=1217.5$
$x=14.1$ cm
hypotenuse $=\sqrt{(14.1)^2+(25.3)^2}=29.0$ cm

29. $Z^2=R^2+(\omega L-1/\omega C)^2$
$\omega L-1/\omega C=\pm\sqrt{Z^2-R^2}$
$\omega L\pm\sqrt{Z^2-R^2}=1/\omega C$
$C=1/\left[\omega^2 L\pm\omega\sqrt{Z^2-R^2}\right]$

Chapter 13 Review Problems

1. $\sqrt{52}=2\sqrt{13}$

5. $\sqrt[6]{4}=\sqrt[3]{\sqrt{4}}=\sqrt[3]{2}$

9. $\sqrt[3]{(a-b)^5 x^4}=(a-b)x\sqrt[3]{(a-b)^2 x}$

13. $\sqrt{x}\sqrt{x^3-x^4 y}=\sqrt{x^4-x^5 y}=\sqrt{x^4(1-xy)}$
$\qquad =x^2\sqrt{1-xy}$

17. $\dfrac{x+\sqrt{x^2-y^2}}{x-\sqrt{x^2-y^2}} \cdot \dfrac{x+\sqrt{x^2-y^2}}{x+\sqrt{x^2-y^2}}$

$\qquad = \dfrac{x^2+2x\sqrt{x^2-y^2}+(x^2-y^2)}{x^2-(x^2-y^2)}$

$\qquad = \dfrac{2x^2+2x\sqrt{x^2-y^2}-y^2}{y^2}$

21. $\sqrt{a}\sqrt[3]{b}=a^{1/2}b^{1/3}=a^{3/6}b^{2/6}=\sqrt[6]{a^3 b^2}$

25. $3\sqrt{50}-2\sqrt{32}=3\cdot5\sqrt{2}-2\cdot4\sqrt{2}=7\sqrt{2}$

29. $\dfrac{\sqrt{2ab}}{\sqrt{4ab^2}} \cdot \dfrac{\sqrt{4ab^2}}{\sqrt{4ab^2}} = \dfrac{\sqrt{8a^2 b^3}}{4ab^2} = \dfrac{2ab\sqrt{2b}}{4ab^2} = \dfrac{\sqrt{2b}}{2b}$

33. $\sqrt{x+6}=4$
$x+6=16$
$x=10$

37. $\sqrt{x}+\sqrt{x-9.75}=6.23$
$\sqrt{x-9.75}=6.23-\sqrt{x}$
$x-9.75=38.8-12.46\sqrt{x}+x$
$12.46\sqrt{x}=48.6$
$\sqrt{x}=3.90$
$x=15.2$

41. $\left(\dfrac{9x^4y^3}{6x^3y}\right)^3 = \left(\dfrac{3xy^2}{2}\right)^3 = \dfrac{27x^3y^6}{8}$

45. $3w^{-2} = \dfrac{3}{w^2}$

49. $x^{-1} - 2y^{-2} = \dfrac{1}{x} - \dfrac{2}{y^2}$

53. $(p^{a-1} + q^{a-2})(p^a + q^{a-1})$
$= p^{2a-1} + (pq)^{a-1} + p^a q^{a-2} + q^{2a-3}$

57. $r^2 s^{-3} = \dfrac{r^2}{s^3}$

61. $V = \dfrac{4}{3}\pi(3r)^3 = \dfrac{4}{3}\pi(27r^3) = 36\pi r^3$

CHAPTER 14: QUADRATIC EQUATIONS

Exercise 1: Solving Quadratics by Factoring

1. $2x = 5x^2$
$5x^2 - 2x = 0$
$x(5x - 2) = 0$
$x = 0$
$5x - 2 = 0$
$5x = 2$
$x = \frac{2}{5}$

5. $5x^2 - 3 = 2x^2 + 24$
$3x^2 = 27$
$x^2 = 9$
$x = \pm \sqrt{9}$
$x = \pm 3$

9. $8.25x^2 - 2.93x = 0$
$x(8.25x - 2.93) = 0$
$x = 0, \quad x = \frac{2.93}{8.25} = 0.355$

13. $x^2 - x - 20 = 0$
$(x + 4)(x - 5) = 0$
$x + 4 = 0$
$x = -4$
$x - 5 = 0$
$x = 5$

17. $2x^2 - 3x - 5 = 0$
$(2x - 5)(x + 1) = 0$
$2x - 5 = 0$
$x = \frac{5}{2}$
$x + 1 = 0$
$x = -1$

21. $5x^2 + 14x - 3 = 0$
$(5x - 1)(x + 3) = 0$
$5x - 1 = 0$
$5x = 1$
$x = \frac{1}{5}$
$x + 3 = 0$
$x = -3$

25. $x(x - 5) = 36$
$x^2 - 5x = 36$
$x^2 - 5x - 36 = 0$
$(x - 9)(x + 4) = 0$
$x = 9$
$x = -4$

29. $4x^2 + 16ax + 12a^2 = 0$
$(2x + 6a)(2x + 2a) = 0$
$x = -3a$
$x = -a$

33. $x = 4, \ x = 7$
$x - 4 = 0, \ x - 7 = 0$
$(x - 4)(x - 7) = 0$
$x^2 - 11x + 28 = 0$

37. $\sqrt{5x^2 - 3x - 41} = 3x - 7$
$5x^2 - 3x - 41 = 9x^2 - 42x + 49$
$4x^2 - 39x + 90 = 0$
$(4x - 15)(x - 6) = 0$
$x = \frac{15}{4}$
$x = 6$

Exercise 2: Solving Quadratics by Completing the Square

1. $x^2 - 8x + 2 = 0$
$x^2 - 8x = -2$
$x^2 - 8x + 16 = -2 + 16$
$(x - 4)^2 = 14$
$x = 4 \pm \sqrt{14}$

5. $4x^2 - 3x - 5 = 0$
$x^2 - \frac{3}{4}x = \frac{5}{4}$
$x^2 - \frac{3}{4}x + \left(\frac{3}{8}\right)^2 = \frac{5}{4} + \left(\frac{3}{8}\right)^2$
$\left(x + \frac{3}{8}\right)^2 = \frac{80}{64} + \frac{9}{64}$
$x = \frac{3 \pm \sqrt{89}}{8}$

9. $4x^2 + 7x - 5 = 0$
$x^2 + \frac{7}{4}x = \frac{5}{4}$
$x^2 + \frac{7}{4}x + \left(\frac{7}{8}\right)^2 = \frac{5}{4} + \left(\frac{7}{8}\right)^2$
$\left(x + \frac{7}{8}\right)^2 = \frac{80}{64} + \frac{49}{64}$
$x = \frac{-7 \pm \sqrt{129}}{8}$

Exercise 3: Solving Quadratics by Formula

1. $x^2 - 12x + 28 = 0$
   $a = 1, \ b = -12, \ c = 28$

   $x = \dfrac{-(-12) \pm \sqrt{(-12)^2 - 4(1)(28)}}{2(1)}$
   $= 8.83, \ \ 3.17$

5. $3x^2 + 12x - 35 = 0$
   $a = 3, \ b = 12, \ c = -35$

   $x = \dfrac{-12 \pm \sqrt{(12)^2 - 4(3)(-35)}}{2(3)}$
   $= 1.96, \ \ -5.96$

9. $49x^2 + 21x - 5 = 0$
   $a = 49, \ b = 21, \ c = -5$

   $x = \dfrac{-21 \pm \sqrt{(21)^2 - 4(49)(-5)}}{2(49)}$
   $= 0.170, \ \ -0.599$

13. $3x^2 + 5x - 7 = 0$
    $a = 3, \ b = 5, \ c = -7$

    $x = \dfrac{-5 \pm \sqrt{(5)^2 - 4(3)(-7)}}{2(3)}$
    $= 0.907, \ \ -2.57$

17. $3x^2 + 6x - 505 = 0$
    $a = 3, \ b = 6, \ c = -505$

    $x = \dfrac{-6 \pm \sqrt{(6)^2 - 4(3)(-505)}}{2(3)}$
    $= 12.0, \ \ -14.0$

21. $4.26x + 5.74 = 1.27x^2 + 2.73x$
    $1.27x^2 - 1.53x - 5.74 = 0$
    $a = 1.27, b = -1.53, c = -5.74$

    $x = \dfrac{1.53 \pm \sqrt{(-1.53)^2 - 4(1.27)(-5.74)}}{2(1.27)}$
    $= 2.81, \ \ -1.61$

25. $2x^2 - 3x = 3x^2 + 12x - 2$
    $x^2 + 15x - 2 = 0$
    $a = 1, \ b = 15, \ c = -2$

    $x = \dfrac{-15 \pm \sqrt{(15)^2 - 4(1)(-2)}}{2(1)}$
    $= 0.132, \ \ -15.1$

29. $x^2 - 5x - 11 = 0$
    $b^2 - 4ac = 25 - 4(1)(-11) = 25 + 44 = 69$
    real and unequal

33. $3x^2 - 3x + 5 = 0$
    $b^2 - 4ac = 9 - 4(3)(5) = 9 - 60 = -51$
    not real

Exercise 4: Applications and Word Problems

1. Let x = the number
   $\frac{1}{x}$ = number's reciprocal
   $x + \frac{1}{x} = 2\frac{1}{6}$
   $x^2 + 1 = \frac{13x}{6}$
   $x^2 - \frac{13x}{6} + 1 = 0$
   $6x^2 - 13x + 6 = 0$
   $(3x - 2)(2x - 3) = 0$
   $x = \frac{2}{3}$
   $x = \frac{3}{2}$

5. Let c = first number
   $c + 10$ = second number
   $c^2 + (c + 10)^2 = 250$
   $c^2 + c^2 + 20c + 100 = 250$
   $2c^2 + 20c - 150 = 0$
   $c^2 + 10c - 75 = 0$
   $(c + 15)(c - 5) = 0$
   $c \neq -15$
   $c = 5$
   $c + 10 = 15$

9. $l = 2w, \ (l - 6)(w - 6)(3) = 648$
   $(2w - 6)(w - 6)(3) = 648$
   $(2w^2 - 18w + 36)3 = 648$
   $w^2 - 9w + 18 = 108$
   $w^2 - 9w - 90 = 0$
   $(w + 6)(w - 15) = 0$
   $w \neq -6$
   $w = 15$
   $l = 2w = 30$
   15 cm $\times$ 30 cm

13. Let x = original side
    then x^3 = original volume
    Thus, $(x - 0.175)^3 = x^3 - 2.48$
    $x^3 - 0.525x^2 + 0.09188x - 0.005359$
    $\quad = x^3 - 2.48$
    $-0.525x^2 + 0.09188x + 2.475 = 0$
    $x^2 - 0.1750x - 4.714 = 0$

    $x = \dfrac{0.1750 \pm \sqrt{(-0.1750)^2 - 4(1)(-4.714)}}{2(1)}$

    $x = \frac{0.1750 \pm 4.346}{2} = -2.09$ or 2.26
    discarding the -2.09, $x = 2.26$ in.

17. Let A = rate of A, B = rate of B

$\frac{355}{A} + \frac{488}{B} = 5.20$

$355B + 448A = 5.20AB$

$B = A + 15.8$

$355(A + 15.8) + 448A = 5.20A(A + 15.8)$

$355A + 5609 + 448A = 5.20A^2 + 82.16A$

$5.20A^2 - 720.8A - 5609 = 0$

$A = \dfrac{720.8 \pm \sqrt{(-720.8)^2 - 4(5.20)(-5609)}}{2(5.20)}$

$= \dfrac{720.8 \pm 797.6}{10.4} = 146 \text{ mi/h}$

21. Let x = newer machine's rate, y = slower machine's rate

$\frac{5}{y} + \frac{3}{x} + \frac{3}{y} = 1$

$8x + 3y = xy$

$y = x + 3$

$8x + 3(x + 3) = x(x + 3)$

$8x + 3x + 9 = x^2 + 3x$

$x^2 - 8x - 9 = 0$

$(x + 1)(x - 9) = 0$

$x \neq -1$

$x = 9 \text{ h/box}$

$y = x + 3 = 12 \text{ h/box}$

25. $M = \frac{1}{2}wlx - \frac{1}{2}wx^2$

$M = 0, \ lx = x^2$

$x = l \ \text{ or } \ x = 0$

Thus the bending moment is zero at each end.

29. $\frac{1}{R_1} + \frac{1}{R_2} = \frac{1}{105}$

$105R_2 + 105R_1 = R_1R_2$

$R_1 + R_2 = 780$

$R_1 = 780 - R_2$

$105R_2 + 105(780 - R_2) = (780 - R_2)R_2$

$105R_2 + 81900 - 105R_2 = 780R_2 - R_2^2$

$R_2^2 - 780R_2 + 81900 = 0$

$(R_2 - 125)(R_2 - 655) = 0$

$R_2 = 125\,\Omega$

$R_2 = 655\,\Omega$

$R_1 = 780 - R_2 = 655\,\Omega \text{ (for } R_2 = 125)$

$R_1 = 780 - R_2 = 125\,\Omega \text{ (for } R_2 = 655)$

33. $P = (I_1 + I_2)^2 R$

$RI_1^2 + 2RI_2I_1 + RI_2^2 - P = 0$

$R = 100\,\Omega$

$I_2 = 0.2\,A$

$P = 9.0\,W$

$I_1 = \dfrac{-2RI_2 \pm \sqrt{(2RI_2)^2 - 4(R)(RI_2^2 - P)}}{2R}$

$= \dfrac{-2(100)(0.2) \pm \sqrt{40^2 - 4(100)(-5)}}{2(100)}$

$= \dfrac{-40 \pm 60}{200} = -0.5\,A, \text{ and } 0.1\,A$

Exercise 5: Graphing the Quadratic Function

1.

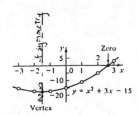

9. $x^2 = 125y$

when $x = 50$,

$(50)^2 = 125y$

$y = \frac{(50)^2}{125} = 20\text{cm} = d$

Exercise 6: Equations of Quadratic Type

1. $x^6 - 6x^3 + 8 = 0$ let $w = x^3$

$w^2 - 6w + 8 = 0$

$(w - 4)(w - 2) = 0$

$w = 2$ and $w = 4$

thus $x^3 = 2$ and $x^3 = 4$

$x = \sqrt[3]{2}$ $\qquad x = \sqrt[3]{4}$

5. $x^{-2/3} - x^{-1/3} = 0$ let $w = x^{-1/3}$

$w^2 - w = 0$

$w(w - 1) = 0$

$w = 0$ and 1

thus $x^{-1/3} = 0$, $x = 0$ doesn't check

and $x^{-1/3} = 1$, $x = 1$

9. $x^{2/3} + 3x^{1/3} = 4$ let $w = x^{1/3}$

$w^2 + 3w - 4 = 0$

$(w + 4)(w - 1) = 0$

$w = -4$ and $w = 1$

thus $x^{1/3} = -4$ and $x^{1/3} = 1$

$x = -64$ $\qquad x = 1$

13. $x^{1/2} - x^{1/4} = 20$ let $w = x^{1/4}$

$w^2 - w - 20 = 0$

$(w - 5)(w + 4) = 0$

$w = 5$ and $w = -4$

thus $x^{1/4} = 5$ and $x^{1/4} = -4$

$x = 625$ $\qquad x = 256$ (doesn't check)

Exercise 7: Simple Equations of Higher Degree

1. $x^3 + x^2 - 17x + 15 = 0$

Since $x = 1$ is a root, $(x - 1)$ is a factor

$$
\begin{array}{r}
x^2 + 2x - 15 \\
x - 1 \overline{\smash{\big)}\ x^3 + x^2 - 17x + 15} \\
\underline{x^3 - x^2} \\
2x^2 - 17x \\
\underline{2x^2 - 2x} \\
-15x + 15 \\
\underline{-15x + 15} \\
0
\end{array}
$$

So, $x^2 + 2x - 15 = 0$

$(x + 5)(x - 3) = 0$

$x = -5$

$x = 3$

5. $2x^3 + 7x^2 - 7x - 12 = 0$

Since $x = -1$ is a root, $(x + 1)$ is a factor.

$$
\begin{array}{r}
2x^2 + 5x - 12 \\
x + 1 \overline{\smash{\big)}\ 2x^3 + 7x^2 - 7x - 12} \\
\underline{2x^3 + 2x^2} \\
5x^2 - 7x \\
\underline{5x^2 + 5x} \\
-12x - 12 \\
\underline{-12x - 12} \\
0
\end{array}
$$

So, $2x^2 + 5x - 12 = 0$

$(2x - 3)(x + 4) = 0$

$x = \frac{3}{2}$

$x = -4$

13. $15.2x^3 - 11.3x + 1.72 = 0$

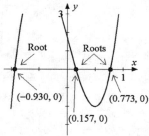

So, the roots are: $-0.930, 0.157$ and 0.773

17. $4.82x^4 - 16.1x^3 + 3.25 = 0$

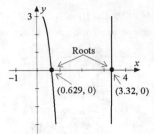

So, the roots are: 0.629 and 3.32

Exercise 8: Systems of Quadratic Equations

1. $x^2 - y^2 = 32$

   $x - y = 4$
   $x = 4 + y$

   $(4 + y)^2 - y^2 = 32$
   $16 + 8y + y^2 - y^2 = 32$
   $8y = 16$
   $y = 2$
   $x = 4 + (2) = 6$ $(6, 2)$

5. $y^2 - 2y + 4x = 11$

   $x + 4y = 14$
   $x = 14 - 4y$

   $y^2 - 2y + 4(14 - 4y) = 11$
   $y^2 - 2y + 56 - 16y = 11$
   $y^2 - 18y + 45 = 0$
   $(y - 3)(y - 15) = 0$
   $y - 3 = 0$
   $y = 3$
   $y - 15 = 0$
   $y = 15$
   $x = 14 - 4(3) = 2$
   $x = 14 - 4(15) = -46$ $(2, 3), (-46, 15)$

9. $x + 3y^2 = 7$
   $2x - y^2 = 9$

   $\quad x + 3y^2 = 7$
   $\underline{\quad 6x - 3y^2 = 27}$
   $\quad 7x \qquad = 34$

   $x = \frac{34}{7} = 4.86$
   $y^2 = 2x - 9 = \left(\frac{68}{7}\right) - \left(\frac{63}{7}\right) = \frac{5}{7}$
   $y = \pm\sqrt{\frac{5}{7}} = \pm 0.845$
   $(4.86, 0.845), (4.86, -0.845)$

13. $3.52x^2 + 4.82y^2 = 34.2$
    $5.92x - 3.72y = 4.58$

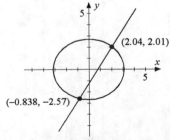

So, approximate solution is: $(2.04, 2.01), (-0.838, -2.57)$

Chapter 14: Review Problems

1. $y^2 - 5y - 6 = 0$
$(y-6)(y+1) = 0$
$y - 6 = 0$
$y = 6$
$y + 1 = 0$
$y = -1$

5. $w^2 - 5w = 0$
$w(w-5) = 0$
$w = 0$
$w - 5 = 0$
$w = 5$

9. $3t^2 - 13t - 10 = 0$
$(3t+2)(t-5) = 0$
$3t + 2 = 0$
$t = -\frac{2}{3}$
$t - 5 = 0$
$t = 5$

13. $9 - x^2 = 0$
$x^2 = 9$
$x = \pm 3$

17. $y + 6 = 5\sqrt{y}$
$y^2 + 12y + 36 = 25y$
$y^2 - 13y + 36 = 0$
$(y-9)(y-4) = 0$
$y - 9 = 0$
$y = 9$
$y - 4 = 0$
$y = 4$

21. $\frac{z}{2} = \frac{5}{z}$
$z^2 = 10$
$z = \pm\sqrt{10}$

25. $3x^2 - 27x = 0$
$3x(x-9) = 0$
$3x = 0$
$x = 0$
$x - 9 = 0$
$x = 9$

29. $5y^2 = 125$
$y^2 = 25$
$y = \pm 5$

33. $5.12y^2 + 8.76y - 9.89 = 0$
$$y = \frac{-8.76 \pm \sqrt{(8.76)^2 - 4(5.12)(-9.89)}}{2(5.12)}$$
$$= \frac{-8.76 \pm \sqrt{279.28}}{10.24}$$
$y = 0.777, \quad -2.49$

37. $3.21z^6 - 21.3 = 4.23z^3$
Let $x = z^3$
$3.21x^2 - 4.23x - 21.3 = 0$
$$x = \frac{-(-4.23) \pm \sqrt{(-4.23)^2 - 4(3.21)(-21.3)}}{2(3.21)}$$
$$= \frac{4.23 \pm \sqrt{291.38}}{6.42}$$
$x = 3.318, \quad -2.000$
So, $z^3 = 3.318$ and $z^3 = -2.000$
$\quad\quad z = 1.49 \quad\quad\quad z = -1.26$

41. $x^4 - 2x^3 - 13x^2 + 14x + 24 = 0$
Since $x = -1$ is a root, $(x+1)$ is a factor.

$$
\begin{array}{r}
x^3 - 3x^2 - 10x + 24 \\
x+1 \overline{)\ x^4 - 2x^3 - 13x^2 + 14x + 24} \\
\underline{x^4 + x^2} \\
-3x^3 - 13x^2 \\
\underline{-3x^3 - 3x^2} \\
-10x^2 + 14x \\
\underline{-10x^2 - 10x} \\
+24x + 24 \\
\underline{24x + 24} \\
0
\end{array}
$$

So, $x^3 - 3x^2 - 10x + 24 = 0$
Try $x = 2$ as a root, so $(x-2)$ will be a factor.

$$
\begin{array}{r}
x^2 - x - 12 \\
x-2 \overline{)\ x^3 - 3x^2 - 10x + 24} \\
\underline{x^3 - 2x^2} \\
-x^2 - 10x \\
\underline{-x^2 + 2x} \\
-12x + 24 \\
\underline{-12x + 24} \\
0
\end{array}
$$

$x = 2$ works!
So, $x^2 - x - 12 = 0$
$\quad (x-4)(x+3) = 0$
$\quad\quad x = 4$
$\quad\quad x = -3$
Thus the roots are
$x = -1, 2, 4, -3$

45. Let $x =$ bags bought, $p =$ price per bag

$xp = 1000$ or $p = \frac{1000}{x}$

$(x + 5)(p - 0.12) = 1000$

$xp + 5p - 0.12x - 0.6 = 1000$

Substituting:

$x\left(\frac{1000}{x}\right) + 5\left(\frac{1000}{x}\right) = 0.12x + 1000.6$

$1000x + 5000 = 0.12x^2 + 1000.6x$

$0.12x + 0.6x - 5000 = 0$

By quadratic formula:

$x = \frac{-0.6 \pm \sqrt{0.6^2 - 4(0.12)(-5000)}}{2(0.12)}$

$x = 202$ bags (closest whole bag)

49. Let $x =$ his original speed

$10 \text{ min} = \frac{1}{6} \text{ h}$

Let $t =$ his original predicted time

$xt = 3$

$t = \frac{3}{x}$

$\frac{1 \text{ mile}}{x} + \frac{1}{6} + \frac{2 \text{ miles}}{x+1} = \frac{3}{x}$

$6(x+1) + x(x+1) + 2(6x) = \frac{3}{x}[6x(x+1)]$

$6x + 6 + x^2 + x + 12x = 18x + 18$

$x^2 + x - 12 = 0$

$(x + 4)(x - 3) = 0$

$x \neq -4$

$x = 3 \text{ mi/h}$

53. $w = 26 - 2d, \quad wd = 80$

$(26 - 2d)d = 80$

$26d - 2d^2 = 80$

$-2d^2 + 26d - 80 = 0$

$d^2 - 13d + 40 = 0$

$(d - 5.0)(d - 8.0) = 0$

$d = 5.0$

$d = 8.0$

$w = \frac{80}{d} = 16$ (for $d = 5.0$)

$w = \frac{80}{d} = 10$ (for $d = 8.0$)

5.0 in. $\times$ 16 in. or

8.0 in. $\times$ $1\overline{0}$ in.

57. $2x^2 = (x + 6.0)(x + 4.0)$

$2x^2 = x^2 + 10x + 24$

$x^2 - 10x - 24 = 0$

$(x + 2)(x - 12) = 0$

$x \neq -2$

$x = 12 \text{ ft}$

dimensions $= 12 \text{ ft} \times 12 \text{ ft}$

CHAPTER 15: OBLIQUE TRIANGLES AND VECTORS

Exercise 1: Trigonometric Functions of Any Angle

1. II

5. I

9. II or III

13. III

17. pos

21. neg

25. sine negative, cosine positive, tangent negative

29. | r | 12.65 |
    |---|---|
    | sin | 0.949 |
    | cos | −0.316 |
    | tan | −3.00 |
    | cot | −0.333 |
    | sec | −3.16 |
    | csc | 1.05 |

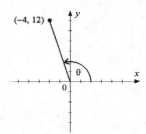

33. | r | 5.00 |
    |---|---|
    | sin | 0 |
    | cos | 1.00 |
    | tan | 0 |
    | cot | — |
    | sec | 1.00 |
    | csc | — |

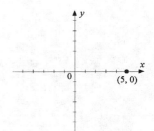

37. For pos tan quadrant I or III, not I, therefore III x and y are neg.
    $y = -2$, $x = -1$, $r = \sqrt{1^2 + 2^2} = \sqrt{5}$
    $\sin = \frac{-2}{\sqrt{5}}$, $\cos = \frac{-1}{\sqrt{5}}$, $\tan = 2$

41. 0.9816, -0.1908, -5.145

45. -0.8898, 0.4563, -1.950

49. 0.9108, -0.4128, -2.206

53. -0.1959, 0.9806, -0.1998

57. 0.8192

61. 1.711

65. tan neg II and IV $\theta = \arctan -1 = -45$
    $360 - 45 = 315°$ $180 - 45 = 135°$

69. tan pos I and III
    $\theta = \arctan 6.372 = 81.1°$
    $180 + 81.1 = 261.1°$

73. $\sin \theta = -0.6358$ sin neg III and IV
    $\theta = \arcsin -0.6358 = -39.48°$
    $180 + 39.48 = 219.5°$
    $360 - 39.48 = 320.5°$

77. $227.4°$, $312.6°$

81. $\arctan (-4.48) = -77.4° = 282.6°$ and
    $180 - 77.4 = 102.6°$

85. $\tan 180° = 0$

89. $\sin 180° = 0$

93. $\cos 180° = -1$

97. $\cos 60° = \frac{1}{2}$

101. $\cos 90° = 0$

Exercise 2: Law of Sines

1. $\dfrac{\sin A}{228} = \dfrac{\sin 46.3}{304}$ $A = 32.8°$
    $C = 180 - 46.3 - 32.8 = 100.9$ $C = 101°$
    $\dfrac{c}{\sin 100.9°} = \dfrac{304}{\sin 46.3°}$ $c = 413$

5. $B = 180 - 61.9 - 47.0$ $\qquad$ $B = 71.1°$

   $\dfrac{b}{\sin 71.1°} = \dfrac{7.65}{\sin 61.9°}$ $\qquad$ $b = 8.20$

   $\dfrac{c}{\sin 47.0°} = \dfrac{7.65}{\sin 61.9°}$ $\qquad$ $c = 6.34$

9. $A = 180 - 125 - 32.0$ $\qquad$ $A = 23.0°$

   $\dfrac{a}{\sin 23.0°} = \dfrac{58.0}{\sin 32.0°}$ $\qquad$ $a = 42.8$

   $\dfrac{b}{\sin 125°} = \dfrac{58.0}{\sin 32.0°}$ $\qquad$ $b = 89.7$

13. $C = 180 - 18.0 - 12.0$ $\qquad$ $C = 15\overline{0}°$

   $\dfrac{b}{\sin 12.0°} = \dfrac{50.7}{\sin 18.0°}$ $\qquad$ $b = 34.1$

   $\dfrac{c}{\sin 150°} = \dfrac{50.7}{\sin 18.0°}$ $\qquad$ $c = 82.0$

Exercise 3: Law of Cosines

1. $c = \sqrt{15.7^2 + 11.2^2 - 2(15.7)(11.2)\cos 106.0°}$
   $\qquad = 21.65 = 21.7$

   $\dfrac{11.2}{\sin B} = \dfrac{21.65}{\sin 106.0°}$

   $B = 29.8°$

   $A = 180 - 106.0 - 29.8 = 44.2°$

5. $c = \sqrt{128^2 + 152^2 - 2(128)(152)\cos 27.3°}$
   $\qquad = 70.1$

   $\dfrac{128}{\sin A} = \dfrac{70.1}{\sin 27.3°}$

   $A = 56.9°$

   $B = 180 - 27.3 - 56.9 = 95.8°$

9. $\cos A = \dfrac{11.3^2 - 15.6^2 - 12.8^2}{-2(15.6)(12.8)}$

   $A = 45.6°$

   $\dfrac{15.6}{\sin B} = \dfrac{11.3}{\sin 45.6°}$

   $B = 80.4°$

   $C = 180 - 45.6 - 80.4 = 54.0°$

13. $c = \sqrt{9.08^2 + 6.75^2 - 2(9.08)(6.75)\cos 67.0°}$

   $c = 8.95$

   $\dfrac{9.08}{\sin A} = \dfrac{8.95}{\sin 67.0°}$

   $A = 69.0°$

   $B = 180 - 67.0 - 69.0$

   $B = 44.0°$

17. $c = \sqrt{1445^2 + 1502^2 - 2(1445)(1502)\cos 41.77°}$
   $\qquad = 1052$

   $\dfrac{1445}{\sin A} = \dfrac{1052}{\sin 41.77°}$

   $A = 66.21°$

   $B = 180 - 41.77 - 66.21 = 72.02°$

21.
```
10 '      COS-LAW
20 '
30 '      This program uses the law of cosines
40 '      to solve an oblique triangle.
50 '
60 INPUT "Which method do you wish to use (SAS, SSS)"; A$
70 IF A$ = "SSS" THEN GOTO 240
80 INPUT "What is the length of the side opposite the acute angle (a)"; A
90 INPUT "What is the length of the other side (b)"; B
100 INPUT "What is angle C, in degrees"; A3
110 R3 = A3 * (3.1416 / 180)
120 C = (A^2 + B^2 - 2 * A * B * COS(R3))^.5
130 S = (A * SIN(R3)) / C
140 R1 = ATN(S / (1 - S^2)^.5): 'We must use a trig-identity here.
150 A1 = R1 * (180 / 3.1416)
160 R2 = 3.1416 - R1 - R3
170 A2 = R2 * (180 / 3.1416)
180 PRINT " Angle A", " Angle B", " Angle C"
190 PRINT A1, A2, A3
200 PRINT
210 PRINT " Side a", " Side b", "Side c"
220 PRINT A, B, C
230 END
240 INPUT "What is one side opposite an acute angle (a)"; A
250 INPUT "What is the other side opposite an acute angle (b)"; B
260 INPUT "What is the final side (c)"; C
270 S = (B^2 + C^2 - A^2) / (2 * B * C)
280 R1 = ATN((1 - S^2)^.5 / S): 'This is another trig-identity.
290 A1 = R1 * (180 / 3.1416)
300 S = (B * SIN(R1)) / A
310 R2 = ATN(S / (1 - S^2)^.5)
320 A2 = R2 * (180 / 3.1416)
330 R3 = 3.1416 - R1 - R2
340 A3 = R3 * (180 / 3.1416)
350 PRINT " Angle A", " Angle B", " Angle C"
360 PRINT A1, A2, A3
370 PRINT
380 PRINT " Side a", " Side b", "Side c"
390 PRINT A, B, C
400 END
```

Exercise 4: Applications

1. $B = 180 - 85.4 - 74.3 = 20.3°$

$$\frac{\sin B}{b} = \frac{\sin C}{88.6\,\text{m}}$$

$$b = \frac{(\sin 20.3°)(88.6\,\text{m})}{\sin 85.4°} = 30.8\,\text{m}$$

$$\frac{\sin A}{a} = \frac{\sin C}{88.6\,\text{m}}$$

$$a = \frac{(\sin 74.3°)(88.6\,\text{m})}{\sin 85.4°} = 85.6\,\text{m}$$

5. $A = \arcsin\frac{15.0(\sin 105°)}{22.0} = 41.2°$

$90.2 - 41.2 = 48.8 = \text{N48.8° W}$

9.

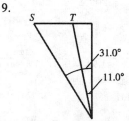

$r = (625\,\text{km/h})(20\,\text{min})(1\,\text{h}/60\,\text{min}) = 208\,\text{km}$

$R = 31.0 - 11.0 = 20.0°$

$T = 180 - (90 - 31.0) - 20.0 = 101°$

$$t = \frac{(208\,\text{km})(\sin 101°)}{\sin 20.0°} = 598\,\text{km}$$

13.

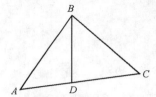

$BDC = 90 - 15.6 = 74.4°$

$C = 180 - 42.5 - 74.4 = 63.1°$

$BC = \dfrac{(71.6\,\text{m})(\sin 74.4°)}{\sin 63.1°} = 77.3\,\text{m}$

$BDA = 180 - 74.4 = 105.6°$

$A = 180 - 42.5 - 105.6 = 31.9°$

$AB = \dfrac{(71.6\,\text{m})(\sin 105.6°)}{\sin 31.9°} = 131\,\text{m}$

17.

$B = \arcsin \dfrac{(112\,\text{mm})(\sin 35.7°)}{255\,\text{mm}} = 14.85°$

$a = 180 - 35.7 - 14.85 = 129.45°$

$x = \dfrac{(255\,\text{mm})(\sin 129.45°)}{\sin 35.7°} = 337\,\text{mm}$

21. $AC = \sqrt{72.1^2 + 105^2 - 2(72.1)(105)\cos 63.0°} = 96.69\,\text{in.}$

$ACD = \arcsin \dfrac{105(\sin 63.0°)}{96.69} = 75.37°$

$ACB = 121.0 - 75.36 = 45.63°$

$AB = \sqrt{43.0^2 + 96.69^2 - 2(43.0)(96.69)\cos 45.63°} = 73.4\,\text{in.}$

25. $ADC = \arcsin \dfrac{112(\sin 62.3°)}{186} = 32.22°$

$DCA = 180 - 62.3 - 32.22 = 85.48°$

$AD = \dfrac{186(\sin 85.48°)}{\sin 62.3°} = 209.4$

$c = 2(AD) = 419$

29. $\dfrac{(8)}{\sin[180-(71.5+90-41.6+41.6)]} = \dfrac{(x)}{\sin(71.5+90-41.6)}$; $\dfrac{8}{\sin\ 18.5} = \dfrac{x}{\sin\ 119.9}$; $x = 21.9\,\text{ft}$

Exercise 5: Addition of Vectors

1. $\theta = 180 - 21.8 = 158.2°$

$R^2 = 244^2 + 287^2 - 2(244)(287)\cos 158.2°$

$R = 521$

$\dfrac{\sin \phi}{244} = \dfrac{\sin 158.2°}{521}$

$\sin \phi = \dfrac{244 \sin 158.2°}{521} = 0.1739$

$\phi = 10.0°$

5. $\theta = 180 - 100.0 = 80.0°$

$R^2 = 4483^2 + 5829^2 - 2(4483)(5829)\cos 80.0°$

$R = 6708$

$\dfrac{\sin \phi}{4483} = \dfrac{\sin 80.0°}{6708}$

$\sin \phi = \dfrac{4483 \sin 80.0°}{6708} = 0.65815$

$\phi = 41.16°$

9.

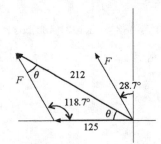

$$\frac{\sin\phi}{125} = \frac{\sin 118.7^\circ}{212}$$
$$\sin\phi = 0.51719$$
$$\phi = 31.144^\circ$$
$$\theta = 180 - 118.7 - 31.14 = 30.16^\circ$$

Direction of the resultant $= 90 - 30.16$
$$= 59.84 = \text{N } 59.8^\circ \text{ W}$$

Second force $= F$

Thus, $\dfrac{\sin 30.16^\circ}{F} = \dfrac{\sin 118.7^\circ}{212}$
$$F = 121 \text{ N}$$

13.
$$2870^2 = 1120^2 + 2210^2 - 2(1120)(2210)\cos\theta$$
$$\cos\theta = -0.42388$$
$$\theta = 115.08^\circ$$
$$\frac{\sin\phi}{1120} = \frac{\sin 115.08^\circ}{2870}$$
$$\sin\phi = 0.35345$$
$$\phi = 20.698^\circ = 20.7^\circ$$
$$180 - 115.08 - 20.698 = 44.2^\circ$$

17.
V_A = velocity relative to air
V_G = velocity relative to ground
V_W = wind velocity = 36.0 km/h
$$\frac{\sin W}{36.0} = \frac{\sin 45^\circ}{388}$$
$$W = 3.76^\circ$$
$$G = 180 - 45 - 3.76 = 131.24^\circ$$
$$V_G^2 = 36^2 + 388^2 - 2(36)(388)\cos 131.24$$
$$V_G = 413 \text{ km/h}$$
Heading $= 131.24 - 90 = \text{N } 41.2^\circ \text{ E}$

21.
$$R = \sqrt{11.3^2 + 18.4^2 - 2(11.3)(18.4)\cos 128.5^\circ}$$
$$= 26.93 \text{ A} = 26.9 \text{ A}$$
$$\frac{\sin\phi}{18.4} = \frac{\sin 128.5^\circ}{26.93}$$
$$\phi = 32.3^\circ$$

25.
```
10  '    VECTORS
20  '
30  '    This program accepts the direc-
40  '    tion and magnitude of any number
50  '    of vectors and computes the
60  '    magnitude and direction of the
70  '    resultant.
80  '
90  '    Enter the vectors in the data
100 '    statements, first magnitude,
110 '    then direction in degrees
120 '
130 READ M, T1
140 IF M = 0 THEN 210
150 LET T = T1 * (3.1416 / 180)
160 LET X = M * COS(T)
170 LET Y = M * SIN(T)
180 LET X1 = X1 + X
190 LET Y1 = Y1 + Y
200 GOTO 130
210 LET R = SQR(X1^2 +Y1^2)
220 LET T2 = ATN(Y1 / X1)
230 LET T3 = T2 * (180 / 3.1416)
240 PRINT R; " at"; T3; "degrees."
250 DATA 273, 34
260 DATA 179, 143
270 DATA 203, 225
280 DATA 138, 314
290 DATA 0, 0
300 END
```

1. $c = \sqrt{44.9^2 + 39.1^2 - 2(39.1)(44.9)\cos 135°}$
    $= 77.6$

    $A = \arcsin \dfrac{44.9(\sin 135°)}{77.6} = 24.1°$
    $B = 180 - 135 - 24.1 = 20.9°$

5. $a = \sqrt{38.2^2 + 51.8^2 - 2(38.2)(51.8)\cos 132°}$
    $= 82.4$

    $C = \arcsin \dfrac{51.8(\sin 132°)}{82.4} = 27.8°$
    $B = 180 - 132 - 27.8 = 20.2°$

9. II

13. quadrant III cos neg

17. $r^2 = (-3)^2 + (-4)^2 = 25, r = 5$
    $\sin \theta = \dfrac{-4}{5} = -0.800$
    $\cos \theta = \dfrac{-3}{5} = -0.600$
    $\tan \theta = \dfrac{-4}{-3} = 1.33$

21. $\phi = 180 - 155 = 25°$
    $R^2 = 44.9^2 + 29.4^2 - 2(44.9)(29.4)\cos 25°$
    $R = 22.08 = 22.1$
    $\dfrac{\sin \alpha}{44.9} = \dfrac{\sin 25°}{22.08}$ $\qquad \alpha = 59.2°$ or $121°$
    $\dfrac{\sin \beta}{29.4} = \dfrac{\sin 25°}{22.08}$ $\qquad \beta = 34.2$
    So we drop 59.2° value and our angle is 121°.

	sin	cos	tan
25.	0.0872	−0.9962	−0.0875

29. $\sin 35°\cos 35° = (0.5736)(0.8192) = 0.4698$

33.

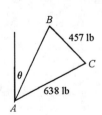

$C = (90 - 28.0) + (90 - 37.0) = 115°$
$BAC = \arcsin \dfrac{457 \text{ lb}(\sin 115°)}{638 \text{ lb}} = 40.5°$
$\theta = 37.0 - 40.5 = -3.5° = \text{S } 3.5° \text{ E}$

37. 47.5°, 132.5°

41. $R^2 = 273^2 + 483^2 - 2(123)(483)\cos 131.8°$
    $= 483,600$
    $R = 695 \text{ lb}$
    $\dfrac{\sin A}{273} = \dfrac{\sin 131.8}{695}$
    $\sin A = 0.293$
    $A = 17.0°$

CHAPTER 16: RADIAN MEASURE, ARC LENGTH, AND ROTATION

Exercise 1: Radian Measure

1. 0.834 rad

5. 9.74 rad

9. 0.497 rev

13. 162°

17. 65.3°

21. $\frac{11\pi}{30}$

25. $\frac{13\pi}{30}$

29. $\frac{9\pi}{20}$

33. 147°

37. 158°

41. 15°

45. 0.4863

49. 1.067

53. −0.8090

57. 0.5854

61. 1.337

65. 1.309

69. 0.5000

73. $62.5°\left(\frac{\pi\,\text{rad}}{180°}\right) = 1.091\,\text{rad}$
    Area $= (5.92\,\text{in.})^2\frac{1.091}{2} = 19.1\,\text{in.}^2$

77. $y = 4\cos\left[25(2.00)\right] = 4\cos 50.0$
    $= 4(0.965) = 3.86\,\text{in.}$

Exercise 2: Arc Length

1. $s = (\theta\ \text{in radians})\ r = 4.83\left(\frac{2\pi}{5}\right) = 6.07\,\text{in.}$

5. $64.8(38.5)\left(\frac{\pi}{180}\right) = 43.5\,\text{in.}$

9. $\left(\frac{15.8\,\text{ft}}{3.87\,\text{m}}\right)\left(\frac{\text{m}}{3.281\,\text{ft}}\right) = 1.24\,\text{rad}$

13. $\frac{28.2}{(\pi/180)(12+55/60)} = 125\,\text{ft}$

17. $r = \frac{s}{\theta} = \frac{180}{17.5\left(\frac{\pi}{180}\right)} = 589\,\text{m}$

21. $s = 125\,\text{days}\left(\frac{360°}{365\,\text{days}}\right)\left(\frac{\pi}{180}\right)(93\times10^6)$
    $= 2.0\times10^8\,\text{mi}$

25. $\tan\beta = \frac{(240-120)/2}{350} = 0.1714$
    $\tan\beta = \frac{120}{h} \qquad h = \frac{120}{0.1714}$
    $\qquad\qquad\qquad h = 700\,\text{mm}$
    $R^2 = 700^2 + 120^2$
    $R = 71\overline{0}\,\text{mm}$
    $k = \sqrt{350^2 + 60^2} = 355$
    $r = R - k = 710 - 355 = 355\,\text{mm}$
    $s = \text{circumference} = \pi d = \pi(240) = 754$
    $\theta = \frac{s}{R} = 754/710 = 1.062\,\text{rad} = 1.062\left(\frac{180}{\pi}\right)$
    $\qquad = 60.8°$

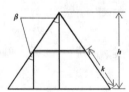

29. $s = \theta\,r = 28.3\left(\frac{\pi}{180}\right)(28.5) = 14.1\,\text{cm}$

33. Given: $A = \frac{r^2\theta}{2}$ and $\theta = \frac{s}{r}$
    Show: Area of sector $= \frac{rs}{2}$
    $A = \frac{r^2\theta}{2} = \frac{r^2}{2}\left(\frac{s}{r}\right) = \frac{rs}{2}$

Exercise 3: Uniform Circular Motion

1. $1850\frac{\text{rev}}{\text{min}}\left(\frac{2\pi\,\text{rad}}{\text{rev}}\right)\left(\frac{\text{min}}{60\,\text{s}}\right) = 194\,\text{rad/s}$
   $1850\frac{\text{rev}}{\text{min}}\left(\frac{360°}{\text{rev}}\right)\left(\frac{\text{min}}{60\,\text{s}}\right) = 11,100\,\text{deg/s}$

5. $48.1\frac{\text{deg}}{\text{s}}\left(\frac{\text{rev}}{360°}\right)\left(\frac{60\,\text{s}}{\text{min}}\right) = 8.02\,\text{rev/min}$
   $48.1\frac{\text{deg}}{\text{s}}\left(\frac{2\pi\,\text{rad}}{360°}\right) = 0.840\,\text{rad/s}$

9. $\frac{2.00\,\text{rad}}{2550\,\text{rev/min}\ (2\pi\,\text{rad/rev})\,(\text{min}/60\,\text{s})} = 0.00749\,\text{s}$

13. $\omega = \dfrac{155 \text{ deg}}{1.25 \text{ s}} = 124 \text{ deg/s}$

 $r = 0.750 \text{ m}$

 $V = \omega r = \dfrac{124 \text{ deg}}{\text{s}}(0.750 \text{ m})\left(\dfrac{\pi \text{ rad}}{180 \text{ deg}}\right)$

 $= 1.62 \text{ m/s}$

17. $r = 31.6 \text{ cm}$

 $V = 65.5 \text{ km/h}$

 $\omega = \dfrac{V}{r} = \dfrac{65.5 \text{ km/h}}{31.6 \text{ cm}}\left(\dfrac{100 \text{ cm}}{\text{m}}\right)\left(\dfrac{1000 \text{ m}}{\text{km}}\right)$

 $= 207,278 \text{ rad/h}$

 $207278 \dfrac{\text{rad}}{\text{h}}\left(\dfrac{\text{h}}{3600 \text{ s}}\right) = 57.6 \text{ rad/s}$

Chapter 16 Review Problems

1. $\dfrac{3\pi}{7}\left(\dfrac{180°}{\pi}\right) = 77.1°$

5. $165°$

9. $300°\left(\dfrac{\pi}{180°}\right) = \dfrac{30\pi}{18} = \dfrac{5\pi}{3}$

13. 0.3420

17. -0.01827

21. $V = \dfrac{2\pi \text{ rad}}{3 \cdot 24 + 7 + 35/60}(4250) = 336 \text{ mi/h}$

25. -0.8812

29. 0.9777

33. $r = \dfrac{s}{\theta} = \dfrac{384 \text{ mm}}{1.73 \text{ rad}} = 222 \text{ mm}$

CHAPTER 17: GRAPHS OF THE TRIGONOMETRIC FUNCTIONS

Exercise 1: The Sine Curve

1.   period = 2s, frequency = $\frac{1}{2}$ cycle/s,
   amplitude = 7

5.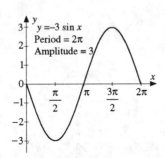

$y = -3 \sin x$
Period = 2π
Amplitude = 3

9.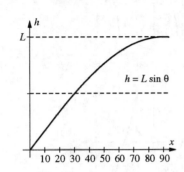

$h = L \sin \theta$

Exercise 2: The General Sine Wave

1.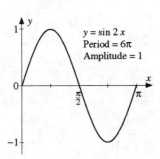

$y = \sin 2x$
Period = 6π
Amplitude = 1

5.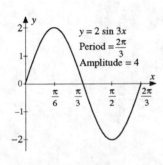

$y = 2 \sin 3x$
Period = $\frac{2\pi}{3}$
Amplitude = 4

9.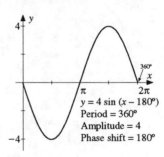

$y = 4 \sin (x - 180°)$
Period = $360°$
Amplitude = 4
Phase shift = $180°$

13.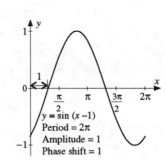

$y = \sin (x - 1)$
Period = 2π
Amplitude = 1
Phase shift = 1

17.   a.   period = $\pi = \frac{2\pi}{b}$
   $b = 2$
   phase disp. = 0
   $c = 0$
   amp = 1
   $y = \sin 2x$

    b.   period = $2\pi = \frac{2\pi}{b}$
   $b = 1$
   phase disp. = (+) shifted right
   $c = -\frac{\pi}{2}$
   amp = 1
   $y = \sin\left(x - \frac{\pi}{2}\right)$

    c. $y = 2 \sin(x - \pi)$

21.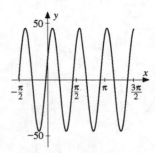

$y = 45.3 \sin(4.22x + 0.372)$

25.

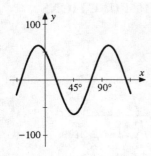

$$y = -61.3 \sin(3.27x - 58.2°)$$

29.

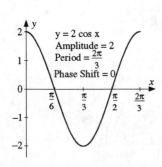

Exercise 4: Graphs of More Trigonometric Functions

1.

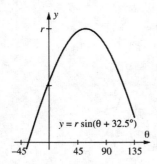

$y = 3 \cos x$
Amplitude = 3
Period = 2π
Phase shift = 0

5.

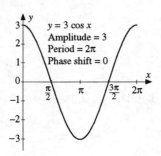

$y = 2 \cos x$
Amplitude = 2
Period = $\frac{2\pi}{3}$
Phase Shift = 0

9.

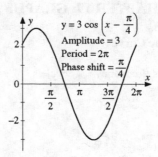

$y = 3 \cos\left(x - \frac{\pi}{4}\right)$
Amplitude = 3
Period = 2π
Phase shift = $\frac{\pi}{4}$

13.

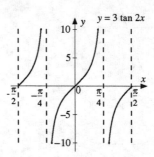

$y = 3 \tan 2x$

17.

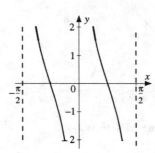

21.

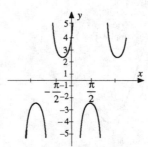

25.

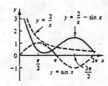

29.

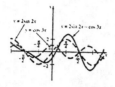

33.

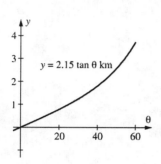

$y = 2.15 \tan \theta$ km

37.

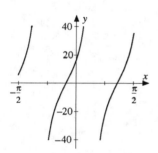

$y = 23.1 \tan(2.24x + 35.2°)$

Exercise 5: The Sine Wave as a Function of Time

1. $P = \frac{2\pi}{\omega}$

 $P = \frac{1}{f} = \frac{1}{68} = 0.0147$ s

 $\omega = 2\pi f = 2\pi(68) = 427$ rad/s

5. $f = \frac{1}{P} = 8$ Hz

 $\omega = 2\pi f = 2\pi(8) = 50.3$ rad/s

9. $\omega = 455$ rad/s

$P = \frac{2\pi}{\omega} = \frac{2\pi}{455} = 0.0138$ s

$f = \frac{1}{P} = \frac{1}{0.0138} = 72.4$ Hz

13. period = 400 ms, amplitude = 10, phase angle = 1.1 rad

17.

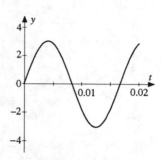

21. $y = 18.3 \sin \omega t + 26.3 \sin(\omega t + 75°)$

 $R = \sqrt{18.3^2 + 26.3^2 - 2(18.3)(26.3) \cos 105°}$
 $= 35.7$

 Then, $\frac{\sin \phi}{26.3} = \frac{\sin 105°}{35.7}$ so $\phi = 45.4°$

 Thus, $y = 35.7 \sin(\omega t + 45.4°)$

25. $y = 7.37 \cos \omega t + 5.83 \cos \omega t$

 $R = \sqrt{5.83^2 + 7.37^2} = 9.40$

 $\phi = \arctan \frac{7.37}{5.83} = 51.7°$

 Thus, $y = 9.40 \sin(\omega t + 51.7°)$

29. $\omega = 2.55$ rev/s $(2\pi$ rad/rev$) = 16.0$ rad/s

 $y = R \sin 16.0 t$

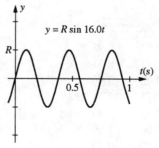

33. $\omega = 2\pi f = 2\pi(35) = 220$

 $i = 49.2 \sin(220t + 63.2°)$ mA

Exercise 6: Polar Coordinates

1-12.

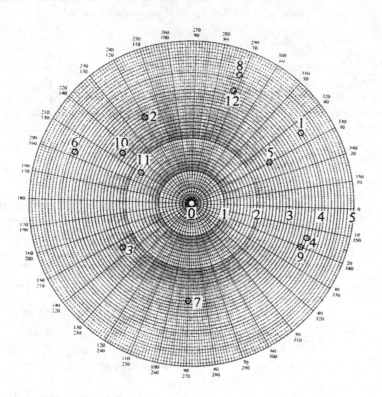

13.

Theta	r	
0	1.00,	−1.00
15	0.93,	−0.93
30	0.71,	−0.71
135	0.00,	−0.00
150	0.71,	−0.71
165	0.93,	−0.93
180	1.00,	−1.00
195	0.93,	−0.93
210	0.71,	−0.71
225	0.00,	−0.00
330	0.71,	−0.71
345	0.93,	−0.93
360	1.00,	−1.00

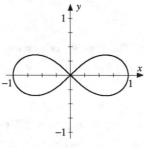

Lemniscate of Bernoulli
$$r = a\sqrt{\cos 2\theta}$$

17.	Theta	r
	0	0.00
	30	0.13
	60	0.50
	90	1.00
	120	1.50
	150	1.87
	180	2.00
	210	1.87
	240	1.50
	270	1.00
	300	0.50
	330	0.13
	360	0.00

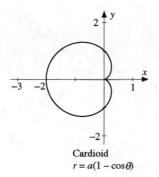

Cardioid
$r = a(1 - \cos\theta)$

21.	Theta	r
	0	––
	30	5.00
	60	4.15
	90	4.00
	120	4.15
	150	5.00
	180	––
	210	1.00
	240	1.85
	270	2.00
	300	1.85
	330	1.00
	360	––

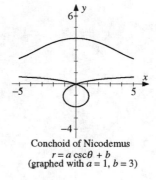

Conchoid of Nicodemus
$r = a\csc\theta + b$
(graphed with $a = 1$, $b = 3$)

25.

Theta	r
0	1.00
30	1.05
60	1.11
90	1.17
120	1.23
150	1.30
180	1.37
210	1.44
240	1.52
270	1.60
300	1.69
330	1.78
360	1.87
390	1.98
420	2.08
450	2.19
480	2.31
510	2.44
540	2.57
570	2.70
600	2.85
630	3.00
660	3.16
690	3.33
720	3.51
750	3.70
780	3.90
810	4.11
840	4.33
870	4.57
900	4.81
930	5.07
960	5.34
990	5.63
1020	5.93
1050	6.25
1080	6.59

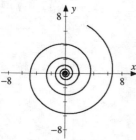

Logarithmic spiral
$r = e^{a\theta}$
(graphed with $a = 0.1$)

29. $r = 2\cos\theta - 1$

33. $r = 2 + \cos 3\theta$

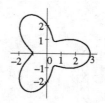

37. $r = \sqrt{4.00^2 + 3.00^2} = 5.00$
$\alpha = \arctan\frac{3.00}{4.00} = 36.9°$
$(5.00, 36.9°)$

41. $r = \sqrt{(-312)^2 + (-509)^2} = 597$
$\alpha = \arctan\frac{-509}{-312} = 58.49°$ in
$\qquad\qquad\qquad\qquad$ Quad III
$\phi = 180° + 58.49° = 238°$
$(597, 238°)$

45. $x = 445\cos 312° = 298$
$y = 445\sin 312° = -331$
$(298, -331)$

49. $x = 15.0\cos(-35.0°) = 12.3$
$y = 15.0\sin(-35.0°) = -8.60$
$(12.3, -8.60)$

53. $r = 2\sin\theta$
$r^2 = 2r\sin\theta$
$x^2 + y^2 = 2y$

57. $r^2 = 4 - r\cos\theta$
$x^2 + y^2 = 4 - x$

61. $x^2 + y^2 = 1$
$r^2 = 1$
$r = 1$

65. $r = 2.53 \sin 2\theta + 4.23$

Exercise 7: Graphing Parametric Equations

1.

Theta	x	y
0	0.00	0.00
30	0.02	0.13
60	0.18	0.50
90	0.57	1.00
120	1.23	1.50
150	2.12	1.87
180	3.14	2.00
210	4.17	1.87
240	5.05	1.50
270	5.71	1.00
300	6.10	0.50
330	6.26	0.13
360	6.28	0.00

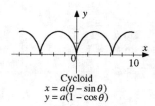

Cycloid
$x = a(\theta - \sin\theta)$
$y = a(1 - \cos\theta)$

5.

Theta	x	y
0	--	--
30	3.46	0.50
60	1.15	1.50
90	0.00	2.00
120	-1.15	1.50
150	-3.46	0.50
180	--	--
210	3.46	0.50
240	1.15	1.50
270	0.00	2.00
300	-1.15	1.50
330	-3.46	0.50
360	--	--

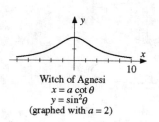

Witch of Agnesi
$x = a \cot\theta$
$y = \sin^2\theta$
(graphed with $a = 2$)

9. $x = \sin\theta$
       $y = \sin 2\theta$

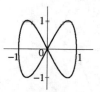

13. $r = 3\cos\theta$
       $x = r\cos\theta = 3\cos\theta\cos\theta = 3\cos^2\theta$
       $y = r\sin\theta = 3\cos\theta\sin\theta$

17. $r = 3\sin 3\theta - 2$
$x = r\cos \theta = (3\sin 3\theta - 2)\cos \theta$
$y = r\sin \theta = (3\sin 3\theta - 2)\sin \theta$

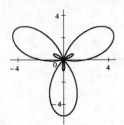

21. $x = 5.83 \sin 2\theta$
$y = 4.24 \sin \theta$

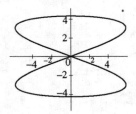

9.

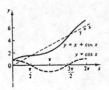

13.

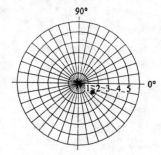

Chapter 17 Review Problems

1.

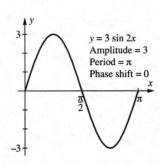

$y = 3\sin 2x$
Amplitude = 3
Period = π
Phase shift = 0

5.

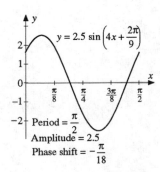

$y = 2.5\sin\left(4x + \dfrac{2\pi}{9}\right)$

Period = $\dfrac{\pi}{2}$
Amplitude = 2.5
Phase shift = $-\dfrac{\pi}{18}$

17. $r = \sqrt{x^2 + y^2} = \sqrt{(-5.30)^2 + 3.80^2} = 6.52$
$\theta = \arctan \dfrac{y}{x} = \arctan \dfrac{3.80}{-5.30}$
$= -36 + 180 = 144°$

21. $x = r\cos \theta = 3.80 \cos - 44.0° = 2.73$
$y = r\sin \theta = 3.80 \cos - 44.0° = -2.64$

25. $5x + 2y = 1$
$5r\cos \theta + 2r\sin \theta = 1$
$r(5\cos \theta + 2\sin \theta) = 1$
$r = \dfrac{1}{5\cos \theta + 2\sin \theta}$

29. 100 cycles $\dfrac{s}{30 \text{ cycles}} = 3.33$ s

33. a. $R = \sqrt{63.7^2 + 42.9^2 - 2(63.7)(42.9)\cos 142°}$
$= 101$
Then, $\dfrac{\sin \phi}{42.9} = \dfrac{\sin 142°}{101}$
so $\phi = 15.2°$ in quad IV
or $-15.2°$. Thus, $y = 101\sin(\omega t - 15.2°)$.

b. $R = \sqrt{1.73^2 + 2.64^2} = 3.16$
$\phi = \arctan \dfrac{2.64}{1.73} = 56.8°$
$y = 3.16\sin(\omega t + 56.8°)$

37.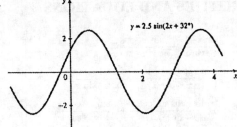

$y = 2.5 \sin(2x + 32°)$

CHAPTER 18: TRIGONOMETRIC IDENTITIES AND EQUATIONS

Exercises 1: Fundamental Identities

1. $\tan x - \sec x = \dfrac{\sin x}{\cos x} - \dfrac{1}{\cos x} = \dfrac{\sin x - 1}{\cos x}$

5. $\dfrac{\tan \theta}{\csc \theta} + \dfrac{\sin \theta}{\tan \theta} = \dfrac{\sin \theta / \cos \theta}{1/\sin \theta} + \dfrac{\sin \theta}{\sin \theta / \cos \theta}$

$= \dfrac{\sin^2 \theta}{\cos \theta} + \cos \theta = \dfrac{\sin^2 \theta + \cos^2 \theta}{\cos \theta} = \dfrac{1}{\cos \theta}$

9. $\dfrac{\cos \theta}{\cot \theta} = \dfrac{\cos \theta}{\cos \theta / \sin \theta} = \sin \theta$

13. $\sec x \sin x = \dfrac{1}{\cos x} \cdot \sin x = \tan x$

17. $\cot \theta \tan^2 \theta \cos \theta = \dfrac{\cos \theta}{\sin \theta} \left(\dfrac{\sin \theta}{\cos \theta} \right)^2 \cos \theta = \sin \theta$

21. $\dfrac{1}{\sec^2 x} + \dfrac{1}{\csc^2 x} = \cos^2 x + \sin^2 x = 1$

25. $\dfrac{\sec x - \csc x}{1 - \cot x} = \dfrac{1/\cos x - 1/\sin x}{1 - \cos x / \sin x}$

$= \dfrac{(\sin x - \cos x)/\cos x \sin x}{(\sin x - \cos x)/\sin x} = \dfrac{1}{\cos x} = \sec x$

29. $\cos \theta \sec \theta - \dfrac{\sec \theta}{\cos \theta} = \cos \theta \cdot \dfrac{1}{\cos \theta} - \dfrac{1}{\cos \theta \cos \theta}$

$= 1 - \sec^2 \theta = -\tan^2 \theta$

33. $\dfrac{\sin x}{\csc x} + \dfrac{\cos x}{\sec x} = 1$

$\dfrac{\sin x}{1/\sin x} + \dfrac{\cos x}{1/\cos x} = 1$

$\sin^2 x + \cos^2 x = 1$

$1 = 1$

37. $\dfrac{\csc \theta}{\sec \theta} = \cot \theta$

$\dfrac{1/\sin \theta}{1/\cos \theta} = \cot \theta$

$\dfrac{\cos \theta}{\sin \theta} = \cot \theta$

$\cot \theta = \cot \theta$

41. $\cot^2 x - \cos^2 x = \cos^2 x \cot^2 x$

$\left(\dfrac{\cos x}{\sin x} \right)^2 - \cos^2 x = \cos^2 x \cot^2 x$

$\dfrac{\cos^2 x - \sin^2 x \cos^2 x}{\sin^2 x} = \cos^2 x \cot^2 x$

$\dfrac{\cos^2 x}{\sin^2 x} (1 - \sin^2 x) = \cos^2 x \cot^2 x$

$\cot^2 x (\cos^2 x) = \cos^2 x (\cot^2 x)$

45. $\dfrac{\tan x + 1}{1 - \tan x} = \dfrac{\sin x + \cos x}{\cos x - \sin x}$

$\dfrac{(\sin x / \cos x) + 1}{1 - (\sin x / \cos x)} = \dfrac{\sin x + \cos x}{\cos x - \sin x}$

$\dfrac{(\sin x + \cos x)/\cos x}{(\cos x - \sin x)/\cos x} = \dfrac{\sin x + \cos x}{\cos x - \sin x}$

$\dfrac{\sin x + \cos x}{\cos x - \sin x} = \dfrac{\sin x + \cos x}{\cos x - \sin x}$

49. $(\sec \theta - \tan \theta)(\tan \theta + \sec \theta) = 1$

$\sec^2 \theta - \tan^2 \theta = 1$

$(1 + \tan^2 \theta) - \tan^2 \theta = 1$

$1 = 1$

Exercise 2: Sum or Difference of Two Angles

1. $\sin(\theta + 30°) = \sin \theta \cos 30° + \cos \theta \sin 30°$

$= \dfrac{\sqrt{3}}{2} \sin \theta + \dfrac{1}{2} \cos \theta$

$= \dfrac{1}{2} \left(\sqrt{3} \sin \theta + \cos \theta \right)$

5. $\cos\left(x + \dfrac{\pi}{2} \right) = \cos x \cos \dfrac{\pi}{2} - \sin x \sin \dfrac{\pi}{2} = -\sin x$

$\left(\text{since } \cos \dfrac{\pi}{2} = 0 \text{ and } \sin \dfrac{\pi}{2} = 1 \right)$

9. $\tan(2\theta - 3\alpha) = \dfrac{\tan 2\theta - \tan 3\alpha}{1 + \tan 2\theta \tan 3\alpha}$

13. $\sin\left(\dfrac{\pi}{3} - x \right) - \cos\left(\dfrac{\pi}{6} - x \right)$

$= \sin \dfrac{\pi}{3} \cos x - \cos \dfrac{\pi}{3} \sin x$

$\qquad - \left(\cos \dfrac{\pi}{6} \cos x + \sin \dfrac{\pi}{6} \sin x \right)$

$= \dfrac{\sqrt{3}}{2} \cos x - \dfrac{\sin x}{2} - \dfrac{\sqrt{3}}{2} \cos x - \dfrac{\sin x}{2}$

$= -\sin x$

17. $\sin(\theta + 60°) + \cos(\theta + 30°) = \sqrt{3} \cos \theta$

$\sin \theta \cos 60° + \cos \theta \sin 60° + \cos \theta \cos 30°$

$\qquad - \sin \theta \sin 30°$

$= \sqrt{3} \cos \theta$

$\dfrac{\sin \theta}{2} + \dfrac{\sqrt{3}}{2} \cos \theta + \dfrac{\sqrt{3}}{2} \cos \theta - \dfrac{\sin \theta}{2} = \sqrt{3} \cos \theta$

$\dfrac{2\sqrt{3}}{2} \cos \theta = \sqrt{3} \cos \theta$

$\sqrt{3} \cos \theta = \sqrt{3} \cos \theta$

21. $\cos x = \sin\left(x + \frac{\pi}{6}\right) - \sin\left(x - \frac{\pi}{6}\right)$

$\cos x = \sin x \cos\frac{\pi}{6} + \cos x \sin\frac{\pi}{6}$

$\qquad - \left(\sin x \cos\frac{\pi}{6} - \cos x \sin\frac{\pi}{6}\right)$

$\cos x = \dfrac{\cos x}{2} + \dfrac{\cos x}{2}$

$\cos x = \cos x$

25. $\dfrac{\sin(x-y)}{\sin(x+y)} = \dfrac{\tan x - \tan y}{\tan x + \tan y}$

$\dfrac{\sin x \cos y - \cos x \sin y}{\sin x \cos y + \cos x \sin y} = \dfrac{\tan x - \tan y}{\tan x + \tan y}$

$\dfrac{(\sin x \cos y)/(\cos x \cos y) - (\cos x \sin y)/(\cos x \cos y)}{(\sin x \cos y)/(\cos x \cos y) + (\cos x \sin y)/(\cos x \cos y)}$

$\qquad = \dfrac{\tan x - \tan y}{\tan x + \tan y}$

$\dfrac{\tan x - \tan y}{\tan x + \tan y} = \dfrac{\tan x - \tan y}{\tan x + \tan y}$

29. $\tan(\alpha + 45°) = \dfrac{\cos\alpha + \sin\alpha}{\cos\alpha - \sin\alpha}$

$\dfrac{\tan\alpha + \tan 45°}{1 - \tan\alpha\tan 45°} = \dfrac{\cos\alpha + \sin\alpha}{\cos\alpha - \sin\alpha}$

$\dfrac{\tan\alpha + 1}{1 - \tan\alpha} = \dfrac{\cos\alpha + \sin\alpha}{\cos\alpha - \sin\alpha}$

$\dfrac{(\sin\alpha/\cos\alpha) + 1}{1 - (\sin\alpha/\cos\alpha)} = \dfrac{\cos\alpha + \sin\alpha}{\cos\alpha - \sin\alpha}$

$\dfrac{(\sin\alpha + \cos\alpha)/\cos\alpha}{(\cos\alpha - \sin\alpha)/\cos\alpha} = \dfrac{\cos\alpha + \sin\alpha}{\cos\alpha - \sin\alpha}$

$\dfrac{\cos\alpha + \sin\alpha}{\cos\alpha - \sin\alpha} = \dfrac{\cos\alpha + \sin\alpha}{\cos\alpha - \sin\alpha}$

33. $R = \sqrt{8470^2 + 7360^2} = 11{,}200$

$\phi = \arctan\dfrac{8470}{7360} = 41.0°$

Thus $y = 11{,}200\sin(\omega t + 41.0°)$

Exercise 3: Functions of Double Angles

1. $2\sin^2 x + \cos 2x = 2\sin^2 x + (\cos^2 x - \sin^2 x)$

$\qquad = \cos^2 x + \sin^2 x = 1$

5. $\dfrac{\sin 6x}{\sin 2x} - \dfrac{\cos 6x}{\cos 2x} = \dfrac{\cos 2x \sin 6x - \sin 2x \cos 6x}{\sin 2x \cos 2x}$

$\qquad = \dfrac{\sin(6x - 2x)}{\sin 2x \cos 2x} = \dfrac{\sin 4x}{\sin 2x \cos 2x}$

$\qquad = \dfrac{2\sin 2x \cos 2x}{\sin 2x \cos 2x} = 2$

9. $2\cot 2x = \cot x - \tan x$

$2\cot 2x = \dfrac{\cos x}{\sin x} - \dfrac{\sin x}{\cos x}$

$2\cot 2x = \dfrac{\cos^2 x - \sin^2 x}{\sin x \cos x}$

$2\cot 2x = \dfrac{\cos 2x}{(\sin 2x)/2}$

$2\cot 2x = 2\cot 2x$

13. $1 + \cot x = \dfrac{2\cos 2x}{\sin 2x - 2\sin^2 x}$

$1 + \cot x = \dfrac{2(\cos^2 x - \sin^2 x)}{2\sin x \cos x - 2\sin^2 x}$

$1 + \cot x = \dfrac{(\cos x - \sin x)(\cos x + \sin x)}{\sin x(\cos x - \sin x)}$

$1 + \cot x = \dfrac{\cos x}{\sin x} + \dfrac{\sin x}{\sin x}$

$1 + \cot x = \cot x + 1$

17. $\sin 2A = \sin 2B$

$2\sin A \cos A = 2\sin B \cos B$

$\cos A = \sin B$

$\cos B = \sin A$

$2\sin A \sin B = 2\sin B \sin A$

Exercise 4: Functions of Half-Angles

1. $2\sin^2\frac{\theta}{2} + \cos\theta = 1$

$2\left(\dfrac{1 - \cos\theta}{2}\right) + \cos\theta = 1$

$1 - \cos\theta + \cos\theta = 1$

$1 = 1$

5. $\dfrac{\sin\alpha}{\cos\alpha + 1} = \sin\frac{\alpha}{2}\sec\frac{\alpha}{2} = \sin\frac{\alpha}{2} \cdot \dfrac{1}{\cos(\alpha/2)}$

$\qquad = \tan\frac{\alpha}{2} = \dfrac{1 - \cos\alpha}{\sin\alpha} \cdot \dfrac{1 + \cos\alpha}{1 + \cos\alpha}$

$\qquad = \dfrac{1 - \cos^2\alpha}{\sin\alpha(1 + \cos\alpha)} = \dfrac{\sin^2\alpha}{\sin\alpha(1 + \cos\alpha)}$

$\qquad = \dfrac{\sin\alpha}{\cos\alpha + 1}$

9. $2\sin\alpha + \sin 2\alpha = 4\sin\alpha\cos^2\frac{\alpha}{2}$

$2\sin\alpha + \sin 2\alpha = \dfrac{4\sin\alpha(1 + \cos\alpha)}{2}$

$2\sin\alpha + \sin 2\alpha = 2\sin\alpha + 2\sin\alpha\cos\alpha$

$2\sin\alpha + \sin 2\alpha = 2\sin\alpha + \sin 2\alpha$

13. $\sin\frac{A}{2} = \sqrt{\dfrac{c - b}{2c}}$

$\sqrt{\dfrac{1 - \cos A}{2}} = \sqrt{\dfrac{c - b}{2c}}$

$\sqrt{\dfrac{1 - b/c}{2}} = \sqrt{\dfrac{c - b}{2c}}$

$\sqrt{\dfrac{(c - b)/c}{2}} = \sqrt{\dfrac{c - b}{2c}}$

$\sqrt{\dfrac{c - b}{2c}} = \sqrt{\dfrac{c - b}{2c}}$

Exercise 5: Trigonometric Equations

1. $\sin\theta = \frac{1}{2}$

$\theta = \arcsin\frac{1}{2}$

$\theta = 30°,\ 150°$

5. $4\sin^2\theta = 3$

$\sin^2\theta = \frac{3}{4}$

$\theta = \arcsin\sqrt{\frac{3}{4}}$

$\theta = \arcsin \pm\frac{\sqrt{3}}{2}$

$\theta = 60°,\ 120°,\ 240°,\ 300°$

9. $2\cos^2\theta = 1 + 2\sin^2\theta$

$2(\cos^2\theta - \sin^2\theta) = 1$

$\cos 2\theta = \frac{1}{2}$

$2\theta = \arccos\frac{1}{2}$

$2\theta = 60°,\ 300°,\ 420°,\ 660°$

$\theta = 30°,\ 150°,\ 210°,\ 330°$

13. $1 + \tan\theta = \sec^2\theta$

$1 + \tan\theta = 1 + \tan^2\theta$

$\tan\theta = \tan^2\theta$

$\tan\theta = 0,\ 1$

$\theta = 0°,\ 45°,\ 180°,\ 225°$

17. $3\sin\frac{\theta}{2} - 1 = 2\sin^2\frac{\theta}{2}$

$2\sin^2\frac{\theta}{2} - 3\sin\frac{\theta}{2} + 1 = 0$

$\left(2\sin\frac{\theta}{2} - 1\right)\left(\sin\frac{\theta}{2} - 1\right) = 0$

$\sin\frac{\theta}{2} = \frac{1}{2},\ 1$

$\theta = 60°,\ 180°,\ 300°$

21. $3\tan\theta = 4\sin^2\theta\tan\theta$

$3\tan\theta - 4\sin^2\theta\tan\theta = 0$

$\tan\theta(3 - 4\sin^2\theta) = 0$

$\tan\theta = 0 \qquad \theta = 0°,\ 180°$

or $4\sin^2\theta = 3$

$\sin\theta = \pm\frac{\sqrt{3}}{2}$

$\theta = 60°,\ 120°,\ 240°,\ 300°$

Thus $\theta = 0°,\ 60°,\ 120°,\ 180°,\ 240°,\ 300°$

25. $\sin\theta = 2\sin\theta\cos\theta$

$\sin\theta - 2\sin\theta\cos\theta = 0$

$\sin\theta(1 - 2\cos\theta) = 0$

$\sin\theta = 0$

$\theta = 0°,\ 180°$

$\cos\theta = \frac{1}{2}$

$\theta = 60°,\ 300°$

Thus $\theta = 0°,\ 60°,\ 180°,\ 300°$

Exercise 6: Inverse Trigonometric Functions

1.   22.0°

5.   81.8°

9.   33.2°

13.

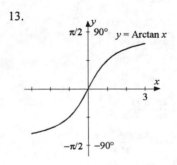

Chapter 18 Review Problems

1. $\sec^2\theta + \tan^2\theta = \sec^4\theta - \tan^4\theta$

$\sec^2\theta + \tan^2\theta = (\sec^2\theta + \tan^2\theta)(\sec^2\theta - \tan^2\theta)$

$\sec^2\theta + \tan^2\theta = (\sec^2\theta + \tan^2\theta)(1)$

$\sec^2\theta + \tan^2\theta = \sec^2\theta + \tan^2\theta$

5. $\sin(45° + \theta) - \sin(45° - \theta) = \sqrt{2}\sin\theta$

$\sin 45°\cos\theta + \cos 45°\sin\theta - (\sin 45°\cos\theta$

$\qquad\qquad\qquad - \cos 45°\sin\theta)$

$\qquad = \sqrt{2}\sin\theta$

$2\cos 45°\sin\theta = \sqrt{2}\sin\theta$

$2(0.707)\sin\theta = \sqrt{2}\sin\theta$

$1.41\sin\theta = 1.41\sin\theta$

9. $\sin\theta\cot\frac{\theta}{2} = \cos\theta + 1$

$\sin\theta\left(\frac{1}{\tan(\theta/2)}\right) = \cos\theta + 1$

$\sin\theta\left(\frac{1+\cos\theta}{\sin\theta}\right) = \cos\theta + 1$

$1 + \cos\theta = \cos\theta + 1$

13. $\dfrac{\sin\theta\sec\theta}{\tan\theta} = \dfrac{\sin\theta/\cos\theta}{\sin\theta/\cos\theta} = 1$

17. $\cos\theta\tan\theta\csc\theta = \cos\theta \cdot \dfrac{\sin\theta}{\cos\theta} \cdot \dfrac{1}{\sin\theta} = 1$

21. $(1-\sin^2\theta)\sec^2\theta = \cos^2\theta\,\dfrac{1}{\cos^2\theta} = 1$

25. $2\cos^2\theta - 5\cos\theta - 3 = 0$
Let $x = \cos\theta$
$2x^2 - 5x - 3 = 0$
$2x^2 - 6x + x - 3 = 0$
$2x(x-3) + (x-3) = 0$
$(2x+1)(x-3) = 0$
$x = \cos\theta = -\dfrac{1}{2}$
$x = \cos\theta = 3$
$\theta = 120°,\ 240°$

29. $\cot 2\theta - 2\cos 2\theta = 0$
$\dfrac{\cos 2\theta}{\sin 2\theta} - 2\cos 2\theta = 0$

$\cos 2\theta\left(\dfrac{1}{\sin 2\theta} - 2\right) = 0$
$\cos 2\theta = 0$
$2\theta = 90°,\ 270°,\ 450°,\ 630°$
$\theta = 45°,\ 135°,\ 225°,\ 315°$
$\dfrac{1}{\sin 2\theta} = 2$
$\sin 2\theta = \dfrac{1}{2}$
$2\theta = 30°,\ 150°,\ 390°,\ 510°$
$\theta = 15°,\ 75°,\ 195°,\ 255°$
Thus $\theta = 15°,\ 45°,\ 75°,\ 135°,\ 195°,\ 225°,$
$255°,\ 315°$

33. $\sin^2\theta = 1 + 6\sin\theta$
$\sin^2\theta - 6\sin\theta - 1 = 0$
$\sin\theta = \dfrac{6\pm\sqrt{36-4(-1)}}{2}$
$\sin\theta = \dfrac{6\pm\sqrt{40}}{2}$
$\sin\theta = -0.16$
$\theta = -9.3°$
$180 + 9.3 = 189.3°$
$360 - 9.3 = 350.7°$

37. Arcsin $0.825 = 55.6°$

41. $y = 47.6\sin\omega t + 62.1\cos\omega t$
$R = \sqrt{47.6^2 + 62.1^2} = 78.2$
$\phi = \arctan\dfrac{62.1}{47.6} = 52.5°$
$y = 78.2\sin(\omega t + 52.5°)$

CHAPTER 19: RATIO, PROPORTION, AND VARIATION

Exercise 1: Ratio and Proportion

1. $3{:}x = 4{:}6$
$4x = 18$
$x = \frac{9}{2}$

5. $x{:}(14 - x) = 4{:}3$
$3x = 56 - 4x$
$7x = 56$
$x = 8$

9. $\frac{x}{3} = \frac{y}{9}$
$9x = 3y$
$y = \frac{9x}{3}$
$y = 3x$

13. $\frac{x+2}{5x} = \frac{y}{5}$
$y = \frac{5(x+2)}{5x}$
$y = \frac{x+2}{x}$

17. $\frac{5}{y} = \frac{y}{45}$
$y = \sqrt{225}$
$y = \pm 15$

21. $4x = 3(56 - x)$
$4x = 168 - 3x$
$7x = 168$
$x = 24$
$56 - x = 32$

25. Let x = smaller number
$20 - x$ = greater number
$\frac{(20-x)+x}{20-x-x} = \frac{10}{1}$
$20 = 200 - 20x$
$20x = 180$
$x = 9$
$20 - x = 11$

Exercise 2: Direct Variation

1. $\frac{y}{x} = \frac{56}{21}$
$\frac{y}{74} = \frac{56}{21}$
$y = 197$

5. $y = kx$
$45 = k(9)$
$k = 5$
$y = 5(11) = 55$
$75 = 5x$
$x = 15$

9. $\frac{828 \text{ km}}{29.5 \text{ cm}} = \frac{x}{15.6 \text{ cm}}$
$x = 438 \text{ km}$

13. $\left(\frac{3.01}{53 \text{ hp}}\right) = \frac{3.81}{x}$
$x = 67 \text{ hp}$

17. $\frac{1530 \text{ cm}^3}{302 \, K} = \frac{v}{358 \text{K}}$
$v = 18\overline{0}0 \text{ cm}^3$

Exercise 3: The Power Function

1. $y = kx^2$
$726 = k(163)^2$
$k = \frac{726}{(163)^2} = 0.02733$
$y = 0.02733(274)^2$
$y = 2050$

5. $y = kx^2$
$\frac{285.0}{y} = \frac{(112.0)^2}{(351.0)^2}$
$y = 2799$

9. $y = kx^4$
$29.7 = k(18.2)^4$
$k = \frac{29.7}{18.2^4} = 2.706 \times 10^{-4}$
$y = k(75.6)^4 = 8840$
$154 = kx^4$
$x^4 = \frac{154}{k}$
$x = 27.5$

13.

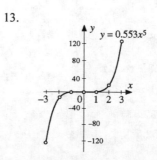

17. $129 = \dfrac{a(2.00)^2}{2}$

    $a = 64.5$

    $525 = \dfrac{64.5(t)^2}{2}$

    $t = 4.03$ s

21. $P = kI^2$

    $\dfrac{P_2}{P_1} = \dfrac{3P_1}{P_1} = \left(\dfrac{I_2}{I_1}\right)^2$

    $\dfrac{I_2}{I_1} = \sqrt{3} = 1.73$

25. $24 = k(22)^2$

    $8 = k(f)^2$

    $\dfrac{8}{24} = \left(\dfrac{f}{22}\right)^2$

    $\dfrac{f}{22} = \sqrt{\dfrac{8}{24}} = 0.577$

    $f = 12.7 \approx$ f13

29. $V = 155(1.15)^3 = 236$ metric tons

33. $\dfrac{500 \text{ gal}}{1.50^2\pi} = \dfrac{750 \text{ gal}}{(\pi d^2)/4}$

    $d = \sqrt{\dfrac{4(750)(\pi)(1.50^2)}{500\pi}} = 3.67$ in.

Exercise 4: Inverse Variation

1. $y = \dfrac{k}{x}$

    $385 = \dfrac{k}{832}$

    $y = \dfrac{k}{226}$

    $\dfrac{y}{385} = \dfrac{832}{226}$

    $y = 1420$

5. $125 = \dfrac{k}{306}$

    $k = 38,250$

    $y = \dfrac{k}{622} = 61.5$

    $418 = \dfrac{k}{x}$

    $x = 91.5$

9. $y = \dfrac{k}{\sqrt{x}}, \quad x_2 = 0.500x_1$

    $\dfrac{y_2}{y_1} = \sqrt{\dfrac{x_1}{0.500x_1}} = \sqrt{2} = 1.414$

    41.4% increase

13. $PV = C$

    $14.7(175) = P(25.0)$

    $P = 7.00(14.7) = 103$ lb/in.^{2}

17. $150(3960)^2 = (3960 + 1500)^2 W$

    $W = 79$ lb

21. $(7.50)^2(426) = x^2(850)$

    $x^2 = 28.2$

    $x = 5.31$ m

25. $CX_c = $ constant

    $CX_c = 0.75 CX_c'$

    $\dfrac{X_c'}{X_c} = \dfrac{1}{0.75} = 1.333$

    $33\frac{1}{3}\%$ increase

Exercise 5: Functions of More Than One Variable

1. $y = kwx$

    $\dfrac{y}{wx} = k$

    $\dfrac{483}{742(383)} = \dfrac{y}{274(756)}$

    $y = 352$

5. $\dfrac{127}{46.2(18.3)} = \dfrac{y}{19.5(41.5)}$

    $y = 121$

    $\dfrac{127}{46.2(18.3)} = \dfrac{155}{8.86w}$

    $w = 116$

    $\dfrac{127}{46.2(18.3)} = \dfrac{79.8}{12.2x}$

    $x = 43.5$

9. $y = k\dfrac{x^{3/2}}{w}$

    $k = \dfrac{284(361)}{(858)^{3/2}} = 4.08$

    $y = 4.08\dfrac{x^{3/2}}{w}$

13. $r = k\dfrac{l}{A} = k\dfrac{l}{\pi(d/2)^2} = k\dfrac{l}{\pi d^2/4}$

    $r' = k\dfrac{3l}{\frac{\pi(3d)^2}{4}} = k\left(\dfrac{3}{9}\right)\left(\dfrac{l}{\frac{\pi d^2}{4}}\right)$

    $r' = \frac{1}{3}r$

    Thus, the resistance will decrease by two-thirds.

17. $\dfrac{75.0}{8.00^2} = \dfrac{x}{12.0^2}$

    $x = 169$ W

21. $\dfrac{5123.73}{3.0(5)} = \$341.58/\text{week}$

    $\dfrac{6148.48}{6} = \$1024.75$

    $\dfrac{\$1024.75}{\$341.58/\text{week}} = 3.0$ weeks

25. $\omega = k\dfrac{\sqrt{T}}{L}$

$k = \dfrac{\omega L}{\sqrt{T}} = \dfrac{325(1.00)}{\sqrt{115}} = \dfrac{\omega(0.750)}{\sqrt{95.0}}$

$\omega = \sqrt{\dfrac{95.0}{115}} \cdot \dfrac{325}{0.750} = 394$ times/s

Chapter 19 Review Problems

1. $736(822) = 583y$
   $y = 1040$

5. $Q = k\sqrt{h}$
   $k = \dfrac{Q}{\sqrt{h}} = \dfrac{225}{\sqrt{3.46}} = \dfrac{Q}{\sqrt{1.00}}$
   $Q = 121$ liters/min

9. $t = kd^{3/2}$
   Earth: $1 = kd^{3/2}$ or $k = \dfrac{1}{d^{3/2}}$

   Saturn: $t = k\left(9\frac{1}{2}d\right)^{3/2}$ or $k = \dfrac{t}{\left(9\frac{1}{2}d\right)^{3/2}}$

   $\dfrac{1}{d^{3/2}} = \dfrac{t}{9\frac{1}{2}^{3/2}d^{3/2}}$

   $t = 9\frac{1}{2}^{3/2} = 29$ yr

13. $F = Av^2$
    $A(35)^2 = A'(12)^2$
    $\dfrac{A'}{A} = \left(\dfrac{35}{12}\right)^2 = 8.5$

17. $\dfrac{82.0mM}{x^2} = \dfrac{mM}{(2.39-x)^2}$
    $82.0(5.71 - 4.78x + x^2) = x^2$
    $468 - 392x + 82.0x^2 = x^2$
    $81x^2 - 392x + 468 = 0$
    $x^2 - 4.84x + 5.78 = 0$
    $x = \dfrac{4.84 \pm \sqrt{23.4 - 23.1}}{2} = \dfrac{4.84 \pm \sqrt{0.3}}{2}$
    $= 2.42 \pm 0.27 = 2.15 = 215,000$ mi

21. $I = k(d)^{3/2}$
    $\dfrac{I_2}{I_1} = \left(\dfrac{d_2}{d_1}\right)^{3/2} = 2^{3/2} = 2.8$

25. $FC = kv^3$
    $1584 = k(15.0)^3$
    $F = k(10.0)^3$
    $\dfrac{F}{1584} = \left(\dfrac{10.0}{15.0}\right)^3$
    $F = 469$ gal

29. $\dfrac{t_1}{t_2} = \dfrac{d_2^2}{d_1^2}$

    $\dfrac{5.0}{9.0} = \dfrac{x^2}{1.5^2}$
    $x = 1.12 = 1.1$ in.

33. $(1.30^3)(5.80) = 12.7$ tons

37. Let $y =$ salary when index is 12.7
    $\dfrac{9.40}{450} = \dfrac{12.7}{y}$
    $9.40y = (12.7)(450)$
    $y = \dfrac{(12.7)(450)}{9.40}$
    $y = \$608$

41. $W_1 = 1.010W$, $L_1 = 1.010L$
    $\dfrac{A}{A_1} = \dfrac{LW}{L_1W_1}$
    $\dfrac{7463}{A_1} = \dfrac{LW}{(1.010L)(1.010W)}$
    $A_1 = 7613$ cm^2

CHAPTER 20: EXPONENTIAL AND LOGARITHMIC FUNCTIONS

Exercise 1: The Exponential Function

1.

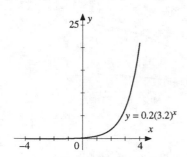

$y = 0.2(3.2)^x$

5.

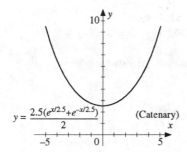

$y = \dfrac{2.5(e^{x/2.5}+e^{-x/2.5})}{2}$ (Catenary)

9. a. $y = a(1+n)^t = 1(1+0.10)^{20} = \$6.73$

   b. $y = 1\left(1+\dfrac{0.10}{12}\right)^{20 \times 12} = \$7.33$

   c. $y = 1\left(1+\dfrac{0.10}{365}\right)^{20 \times 365} = \$7.39$

Exercise 2: Exponential Growth and Decay

1. $y = ae^{nt} = 200e^{0.050(7)} = 284$ units

5. $y = ae^{nt} = 12.0e^{(0.0830)5} = 18.2$ million barrels

9. $i = \dfrac{E}{R}(1 - e^{-Rt/L})$ amperes

   $\dfrac{Rt}{L} = \dfrac{6.25(0.0750)}{186} = 0.00252$

   $i = \dfrac{250}{6.25}(1 - e^{-0.00252}) = 0.101$ amperes

13. $p = 29.92e^{-h/5}$ $h = \dfrac{300\overline{0}0 \text{ ft}}{5280 \text{ ft/mi}} = 5.682$ mi

    $p = 29.92e^{-5.682/5} = 9.604$ in. Hg

17. $\dfrac{y}{a} = e^{-nt} = e^{-0.24(5)} = 0.301 = 3\overline{0}\%$

21.

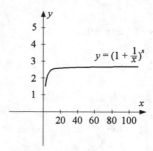

$y = \left(1+\dfrac{1}{x}\right)^x$

As x increases, y will approach 2.718, the value of e.

Exercise 3: Logarithms

1. $\log_3 81 = 4$

5. $\log_x 995 = 5$

9. $\log_5 125 = 3$ $\qquad 5^3 = 125$

13. $\log_5 x = y$ $\qquad 5^y = x$

17. $\log 27.6 = 1.4409$

21. $\log 48.3 = 1.6839$

25. $\log 27.4 = 1.4378$

29. $\log 34970 = 4.5437$

33. $\log x = 5.273$
    $x = 10^{5.273} = 187,000$

37. $\log x = -2.227$
    $x = 10^{-2.227} = 0.00593 = 5.93 \times 10^{-3}$

41. $\log x = -3.972$
    $x = 10^{-3.972} = 0.000107 = 1.07 \times 10^{-4}$

45. $\ln 2365 = 7.7685$

49. $\ln 0.00836 = -4.7843$

53. $\ln 3.84 \times 10^4 = 10.5558$

57. $\ln x = 0.879$
    $x = e^{0.879} = 2.408$

61. $\ln x = 0.936$
    $x = e^{0.936} = 2.550$

65. $\ln x = -18.36$
$x = e^{-18.36} = 1.063 \times 10^{-8}$

Exercise 4: Properties of Logarithms

1. $\log \frac{2}{3} = \log 2 - \log 3$

5. $\log xyz = \log x + \log y + \log z$

9. $\log \frac{1}{2x} = \log 1 - \log 2x = 0 - \log 2x$
$\qquad = -\log 2 - \log x$

13. $\log 3 + \log 4 = \log 3 \cdot 4 = \log 12$

17. $4 \log 2 + 3 \log 3 - 2 \log 4$
$\qquad = \log 2^4 + \log 3^2 - \log 4^2$
$\qquad = \log 16 \cdot \frac{27}{16} = \log 27$

21. $\log \frac{x}{a} + 2 \log \frac{y}{b} + 3 \log \frac{z}{c}$
$\qquad = \log \frac{x}{a} + \log \frac{y^2}{b^2} + \log \frac{z^3}{c^3}$
$\qquad = \log x - \log a + \log y^2$
$\qquad\qquad - \log b^2 + z^3 - \log c^3$
$\qquad = \log \frac{xy^2z^3}{ab^2c^3}$

25. $\log_2 x + 2 \log_2 y = x$
$\log_2 xy^2 = x$
$2^x = xy^2$

29. $\log(p^2 - q^2) - \log(p + q) = 2$
$\log \frac{p^2 - q^2}{p + q} = 2$
$\log(p - q) = 2$
$p - q = 10^2$
$p - q = 100$

33. $\log_3 3^2 = 2$

37. $2^{\log_2 3y} = 3y$

41. $3.97 \times 8.25 \times 9.82 = 10^{\log(3.97 \cdot 8.25 \cdot 9.82)}$
$\qquad = 10^{(\log 3.97 + \log 8.25 + \log 9.82)}$
$\qquad = 10^{2.507} = 322$

45. $83.62^{0.5720} = N$
$\log N = 0.5720 \log 83.62$
$\log N = 1.10$
$N = 12.58$

49. $\log N = \frac{\ln N}{\ln 10} = \frac{8.36}{\ln 10} = 3.63$

53. $\log N = \frac{5.26}{\ln 10} = 2.28$

57. $\ln N = \ln 10^{-3.82} = -8.80$

61.

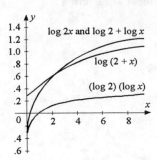

These graphs show us clearly that
$\log 2x = \log 2 + \log x$
$\log(2 + x) \neq \log 2 + \log x$
$(\log 2)(\log x) \neq \log 2 + \log x$

Exercise 5: Exponential Equations

1. $2^x = 7$
$\log_2 2^x = \log_2 7$
$x = \ln \frac{7}{\ln 2}$
$x = 2.81$

5. $(15.4)^{\sqrt{x}} = 72.8$
$\sqrt{x} = \log_{15.4} 72.8$
$x = \left(\frac{\ln 72.8}{\ln 15.4}\right)^2$
$x = 2.46$

9. $e^{2x-1} = 3e^{x+3}$
$\frac{e^{2x-1}}{e^{x+3}} = 3$
$e^{(2x-1)-(x+3)} = 3$
$e^{x-4} = 3$
$x = (\ln 3) + 4$
$x = 5.10$

13. $10^{3x} = 3(10^x)$
$\frac{10^{3x}}{10^x} = 3$
$10^{2x} = 3$
$2x = \log 3$
$x = \frac{\log 3}{2}$
$x = 0.239$

17. $7e^{1.5x} = 2e^{2.4x}$
$\frac{7}{2} = \frac{e^{2.4x}}{e^{1.5x}}$
$e^{0.9x} = 3.5$
$0.9x = \ln 3.5$
$x = \frac{\ln 3.5}{0.9}$
$x = 1.39$

21. $3(4^{3x}) = e^n$

$n \ln e = \ln 3 + 3x \ln 4$

$n = \ln 3 + x \ln 4^3$

$n = \ln 3 + x \ln 64$

$3(4^{3x}) = e^n = e^{\ln 3 + x \ln 64} = 3(e^{x \ln 64}) = 3e^{4.159x}$

25. If $E = 20.3$ V, $R = 4510\,\Omega$, $C = 545\,\mu$F, and $v = 10.1$ V,

$10.1 = 20.3\big(1 - e^{-t/[(4510)(545 \times 10^{-6})]}\big)$

$0.4975 = 1 - e^{-t/2.458}$

$e^{-t/2.458} = 0.5025$

$\frac{-t}{2.458} \ln e = \ln 0.5025$

$t = 1.69$ s

29. $p = 30e^{-kh}$

$p = 10$ in. Hg, $k = 3.83 \times 10^{-5}$

$e^{-kh} = \frac{10}{30}$

$e^{-kh} = \frac{1}{3}$

$h = \frac{\ln(1/3)}{-3.83 \times 10^{-5}}$

$h = 28,700$ ft

33. $a = \frac{y}{(1+n)^t}$

$a = \$50000$, $y = \$70000$, $n = 0.15$

$(1 + 0.15)^t = \frac{70000}{50000}$

$(1 + 0.15)^t = \frac{7}{5}$

$t = \log_{1.15} \frac{7}{5} = \frac{\ln(7/5)}{\ln 1.15} = 2.4$ yr

37. $t = \frac{\ln 2}{N} = \frac{\ln 2}{0.070/\text{yr}} = 9.9$ yr

41.
```
10  '   OIL
20  '
30  '   This program calculates the
40  '   depletion of the world's
50  '   oil reserves.
60  '
70  LPRINT "Year", "Ann'l Cons.", "Oil Left"
80  LET Y = 0: C=17: R = 1700
90  LPRINT Y, C, R
100 FOR Y = 1 TO 100
110 LET C = C*1.05
120 LET R = R - C
130 IF R < 0 THEN 160
140 LPRINT Y, C, R
150 NEXT Y
160 LPRINT "Oil reserves are gone"
170 END
```

Year	Ann'l Cons.	Oil Left
0	17	1700
1	17.85	1682.15
2	18.7425	1663.408
3	19.67962	1643.728
4	20.6636	1623.064
5	21.69678	1601.367
•	•	•
•	•	•
•	•	•
33	85.05404	270.8633
34	89.30673	181.5565
35	93.77206	87.78446

Oil reserves are gone

Exercise 6: Logarithmic Equations

1. $x = \log_3 9 = \dfrac{\log 9}{\log 3} = 2$

5. $x = \log_{27} 9 = \dfrac{\ln 9}{\ln 27} = \dfrac{2}{3}$

9. $\log_x 8 = 3$
$x^3 = 8$
$x^3 = 2^3$
$x = 2$

13. $\log_5 x = 2$
$x = 5^2$
$x = 25$

17. $x = \log_{25} 125$
$\dfrac{\ln 125}{\ln 25} = \dfrac{3}{2}$

21. $\log(2x + x^2) = 2$
$2x + x^2 = 10^2$
$2x + x^2 = 100$
$x^2 + 2x - 100 = 0$
$x = \dfrac{-2 \pm \sqrt{4+400}}{2}$
$x = -1 \pm \sqrt{101}$
$x = -11.0,\ 9.05$

25. $\ln(5x + 2) - \ln(x + 6) = \ln 4$
$\ln \dfrac{5x+2}{x+6} = \ln 4$
$\dfrac{5x+2}{x+6} = 4$
$5x + 2 = 4x + 24$
$x = 22$

29. $2 \log x - \log(1 - x) = 1$
$\log x^2 - \log(1 - x) = 1$
$\log \dfrac{x^2}{1-x} = 1$
$10(1 - x) = x^2$
$x^2 + 10x - 10 = 0$
$x = \dfrac{-10 \pm \sqrt{10^2 + 40}}{2}$
$x = \dfrac{-10 \pm \sqrt{140}}{2}$
$x = 0.916$

33. $\log(x^2 - 1) - 2 = \log(x + 1)$
$\log \dfrac{x^2 - 1}{x + 1} = 2$
$\log(x - 1) = 2$
$x - 1 = 10^2$
$x - 1 = 100$
$x = 101$

37. $t = \dfrac{\log[(ny/R)+1]}{\log(1+n)}$

$n = 0.090/\text{yr},\ R = \$1500,\ y = \$13800$

$t = \dfrac{\log\left[\frac{(0.090)(13800)}{1500} + 1\right]}{\log(1+0.090)} = 7.0\ \text{yr}$

41. $h = 60470 \log\left(\dfrac{B_2}{B_1}\right)$
$B_2 = 28.22$ in. Hg
$h = 815.0'$
$\dfrac{B_2}{B_1} = 10^{h/60470}$
$B_1 = \dfrac{B_2}{10^{h/60470}}$
$B_1 = \dfrac{28.22}{10^{815.0/60470}}$
$B_1 = 27.36$ in. Hg

45. $P_2 = \frac{1}{2} P_1$
$G = 10 \log\left[\dfrac{(1/2)P_1}{P_1}\right] = 10 \log 0.5 = -3.01$ dB

49. $7 = -\log C$
$C = 10^{-7}$

Exercise 7: Graphs on Logarithmic and Semilogarithmic Paper

1.
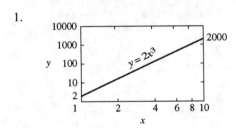

5.
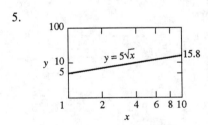

9.
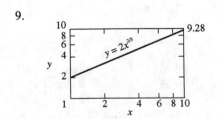

13.

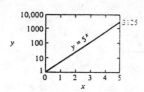

17.

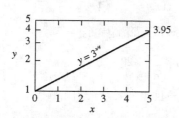

21. y-intercept $= a = 1$

slope $= \dfrac{\ln 97.7 - \ln 1}{5} = 0.92 = \ln b$

$b = e^{0.92} = 2.5$

So, $y = 2.5^x$

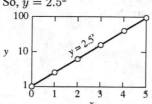

25. When $v = 1$, $p = 460 = a$

slope $= \dfrac{\ln 52.5 - \ln 14.7}{\ln 8 - \ln 26.47} = -1.07$

So, $p = 460v^{-1.07}$

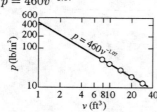

Chapter 20 Review Problems

1. $\log_x 352 = 5.2$

5. $x^{124} = 5.2$

9. $x^{-5/7} = 32$
$x^{-5/7} = 2^5$
$(x^{-5/7})^7 = (2^{35/7})^7$
$(x^{-5})^{1/5} = (2^{35})^{1/5}$
$x^{-1} = 2^7$
$\dfrac{1}{x} = 128$
$x = \dfrac{1}{128}$

13. $\log 10$

17. $\log 364 = 2.5611$

21. $\log x = -0.473$
$x = 0.337$

25. $\ln 84.72 = 4.4394$

29. $\ln x = 1.473$
$x = 4.362$

33. $\log x = 5.837$
$\ln x = 2.3026(5.837) = 13.44$

37. $\log x^3 - 3 = \log x^4$
$\log x^3 - \log x^4 = 3$
$\log \dfrac{1}{x} = 3$
$\dfrac{1}{x} = 10^3$
$x = 1.00 \times 10^{-3}$

41. $y = ae^{-nt}$
$y = 2250e^{-0.0500 \times 20.0} = 2250e^{-1} = 828$ rev/min

45.

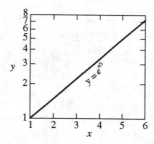

CHAPTER 21: COMPLEX NUMBERS

Exercise 1: Complex Numbers in Rectangular Form

1. $\sqrt{-9} = j3$

5. $4 + \sqrt{-4} = 4 + j2$

9. $\sqrt{-9} + \sqrt{-4} = j3 + j2 = j5$

13. $(a - j3) + (a + j5) = 2a + j2$

17. $(2.28 - j1.46) + (1.75 + j2.66) = 4.03 + j1.20$

21. $j^{21} = j^{20}j = (j^4)^5 j = 1(j) = j$

25. -15

29. -25

33. $8 + j20$

37. $(6 + j3)(3 - j8) = 18 - j48 + j9 - j^2 24$
$$= 18 - j39 + 24 = 42 - j39$$

41. $2 - j3 \rightarrow 2 + j3$

45. $-jm + n = n - jm \rightarrow n + jm$

49. $j12 \div j6 = 2$

53. $\frac{-2+j3}{1-j} \cdot \frac{1+j}{1+j} = \frac{-2-j2+j3+j^2 3}{1-j^2} = \frac{-5+j}{2} = -\frac{5}{2} + \frac{j}{2}$

57. $x = \frac{5 \pm \sqrt{25 - 4(3)(7)}}{2(3)} = \frac{5 \pm \sqrt{-59}}{6} = \frac{5 \pm j7.68}{6}$
$$= 0.833 \pm j1.28$$

61. $x^2 + 9 = (x + j3)(x - j3)$

65.
```
10  '  QUAD-2
20  '
30  '  This program solves a quadratic
40  '  by the quadratic formula.
50  '
60  PRINT "Enter the coefficients"
70  INPUT "A":  A
80  INPUT "B":  B
90  INPUT "C":  C
100 D = B^2 - 4 * A * C
110 IF D > 0 THEN GOTO 140
120 PRINT "The roots are imaginary."
130 GOTO 180
140 X1 = (-B + D^.5) /(2 * A)
150 X2 = (-B - D^.5) /(2 * A)
160 PRINT "The roots are "; X1 ; "and "; X2
170 END
180 R = -B / (2 * A)
190 I = ((-D)^.5) / (2 * A)
200 PRINT "One root is "; R; "+ j"; I
210 PRINT "Other root is "; R : "- j"; I
220 END
```

Exercise 2: Graphing Complex Numbers

1.

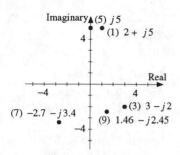

Exercise 3: Complex Numbers in Trigonometric and Polar Forms

1. $5 = j4$
$r = \sqrt{25 + 16} = \sqrt{41} = 6.40$
$\theta = \arctan \frac{4}{5} = 38.7°$
$6.40\angle 38.7° = 6.40(\cos 38.7° + j\sin 38.7°)$

5. $-5 - j2$
$r = \sqrt{25 + 4} = \sqrt{29} = 5.39$
$\theta = \arctan \frac{-2}{-5} = 202°$
$5.39\angle 202° = 5.39(\cos 202° + j\sin 202°)$

9. $-4 - j7$
$r = \sqrt{16 + 49} = \sqrt{65} = 8.06$
$\theta = \arctan \frac{-7}{-4} = 240°$
$8.06\angle 240° = 8.06(\cos 240° + j\sin 240°)$

13. $9(\cos 150° + j\sin 150°) = 9\angle 150°$
$a = 9\cos 150° = -7.79$
$b = 9\sin 150° = 4.5$
$-7.79 + j4.50$

17. $9\underline{/59°} = 9(\cos 59° + j\sin 59°)$
$a = 9\cos 59° = 4.64$
$b = 9\sin 59° = 7.71$
$4.64 + j7.71$

21. $3(\cos 12° + j\sin 12°) \cdot 5(\cos 28° + j\sin 28°)$
$= 15(\cos 40° + j\sin 40°)$

25. $(8\underline{/45°})(7\underline{/15°}) = 56\angle 60°$

29. $\dfrac{58.3(\cos 77.4° + j\sin 77.4°)}{12.4(\cos 27.2° + j\sin 27.2°)}$
$= 4.70(\cos 50.2° + j\sin 50.2°)$

33. $[2(\cos 15° + j\sin 15°)]^3 = 8(\cos 45° + j\sin 45°)$

37. $\sqrt{57\underline{/52°}} = \sqrt{57}\underline{/(52° + 360k)/2}$
$\qquad = 7.55\underline{/26° + 180k}$
$k = 0: \; = 7.55\underline{/26°}$
$k = 1: \; = 7.55\underline{/206°}$

41. $\sqrt{135 + j204}$
$r = \sqrt{(135)^2 + (204)^2} = 244.6$
$\phi = \arctan\left(\dfrac{204}{135}\right) = 56.5°$
Thus, $\sqrt{135 + j204} = \sqrt{244.6\angle 56.5°}$
$\qquad = \sqrt{244.6}\underline{/(56.5° + 360k)/2}$
$\qquad = 15.64\underline{/28.3° + 180k}$
$k = 0: \; = 15.64(\cos 28.3° + j\sin 28.3°)$
$\qquad = 13.8 + j7.41$
$k = 1: \; = 15.64(\cos 208.3° + j\sin 208.3°)$
$\qquad = -13.8 - j7.41$

Exercise 4: Complex Numbers in Exponential Form

1. $2 + j3$
$r = \sqrt{4 + 9} = \sqrt{13} = 3.61$
$\theta = \arctan \frac{3}{2} = 0.983$
$3.61e^{j0.983}$

5. $2.5\underline{/\frac{\pi}{6}} = 2.5e^{j\pi/6}$

9. $5e^{j3} = 5\underline{/3} = 5\underline{/172°}$
$\qquad = 5(\cos 172° + j\sin 172°)$
$a = 5\cos 3 = -4.95$
$b = 5\sin 3 = 0.706$
$-4.95 + j0.706$

13. $9e^{j2} \cdot 2e^{j4} = 18e^{j6}$

17. $1.7e^{j5} \cdot 2.1e^{j2} = 3.6e^{j7}$

21. $\dfrac{55e^{j9}}{5e^{j6}} = 11e^{j3}$

25. $(3e^{j5})^2 = 9e^{j10}$

Exercise 5: Vector Operations Using Complex Numbers

1. polar: $7.00\underline{/49.0°}$
   $a = 7.00\cos 49.0° = 4.59$
   $b = 7.00\sin 49.0° = 5.28$
   rectangular: $4.59 + j5.28$

5. $2.50 \text{ rad} = 143°$
   polar: $39.0\underline{/143°}$
   $a = 39.0\cos 143° = -31.2$
   $b = 39.0\sin 143° = 23.3$
   rectangular: $-31.2 + j23.3$

9. $58\underline{/72°} + 21\underline{/14°}$
   $a_1 = 58\cos 72° = 17.92$
   $b_1 = 58\sin 72° = 55.16$
   $a_2 = 21\cos 14° = 20.38$
   $b_2 = 21\sin 14° = 5.08$
   $(17.9 + j55.2) + (20.4 + j5.08)$
   $38.3 + j60.2$

13. $(7 + j3)(2 - j5) = 14 - j35 + j6 - j^2 15$
    $= 29 - j29$

17. $\dfrac{7.70\underline{/47°}}{2.50\underline{/15°}} = 3.08\underline{/32°}$

Exercise 6: Alternating-Current Applications

1. $i = 250\sin(wt + 25°)$
   $\mathbf{I} = \dfrac{250}{\sqrt{2}}\underline{/25°} = 177\underline{/25°}$

5. $v = 144\sin wt$
   $\mathbf{V} = \dfrac{144}{\sqrt{2}}\underline{/0°} = 102\underline{/0°}$

9. $\mathbf{V} = 300\underline{/-90°}$
   $v = 300\sqrt{2}\sin(wt - 90°) = 424\sin(wt - 90°)$

13. $\mathbf{Z} = 155 + j0 = 155\underline{/0°}$

17. $\mathbf{Z} = 72 - j42 = 83.4\underline{/-30.3°}$

21. $\mathbf{V} = \mathbf{IZ}$

   a. $\mathbf{I} = \dfrac{1.7}{\sqrt{2}}\underline{/25°} = 1.202\underline{/25°}$
      $\mathbf{Z} = 176 + j308$
      $\sqrt{R^2 + X^2} = \sqrt{176^2 + 308^2} = 354.74$
      $\phi = \arctan\left(\dfrac{308}{176}\right) = 60.26°$
      $\mathbf{Z} = 354.7\underline{/60.26°}$
      $\mathbf{V} = (1.202\underline{/25°})(354.74\underline{/60.26°})$
      $\quad = 426.4\underline{/85.26°}$
      $v = 426.4\sqrt{2}\sin(wt + 85.26°)$
      $\quad = 603\sin(wt + 85.3°)$

   b. $\mathbf{I} = \dfrac{43}{\sqrt{2}}\underline{/-30°} = 30.4\underline{/-30°}$
      $\mathbf{Z} = 4.78 + j(2.35 - 7.21)$
      $\quad = 4.78 - j4.86$
      $\sqrt{R^2 + X^2} = \sqrt{4.78^2 + (-4.86)^2}$
      $\quad = 6.817$
      $\phi = \arctan\left(\dfrac{-4.86}{4.78}\right) = -45.48°$
      $\mathbf{Z} = 6.817\underline{/-45.48°}$
      $\mathbf{V} = (30.4\underline{/-30°})(6.82\underline{/-45.5°})$
      $\quad = 207.3\underline{/-75.5°}$
      $v = 207.3\sqrt{2}\sin(wt - 75.5°)$
      $\quad = 293\sin(wt - 75.5°)$

Chapter 21 Review Problems

1. $3 + j2$
   $r = \sqrt{3^2 + 2^2} = 3.61$
   $\theta = \arctan\left(\dfrac{2}{3}\right) = 33.69°$
   $\theta = 0.588 \text{ rad}$
   $3.61\underline{/33.7°}$
   $3.61e^{j0.588}$
   $3.61(\cos 33.7° + j\sin 33.7°)$

5. $j^{17} = jj^{16} = j(j^2)^8 = j(-1)^8 = j$

9. $52\underline{/50°} + 28\underline{/12°}$
   $= (33.42 + j39.83) + (27.39 + j5.822)$
   $= 60.8 + j45.7$

13. $12(\cos 38° + j\sin 38°)$

17. $2(\cos 45° + j\sin 45°)$

21.

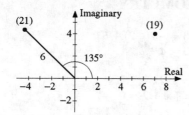

25. $125(\cos 30° + j \sin 30°)$

29.   Period $= P = 0.4$ s $= \dfrac{2\pi}{\omega}$
$\omega = \dfrac{2\pi}{0.4} = 15.7$ rad/s
$\phi = 0.05$ s $\cdot$ $(15.7$ rad/s$) = 0.785$ rad $= 45°$
$i = 1.25 \sin(15.7t - 45°)$
$\mathbf{I} = \dfrac{1.25}{\sqrt{2}} \underline{/-45°} = 0.884\underline{/-45°}$

Exercise 1: The Straight Line

1. $d = \sqrt{(5-2)^2 + (0-0)^2} = \sqrt{9} = 3$

5. $d = \sqrt{(0-0)^2 + (-2.74 - 3.86)^2}$
   $= \sqrt{43.56} = 6.60$

9. $\sqrt{(2.25 - 8.38)^2 + [-4.99 - (-3.95)]^2}$
   $= \sqrt{38.7} = 6.22$

13. $17 - (-9) = 26$

17. $\tan 77.9° = 4.66$

21. $\tan 132.8° = -1.080$

25. $\arctan(-2.75) = -70°$
    $180° - 70° = 11\bar{0}°$

29. $-\frac{1}{2}$

33. $-\frac{1}{-5.372} = 0.1862$

37. $m = -\frac{2}{8} = -\frac{1}{4}$
    $\theta = 166°$

41. The tangent of the angle of intersection between two straight lines is given by:
    $\tan \phi = \frac{m_2 - m_1}{1 + m_1 m_2}$
    Letting $m_2 = 6$ and $m_1 = 1$ and substituting into the above formula, we obtain,
    $\tan \phi = \frac{6-1}{1 + 1(6)} = \frac{5}{7}$
    $\phi = \arctan \frac{5}{7} = 35.5°$

45.

49. $2.25 = \frac{y - (-1.48)}{x - 0}$
    $2.25x = y + 1.48$
    $2.25x - y - 1.48 = 0$

53. $-3 = -2(-2) + b$
    $-3 = 4 + b$
    $b = -7$
    $y = -2x - 7$
    $2x + y + 7 = 0$

57. $(0, -2.3)$ $-2.3 = \left(\frac{2}{3}\right)(0) + b$
    $2x - 3y + 1 = 0$ $b = -2.3$
    $3y = 2x + 1$ $y = \left(\frac{2}{3}\right)x - 2.3$
    $y = \frac{2}{3}x + \frac{1}{3}$ $2x - 3y - 6.9 = 0$
    $m = \frac{2}{3}$

61. $4x - y = -3$
    $y = 4x + 3$
    $m = 4$
    $-1 = 4(4) + b$
    $b = -17$
    $y = 4x - 17$
    $4x - y - 17 = 0$

65. $(0, -3.52)$
    $m = \tan\left(154° + \frac{44}{60}\right) = -0.472$
    $-3.52 = -0.472(0) + b$
    $b = -3.52$
    $y = -0.472x - 3.52$
    $0.472x + y + 3.52 = 0$

69. The line is vertical and 3 units to the left of the origin, so its equation is $x = -3$.

73. $y = -\frac{1}{2}x + 1$
    $2y = -x + 2$
    $x + 2y - 2 = 0$

77. $\cos 12.3° = \frac{x}{2055}$
    $x = 2008$ ft

81. $\sin \theta = \frac{12}{755}$
    $\theta = 0.911°$

85. $F = kx$
    $\frac{F - 0}{L - L_0} = k$
    $F = kL - kL_0$

89. $\alpha = \frac{1}{234.5t_1}$, $t = 75$, $t_1 = 20$
    $R_1 = 148.4 \ \Omega$
    $R = 148.4\left[1 + \frac{1}{234.5 \cdot 20}(75 - 20)\right] = 150.1 \ \Omega$

93. $P = 20.6 + 0.432x$
$x = \frac{30.0 - 20.6}{0.432} = 21.8$ ft

97. y int $= P$, slope $= \frac{(S-P)}{L}$,
so $y = P + t\frac{(S-P)}{L}$
Substituting $P = \$15428$, $t = 15$ years,
$S = \$2264$, and $L = 20$ years,
$y = 15,428 + \frac{15(2264 - 15,426)}{20} = \$5555$

Exercise 2: The Circle

1. $h = 0$, $k = 0$, $r = 7$, so $x^2 + y^2 = 49$

5. $h = 5$, $k = -3$, $r = 4$, so
$(x - 5)^2 + (y + 3)^2 = 16$

9. $C(2, -4)$, $r = 4$

13. $(x^2 - 8x + 16) + y^2 = 0 + 16$
$(x - 4)^2 + y^2 = 16$
$C(4, 0)$, $r = 4$

17. $(x^2 + 6x) + (y^2 - 2y) = 15$
$(x^2 + 6x + 9) + (y^2 - 2y + 1) = 15 + 9 + 1$
$(x + 3)^2 + (y - 1)^2 = 25$
$C(-3, 1)$, $r = 5$

21. $x^2 + y^2 = 25$ at $(4, 3)$
Slope of radius $= \frac{3}{4}$
Slope of tangent $= -\frac{4}{3}$
$\frac{y-3}{x-4} = -\frac{4}{3}$
$3y - 9 = -4x + 16$
$4x + 3y = 25$

25. $x^2 + y^2 - 5x - 7y + 6 = 0$
when $y = 0$, $\quad x^2 - 5x + 6 = 0$
$\quad\quad\quad\quad\quad (x - 3)(x - 2) = 0$
$\quad\quad\quad\quad\quad x = 3$, $x = 2$
when $x = 0$, $\quad y^2 - 7y + 6 = 0$
$\quad\quad\quad\quad\quad (y - 6)(y - 1) = 0$
$\quad\quad\quad\quad\quad y = 6$, $y = 1$
so intercepts are:
$(3, 0)$, $(2, 0)$, $(0, 6)$, $(0, 1)$

29. <u>Slope of line 1</u>: $m_1 = \frac{y}{r+x}$
<u>Slope of line 2</u>: $m_2 = \frac{y}{r-x}$

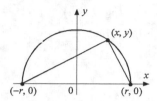

$m_1 = \frac{y}{r+x} \cdot \frac{r-x}{r-x} = \frac{y(r-x)}{r^2 - x^2} = \frac{y(r-x)}{y^2}$
$\quad\quad = \frac{r-x}{y} = -\frac{1}{m_2}$
Therefore, line 1 is perpendicular to line 2.

33. $h = 6.00$, $k = 0$, $r = 10.0$, so
$(x - 6)^2 + y^2 = 100$. When $y = 3.00$,
$x^2 - 12x + 36 + 9 = 100$
$x^2 - 12x - 55 = 0$
$x = 15.5$ ft (drop) or $x = -3.54$ ft
so $w = 2(3.54) = 7.08$ ft

Exercise 3: The Parabola

1. $y^2 = 8x$, $4p = 8$; $p = 2$; $F(2, 0)$, $L = 8$

5. $p = -2$; $x^2 = -8y$; $x^2 + 8y = 0$

9. From the given equation, we see that the axis is
horizontal with $h = 3$, $k = 5$, and $4p = 12$. This
implies the vertex is at $V(3, 5)$. Since $4p = 12$
yields $p = 3$ and the focus is three units to the
right of the vertex its coordinates must be $F(6, 5)$.
The focal width is $|4p| = 12$. The axis is
horizontal and passes through the vertex, thus, its
equation must be $y = 5$.

13. $3x + 2y^2 + 4y - 4 = 0$
$2y^2 + 4y = -3x + 4$
$2(y^2 + 2y + 1) = -3x + 4 + 2$
$(y + 1)^2 = -\frac{3}{2}(x - 2)$
so, $V(2, -1)$, $F\left(\frac{13}{8}, -1\right)$, $L = \frac{3}{2}$, axis: $y = -1$

17. $L = 4p = 8$
$p = 2$
$h = 1$, $k = 2$
$(y - k)^2 = 4p(x - h)$
$(y - 2)^2 = 8(x - 1)$
$y^2 - 8x - 4y + 12 = 0$

21. $k = -1$
     $(y+1)^2 = 4p(x-h)$
     at $(2, 1)$, $4 = 4p(2-h)$
     $p = \frac{1}{2-h}$
     at $(-4, -2)$, $1 = 4p(-4-h)$
     $1 = \frac{4(-4-h)}{2-h}$ from which, $h = -6$
     $p = \frac{1}{2-h} = \frac{1}{8}$
     so, $(y+1)^2 = \frac{1}{2}(x+6)$
     or $2y^2 + 4y - x - 4 = 0$

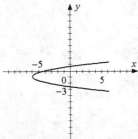

25. Taking the origin at the vertex of the parabola,
     and y-axis downward,
     $x^2 = 4py$
     at $(18, 18)$, $4p = \frac{(18)^2}{18} = 18$
     When $y = 8$ ft, $x^2 = 18(8)$
     $x = 12$ ft

29. $x^2 = 4py = 4y$, since $p = 1.00$
     When $x = 1.50$ m, $y = \frac{(1.50)^2}{4} = 0.563$ m

33. When $x = 10.0$, $(10.0 - 16.0)^2 = 3070\left(y - \frac{1}{12}\right)$
     $y - \frac{1}{12} = -0.0117$
     $y = 0.0716$ ft

Exercise 4: The Ellipse

1. $\frac{x^2}{25} + \frac{y^2}{16} = 1$
     $a = 5$, $b = 4$, $c = \sqrt{25-16} = 3$
     $V(\pm 5, 0)$, $F(\pm 3, 0)$

5. $4x^2 + 3y^2 = 48$
     $\frac{x^2}{12} + \frac{y^2}{16} = 1$
     $a = 4$, $b = \sqrt{12}$, $c = \sqrt{16-12} = 2$
     so, $V(0, \pm 4)$ and $F(0, \pm 2)$

9. $a = 6$, through $(3, \sqrt{3})$
     $\frac{x^2}{36} + \frac{y^2}{b^2} = 1$
     at $(3, \sqrt{3})$, $\frac{9}{36} + \frac{3}{b^2} = 1$
     $\frac{3}{b^2} = 1 - \frac{1}{4} = \frac{3}{4}$
     $b^2 = 4$
     so $\frac{x^2}{36} + \frac{y^2}{4} = 1$

13. $(1, 4)$ and $(-6, 1)$
     $\frac{x^2}{a^2} + \frac{y^2}{b^2} = 1$
     at $(1, 4)$, $\frac{1}{a^2} + \frac{16}{b^2} = 1$ (1)
     at $(-6, 1)$, $\frac{36}{a^2} + \frac{1}{b^2} = 1$
     or $\frac{576}{a^2} + \frac{16}{b^2} = 16$ (2)
     taking $(2) - (1)$, $\frac{576}{a^2} = 15$
     $a^2 = \frac{115}{3}$
     so, $\frac{1}{b^2} = 1 - \frac{36(3)}{115} = \frac{7}{115}$
     $b^2 = \frac{115}{7}$
     so, $\frac{3x^2}{115} + \frac{7y^2}{115} = 1$

17. $5(x^2 + 4x + 4) + 9(y^2 - 6y + 9)$
     $\qquad\qquad = -56 + 20 + 81$
     $5(x+2)^2 + 9(y-3)^2 = 45$
     $\frac{(x+2)^2}{9} + \frac{(y-3)^2}{5} = 1$
     so, $C(-2, 3)$, $V(1, 3)$ and $(-5, 3)$
     $f = \sqrt{9-5} = 2$
     so, $F(0, 3)$ and $(-4, 3)$

21. $25(x^2 + 6x + 9) + 9(y^2 - 4y + 4) = -36 + 225 + 36$
$25(x + 3)^2 + 9(y - 2)^2 = 225$
$\frac{(x+3)^2}{9} + \frac{(y-2)^2}{25} = 1$
$f = \sqrt{25 - 9} = 4$
so, $C(-3, 2)$, $V(-3, 7)$ and $(-3, -3)$, $F(-3, 6)$ and $(-3, -2)$

25. $a = 4$, $f = 2$, $h = -2$, $k = -3$
$b = \sqrt{16 - 4} = \sqrt{12}$
so, $\frac{(x+2)^2}{12} + \frac{(y+3)^2}{16} = 1$

29. The tacks are located at the foci of the ellipse. We know that $a = \frac{84.0}{2} = 42.0$ and $b = \frac{58.0}{2} = 29.0$. We find c using
$c = \sqrt{a^2 - b^2} = \sqrt{42.0^2 - 29.0^2}$
$= \sqrt{923} = 30.4$.
The tacks should be
$2c = 2(30.4) = 60.8$ cm apart.

33. Letting the origin be located at the center of the culvert, we have, $a = 4.0$, $b = 2.0$. Thus, the equation of the ellipse is $\frac{x^2}{16} + \frac{y^2}{4.0} = 1$ \qquad (1)
We seek the x coordinate when $y = -1$ (one unit down from the major axis). Substituting into eq(1), we obtain, $\frac{x^2}{16} + \frac{1}{4.0} = 1$ from which $x = \sqrt{12}$. The width of the stream is thus
$2x = 2\sqrt{12} = 6.9$ ft.

Exercise 5: The Hyperbola

1. The equation is in standard form having horizontal transverse axis, with $a = 4$, and $b = 5$. The vertices, thus, become $V(4, 0)$, $V(-4, 0)$. The distance c, to the foci, is
$c = \sqrt{a^2 + b^2} = \sqrt{16 + 25} = \sqrt{41}$. The foci then are $F(\sqrt{41}, 0)$, $F(-\sqrt{41}, 0)$. The slope of the asymptotes is $\pm \frac{b}{a} = \pm \frac{5}{4}$.

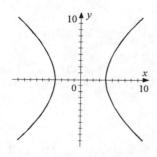

5. $x^2 - 4y^2 = 16$
$\frac{x^2}{16} - \frac{y^2}{4} = 1$
$a = 4$, $b = 2$, $c = \sqrt{16 + 4} = \sqrt{20} = 2\sqrt{5}$
$V(\pm 4, 0)$, $F(\pm 2\sqrt{5}, 0)$, slope $= \pm \frac{1}{2}$

9. $c = 4$, $a = 3$, $b = \sqrt{16 - 9} = \sqrt{7}$
so $\frac{x^2}{9} - \frac{y^2}{7} = 1$

13. We know $2a = 10$, or $a = 5$. Since the axis is vertical, we obtain, $\frac{y^2}{25} - \frac{x^2}{b^2} = 1$. Substituting $(8, 10)$ for (x, y) and solving for b^2 we obtain,
$\frac{100}{25} - \frac{64}{b^2} = 1$
$4b^2 - 64 = b^2$
$b^2 = \frac{64}{3}$
Therefore, the equation is $\frac{y^2}{25} - \frac{3x^2}{64} = 1$.

17. $(16x^2 - 64x) - (9y^2 + 54y) = 161$
$16(x^2 - 4x + 4) - 9(y^2 + 6y + 9)$
$\qquad = 161 + 64 - 81$
$16(x - 2)^2 - 9(y + 3)^2 = 144$
$\frac{(x-2)^2}{9} - \frac{(y+3)^2}{16} = 1$
$a = 3$, $b = 4$
$C(2, -3)$, $V(5, -3)$, $V'(-1, -3)$
$C = \sqrt{9 + 16} = 5$
so, $F(7, -3)$ and $(-3, -3)$
slope $= \pm \frac{4}{3}$

21. $a = 4$, $b = 2$, $h = 3$, $k = 2$, transverse axis is vertical so $\frac{(y-2)^2}{16} - \frac{(x-3)^2}{4} = 1$

25. A hyperbola whose asymptotes are the x- and y-axis has the form, $xy = k$. To find k we simply substitute $x = 6$, $y = 6$, obtaining $k = 36$. Thus, the equation is $xy = 36$.

29. $pv = c$

$c = 25.0(1,000) = 25,000$

so, $pv = 25,000$

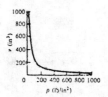

Chapter 22 Review Problems

1. $d = \sqrt{(3-7)^2 + (0-0)^2} = \sqrt{16} = 4$

5. $-\frac{1}{m} = -\frac{1}{1.55} = -0.645$

$\tan \theta = -0.645$

$\arctan(-0.645) = -32.8°$

$\theta = 180° - 32.8° = 147°$

9. $2y - 5 = 3x - 12$

$2y = 3x - 7$

$y = \frac{3}{2}x - \frac{7}{2}$

$m = \frac{3}{2}, b = -\frac{7}{2}$

13. $\frac{y-7}{x+4} = 5$

$y - 7 = 5x + 20$

$5x - y + 27 = 0$

17. Since every point on the line has a y coordinate of 5, the equation of the line is $y = 5$.

21. Since every point on the line has an x coordinate of -3, the equation of the line is $x = -3$ or $x + 3 = 0$.

25. Since the slope of the normal is $-\frac{1}{m}$,

$y - y_1 = -\left(\frac{1}{m}\right)(x - x_1)$ or

$x - x_1 + m(y - y_1) = 0$ $\qquad$ (2)

29. By the Pythagorean theorem,

$(PB)^2 = x_1^2 + (y_1 + OB)^2$

$\qquad = x_1^2 + (y_1 + mx_1 - y_1)^2 = x_1^2 + m^2 x_1^2$

$\qquad = x_1^2(1 + m^2)$

so, $PB = x_1\sqrt{1 + m^2}$

33. By the Pythagorean theorem,

$(PD)^2 = x_1^2 + (OD - y_1)^2 = x_1^2 + \left(y_1\frac{x_1}{m} - y_1\right)^2$

$\qquad = x_1^2 + \frac{x_1^2}{m^2} = x_1^2\frac{(1+m^2)}{m^2}$

$PD = \frac{x_1}{m}\sqrt{1 + m^2}$

37. $25x^2 - 200x + 9y^2 - 90y = 275$

$25(x^2 - 8x + 16) + 9(y^2 - 10y + 25)$

$\qquad = 275 + 400 + 225$

$25(x - 4)^2 + 9(y - 5)^2 = 900$

$\frac{(x-4)^2}{36} + \frac{(y-5)^2}{100} = 1$

Ellipse, major axis vertical, $C(4, 5)$, $a = 10$,

$b = 6, f = \sqrt{100 - 36} = 8$

$F(4, 13)$ and $(4, -3)$, $V(4, 15)$ and $(4, -5)$

41. $x^2 + y^2 + D + Ey + F = 0$

at $(0, 0)$, $F = 0$

at $(8, 0)$, $64 + 8D = 0$

$\qquad\qquad\qquad D = -8$

at $(0, -6)$, $36 - 6E = 0$

$\qquad\qquad\qquad E = 6$

so, $x^2 + y^2 - 8x + 6y = 0$

45. $a = 4, h = k = 0$, so, $\frac{x^2}{16} - \frac{y^2}{b^2} = 1$

at $(10, 25)$, $\frac{100}{16} - \frac{625}{b^2} = 1$

$\frac{84}{16} = \frac{625}{b^2}$

$b^2 = \frac{625(16)}{84} = \frac{625(4)}{21}$

$\frac{x^2}{16} - \frac{84y^2}{16(625)} = 1$

$625x^2 - 84y^2 = 10,000$

49. $x^2 + y^2 + 2x + 2y = 2$ $\qquad$ (1)

$3x^2 + 3y^2 + 5x + 5y = 10$ $\qquad$ (1)

from (1), $\qquad\qquad 3x^2 + 3y^2 + 6x + 6y = 6$

subtracting, $\qquad\qquad x + y = -4$

$\qquad\qquad\qquad\qquad y = -x - 4$

substituting in (1),

$x^2 + (-x - 4)^2 + 2x + 2(-x - 4) = 2$

$x^2 + x^2 + 8x + 16 + 2x - 2x - 8 - 2 = 0$

$2x^2 + 8x + 6 = 0$

$x^2 + 4x + 3 = 0$

$(x + 1)(x + 3) = 0$

$\qquad\qquad x = -1 \mid x = -3$

$y(-1) = 1 - 4 = -3 \mid y(-3) = 3 - 4 = -1$

Curves intersect at $(-1, -3)$ and $(-3, -1)$

53.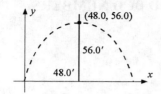

$$(x - 48.0)^2 = 4p(y - 56.0)$$

at $(96.0, 0)$, $\dfrac{48.0^2}{-56.0} = 4p$

$$4p = -41.14$$

when $y = 25.0$,

$$(x - 48.0)^2 = -41.14(25.0 - 56.0) = 1,275$$

$$x - 48.0 = \pm 35.7$$

$$x = 48.0 \pm 35.7 = 12.3 \text{ ft and } 83.7 \text{ ft}$$

CHAPTER 23: BINARY, HEXADECIMAL, OCTAL, AND BCD NUMBERS

Exercise 1: The Binary Number System

1. $10 = 1 \times 2^1 + 0 \times 2^0 = 2 + 0 = 2$

5. $\begin{aligned} 0110 &= 0 \times 2^3 &+& 1 \times 2^2 &+& 1 \times 2^1 &+& 0 \times 2^0 \\ &= 0 &+& 4 &+& 2 &+& 0 \\ &= 6 \end{aligned}$

9. $\begin{aligned} 1100 &= 1 \times 2^3 &+& 1 \times 2^2 &+& 0 \times 2^1 &+& 0 \times 2^0 \\ &= 8 &+& 4 &+& 0 &+& 0 \\ &= 12 \end{aligned}$

13. $0110\ 0111$
$\begin{aligned} &= 0 \times 2^7 &+& 1 \times 2^6 &+& 0 \times 2^5 &+& 0 \times 2^4 &+& 0 \times 2^3 &+& 1 \times 2^2 &+& 1 \times 2^1 &+& 1 \times 2^0 \\ &= 0 &+& 64 &+& 32 &+& 0 &+& 0 &+& 4 &+& 2 &+& 1 \\ &= 103 \end{aligned}$

17.
```
           Remainder
2:  5
2:  2    1
2:  1    0
    0    1     101
```

21.
```
2:  72
2:  36   0
2:  18   0
2:  9    0
2:  4    1
2:  2    0
2:  1    0
    0    1    0100 1000
```

25.
```
2:  274
2:  137  0
2:  68   1
2:  34   0
2:  17   0
2:  8    1
2:  4    0
2:  2    0
2:  1    0
    0    1   0001 0001 0010
```

29.
```
2:  8375
2:  4187  1
2:  2093  1
2:  1046  1
2:  523   0
2:  261   1
2:  130   1
2:  65    0
2:  32    1
2:  16    0
2:  8     0
2:  4     0
2:  2     0
2:  1     0
    0     1  0010 0000 1011 0111
```

33.
```
    0.5
  × 2
  ―――
   1.0   1
   0.0      0.1
```

37.
```
      0.3
    × 2
    ─────
      0.6   0
    × 2
    ─────
      1.2   1
      0.2
    × 2
    ─────
      0.4   0
    × 2
    ─────
      0.8   0
    × 2
    ─────
      1.6   1
      0.6
    × 2
    ─────
      1.2   1
      0.2
    × 2
    ─────
      0.4   0
    × 2
    ─────
      0.8   0   0.0100 1100
```

41.
```
      0.875
    × 2
    ─────
      1.75   1
      0.75
    × 2
    ─────
      1.5   1
      0.5
    × 2
    ─────
      1.0   1
      0.0        0.111
```

45. $0.1 = 1 \times 2^{-1} = 0.5$

49.
$$0.1001 = 1 \times 2^{-1} + 0 \times 2^{-2} + 0 \times 2^{-3} + 1 \times 2^{-4}$$
$$= 0.5 + 0 + 0 + 0.0625 = 0.5625$$

53.
```
2:  4          0.375
2:  2   0    × 2
2:  1   0    ──────
    0   1      0.75   0
             × 2
             ──────
               1.5   1
               0.5
             × 2
             ──────
               1.0   1
               0.0        100.011
```

57.
$$1.1 = 1 \times 2^0 + 1 \times 2^{-1}$$
$$= 1 + 0.5$$
$$= 1.5$$

61.
$$1\,1001.0110\,1 = 1 \times 2^4 + 1 \times 2^3 + 1 \times 2^0 + 1 \times 2^{-2} + 1 \times 2^{-3} + 1 \times 2^{-5}$$
$$= 16 + 8 + 1 + 0.25 + 0.125 + 0.03125$$
$$= 25.40625$$

Exercise 2: The Hexadecimal Number System

1. $1101 = 13_{10} = D$

5. 93

9. 92A6

13. 1.38

17. 0100 1010

21. 0100 0111 1010 0010

25. 1001.1010 1010

29.　　$33 = 0011\ 0011 = 51$

33.　　$F274 = 1111\ 0010\ 0111\ 0100 = 62068$

37.　　$ABC.DE = 1010\ 1011\ 1100.1101\ 1110$
　　　　　　$= 2748.8671875$

41.
16:	921	
16:	57	9
16:	3	9
	0	3	399

45.
16:	1736	
16:	108	8
16:	6	12
	0	6	6C8

13.
16:	8362	
16:	522	10
16:	32	10
16:	2	0
	0	2	20AA

17.　　$100\ 111\ 001\ 011\ 101_8$
　　　　　　$= 0100\ 1110\ 0101\ 1101$

Exercise 3: The Octal Number System

1.　　6

5.　　63

9.　　111

13.　　1 1001 0011

17.　　0110 0110 0110 1000

Exercise 4: BCD Codes

1.　　0110 0010

5.　　0100 0010.1001 0001

9.　　61

Chapter 23 Review Problems

1.
110.	=	6.
0.001	=	0.125
110.001	=	6.125

5.　　$2B4 = 0010\ 1011\ 0100 = 1264$

9.　　$101\ 011\ 100 = 1\ 0101\ 1100_2 = 348$

CHAPTER 24: INEQUALITIES & LINEAR PROGRAMMING

Exercise 1: Definitions

1. $x^2 > -4$
   unconditional, nonlinear

5. $3x + 3 \geq x - 7$
   conditional, linear

9. $x > -11$ and $x \leq 1$

13. $-1 \leq x < 24$

17.

$$x < 3$$

21.

$$x \leq 1$$

25.

$$x > -3$$

29.

$$4 < x \leq 22$$

33.

37.

Exercise 2: Solving Inequalities

1. $2x - 5 > x + 4$
   $x > 9$

5. $2x^2 - 5x + 3 > 0$
   $2x^2 - 2x - 3x + 3 > 0$
   $2x(x - 1) - 3(x - 1) > 0$
   $(2x - 3)(x - 1) > 0$

   $\left.\begin{array}{l} x = \frac{3}{2} \\ x = 1 \end{array}\right\}$ critical values

   $x < 1 \quad$ or $\quad x > \frac{3}{2}$

9. $1 < x + 3 < 8$
   $-2 < x < 5$

13. $|3x| > 9$
    $-9 > 3x > 9$
    $x > 3 \quad$ or $\quad x < -3$

17. $-8 \geq (2 - 3x) \geq 8$
    $-10 \geq -3x \geq 6$
    $x \geq \frac{10}{3}$ or $x \leq -2$

21. $|3 - 5x| \le 4 - 3x$
$-(4 - 3x) \le 3 - 5x \le 4 - 3x$
$-4 + 3x \le 3 - 5x \le 4 - 3x$
$-4 + 8x \le 3 \le 4 + 2x$
thus $\quad 8x \le 7$ and $-1 \le 2x$
$\qquad x \le \frac{7}{8} \qquad -\frac{1}{2} \le x$
thus $\quad -\frac{1}{2} \le x \le \frac{7}{8}$

25. $y^2 = x^2 - x - 6$
$y = \sqrt{x^2 - x - 6}$
$x^2 - x - 6 = 0$
$(x - 3)(x + 2) = 0$
$x = 3$ or $x = -2$

	$(x - 3)$	$(x + 2)$	$(x - 3)(x + 2)$
$x < -2$	–	–	+
$-2 < x < 3$	–	+	–
$x > 3$	+	+	+

Therefore, $x \le -2$ or $x \ge 3$

29. Let x = no. of lenses
$(39.99 - 12.75)x > 2884$
$27.24x > 2884$
$x > 105.9$
thus minimum is 106 lenses per week

33. $4.77x^2 - 7.14 < 4.72 - 2.33x$

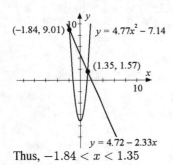

Thus, $-1.84 < x < 1.35$

Exercise 3: Linear Programming

1. Maximize $\quad z = x + 2y$,
   Given: $\qquad x + y \le 5$
   $\qquad\qquad x - y \ge -2$

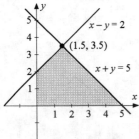

The given lines intersect at $(1.5, 3.5)$ so the corners of the region are $(0, 0)$, $(0, 2)$, $(1.5, 3.5)$, and $(5, 0)$.

At $(0, 0)$, $\qquad z = x + 2y = 0$
At $(0, 2)$, $\qquad z = 0 + 4 = 4$
At $(5, 0)$, $\qquad z = 5 + 0 = 5$
At $(1.5, 3.5)$, $\qquad z = 1.5 + 7 = 8.5$
$\qquad\qquad\qquad$ (optimum solution)

5. Let x = no. of pulleys made
          y = no. of sprockets made
    Profit: $z = 2.10x + 2.35y$
    Time: $3x + 4y \leq 480$
    Cost: $1.25x + 1.30y \leq 175$

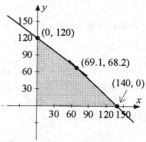

The given lines intersect at $(69.1, 68.2)$ so the
corners of the region are $(0, 0)$,
$(0, 120), (69.1, 68.2),$ and $(140, 0)$.
At $(0,0)$, $z = 2.10x + 2.35y = 0$
At $(0, 120)$, $z = 0 + 282 = 282$
At $(140, 0)$, $z = 294 + 0 = 294$
At $(69.1, 68.2)$, $z = 145 + 160 = 305$
                          (optimum solution)
Thus, for maximum profit make 69 pulleys and 68
sprockets.

Chapter 24: Review Problems

1. $x > -11$ and $x \leq 1$

5.

$$x > 8$$

6 7 8 9 10 11 12 13 14 x

9. $-7 > x - 4 > 7$
     $x < -3$ or $x > 11$

13. $x^2 + 8x + 15 \geq 0$
     $(x + 5)(x + 3) \geq 0$
     Critical values: $x = -5$
                        $x = -3$
     $x \leq -5$ or $x \geq -3$

17. $-4.73 < x - 3 \leq 5.82$
     $-1.73 < x \leq 8.82$

CHAPTER 25: SEQUENCES, SERIES, AND THE BINOMIAL THEOREM

Exercise 1: Sequences and Series

1. $u_n = 3n$. Letting n take on the values $1, 2, 3, 4,$ and 5, and substituting into the formula $u_n = 3n$, we obtain,
$$3 + 6 + 9 + 12 + 15 + ... + 3n + ...$$

5. $2 + 4 + 6 + ...$
Letting n be the nth term of the series it appears that $2n$ is a formula for each individual term. Using this formula we compute the 4th and 5th terms as $8, 10$.

9. $1 + 5 + 9 + ...$
By adding 4 to the previous term we get the next term, $u_n = u_{n-1} + 4$
The next two terms will be $9 + 4 = 13$ and $13 + 4 = 17$.

Program for Problems 13. - 15.
```
10 '          SERIES
20 '
30 '  THIS PROGRAM GENERATES AN INFINITE SERIES
40 '  AND COMPUTES PARTIAL SUMS AND RATIOS
50 '
70 LAST=1
80 A$="###      #.######      #.######          #.######"
90 PRINT " N","TERM","SUM","RATIO"
100   PRINT "_____"
110   FOR N=1 TO 85
120   T=1/N
130   S=S+T
140   R=T/LAST
150   IF N<11 THEN 170
160   IF N/10<>INT(N/10) THEN 180
170   PRINT USING A$;N,T, S, R
180   LAST=T
190   NEXT N
```

13. $1 + \frac{1}{2} + \frac{1}{3} + \frac{1}{4} + \cdots + \frac{1}{n} + \cdots$

N	TERM	SUM	RATIO
1	1.000000	1.000000	1.000000
2	0.500000	1.500000	0.500000
3	0.333333	1.833333	0.666667
4	0.250000	2.083334	0.750000
5	0.200000	2.283334	0.800000
6	0.166667	2.450000	0.833333
7	0.142857	2.592857	0.857143
8	0.125000	2.717857	0.875000
9	0.111111	2.828969	0.888889
10	0.100000	2.928969	0.900000
20	0.050000	3.597740	0.950000
30	0.033333	3.994987	0.966667
40	0.025000	4.278544	0.975000
50	0.020000	4.499206	0.980000
60	0.016667	4.679871	0.983333
70	0.014286	4.832838	0.985714
80	0.012500	4.965480	0.987500

Ans: diverges

Exercise 2: Arithmetic Progressions

1. Find the fifteenth term of an AP with first term 4 and common difference 3.
    Using $a_n = a + (n-1)d$ with $a = 4$, $d = 3$, $n = 15$,
    $$a_n = 4 + (15 - 1)3 = 4 + 143 = 46$$

5. Find the eleventh term of the AP 9, 13, 17, $\cdots$
    Using $a_n = a + (n-1)d$ with $a = 9$, $d = 4$, $n = 11$,
    $$a_n = 9 + (11 - 1)4 = 9 + 40 = 49$$

9. First term is 3 and the 13th term is 55. First we will find the common difference d,
    $$a_n = a + (n-1)d$$
    $$55 = 3 + (13 - 1)d$$
    $$12d = 52$$
    $$d = \frac{52}{12} = 4\frac{1}{3}$$
    Adding $4\frac{1}{3}$ to each term gives 3, $7\frac{1}{3}$, $11\frac{2}{3}$, 16, $20\frac{1}{3}$

13. Find the first term in an AP whose common difference is 3 and whose 7th term is 11.
    Using $a_n = a + (n-1)d$ with $a_7 = 11$, $n = 7$, $d = 3$,
    $$11 = a + (7 - 1)3$$
    $$11 = a + 18$$
    $$a = -7$$

17. Find the sum of the first nine terms of the AP; 5, 10, 15, 20, $\cdots$
    Letting $a = 5$, $d = 5$, and $n = 9$,
    $$S_9 = \left(\frac{9}{2}\right)[2(5) + (9 - 1)5] = \left(\frac{9}{2}\right)[10 + 40] = 225$$

21. Insert two arithmetic means between 5 and 20.
    Our AP will have four terms, with a first term of 5 and a fourth term of 20, thus,
    $$a_n = a + (n-1)d$$
    $$20 = 5(4 - 1)d$$
    $$3d = 15$$
    $$d = 5$$
    The progression is then 5, 10, 15, 20.

25. Find the fourth term of the harmonic progression, $\frac{3}{5}$, $\frac{3}{8}$, $\frac{3}{11}$, $\cdots$
    Since the reciprocals of each term form an AP we will find the 4th term of the AP and then take the reciprocal to find the 4th term in the harmonic progression. Taking the reciprocals
    $$\frac{5}{3}, \frac{8}{3}, \frac{11}{3}$$
    we can see that the fourth term will be $\frac{14}{3}$. Thus $\frac{3}{14}$ is the 4th term in the harmonic progression.

29. A person agrees to repay a loan of $10,000 with an annual payment of $1000 plus 8 percent of the unpaid balance.
(a) Show that the interest payments alone form the AP; $800, $720, $640, $\cdots$.
(b) Find the total amount of interest paid.
The interest each year is
$$10,000 \times 0.08 = 800 \qquad 1^{\text{st}} \text{ year}$$
$$9,000 \times 0.08 = 720 \qquad 2^{\text{nd}} \text{ year}$$
$$8,000 \times 0.08 = 640 \qquad 3^{\text{rd}} \text{ year}$$
We see that this forms an arithmetic progression with first term $800 and common difference (-80). Thus the interest for any given year using $a_n = a + (n-1)d$ is
$$I_n = 800 - 80(n-1)$$
The total amount of interest is calculated using
$$S_n = \left(\tfrac{n}{2}\right)[2a + (n-1)d]$$
$$I_n = \left(\tfrac{10}{2}\right)[2 \cdot 800 + (10-1)(-80)] = 5[1600 - 720] = \$4400$$

33. A freely falling body falls $\frac{g}{2}\text{ft}$ during the first second, $\frac{3g}{2}\text{ft}$ during the next second, $\frac{5g}{2}$ during the third second, and so on, where $g \approx \frac{32.2\text{ ft}}{s^2}$. Find the total distance the body falls during the first $10s$.
The sequence $\frac{g}{2}, \frac{3g}{2}, \frac{5g}{2}, \cdots$ is an arithmetic sequence with $d = g$. To find the sum of the first 10 terms we use $S_n = \left(\tfrac{n}{2}\right)[2a + (n-1)d]$ with $n = 10$, $a = \frac{g}{2}$, and $d = g$.
$$S_n = \left(\tfrac{10}{2}\right)[2\left(\tfrac{g}{2}\right) + (10-1)g] = 5[g + 9g] = 50g$$
$$= 50(32.2) = 1610 \text{ ft}$$

Exercise 3: Geometric Progressions

1. Find the 5th term of a GP with first term 5 and common ratio 2. Using $a_n = ar^{n-1}$ with $a = 5$, $r = 2$, $n = 5$,
$$a_n = 5(2)^{5-1} = 5(2)^4 = 5(16) = 80$$

5. Find the sum of the first 10 terms of the GP in problem 1. Using $S_n = \frac{a(1-r^n)}{1-r}$ with $a = 5$, $r = 2$, $n = 10$
$$S_n = \frac{5(1-2^{10})}{1-2} = \frac{5(-1024)}{-1} = -5(-1023) = 5115$$

9. Insert a geometric mean between 5 and 45. Using $b = \pm\sqrt{ac}$ with $a = 5$, $c = 45$,
$$b = \pm\sqrt{5(45)} = \pm\sqrt{225} = \pm 15$$
Our GP is then: 5, 15, 45 or 5, −15, 45

13. Insert two geometric means between 8 and 216.
Here $a = 8$, $a_4 = 216$ and $n = 4$,
$$a_4 = ar^3 \geq 216 = 8r^3 \geq r^3 = 27 \geq r = 3$$
Our GP is then: 8, 24, 72, 216

17. Using $y = e^{t/2}$, compute values of y for $t = 0, 1, 2, \cdots, 10$. Show that while the values of t form an AP, the values of y form a GP. Find the common ratio.

t	0	1	2	3	4	5	6	7	8	9	10
y	1	1.65	2.72	4.48	7.39	12.2	20.1	33.1	54.6	90.0	148

$$r = \frac{a_n}{a_{n-1}} = \frac{e^{(t+1)/2}}{e^{t/2}} = e^{1/2} \approx 1.649$$

21.   A certain radioactive material decays so that after each year the radioactivity is 8% less than at the start of that year. How many years will it take for its radioactivity to be 50% of its original value?

The amount of radioactive material present after each year forms a GP given by, $a, 0.92a, (0.9)^2a, (0.92)^3a, \cdots$. The nth term is given by $a_n = ar^{n-1}$. We seek the n value that will make $a_n = 0.5a$. Substituting we obtain,
$$0.5a = a(0.92)^{n-1}$$
Dividing by a: $0.5 = (0.92)^{n-1}$
Taking the logarithm of both sides,
$$\log 0.5 = (n-1)\log 0.92$$
$$n - 1 = \frac{\log 0.5}{\log 0.92} = 8.3$$
$$n = 9.3$$
However, since n begins with 1, $n - 1$ will accurately represent the number of years that have gone by. Thus our answer is: $n - 1 = 9.3 - 1 = 8.3$ years.

25.   A person has two parents, and each parent has two parents, and so on. We can write a GP for the number of ancestors as 2, 4, 8, $\cdots$. Find the total number of ancestors in five generations, starting with the parents generation.

We are asked to find the sum of the GP having first term of 2, and a ratio of 2. Substituting $a = 2$, $r = 2$, $n = 5$ into
$$Sn = \frac{a(1-r^n)}{1-r}$$
we obtain,
$$S_n = \frac{2(1-2^5)}{1-2} = \frac{2(-31)}{-1} = 62$$

29.   If the U.S. energy consumption is 7.00% higher each year, by what factor will the energy consumption have increased after 10.0 years?

Let a_0 be the amount present at the start. The amount present at the start of any future year is given by the GP,
$$a_0, 1.07a_0, (1.07)^2a_0, \cdots, (1.07)^{n-1}a_0$$
The amount present at the start of the 11th year is the same as the amount at the end of the 10th year. We calculate the 11th term of the GP having $r = 1.07$, $n = 11$

$a = a_0$ by substituting into the formula $a_n = ar^{n-1}$,
$$a_{11} = a_0(1.07)^{10} = 1.97a_0$$
The factor that energy consumption has increased is
$$\frac{\text{value of 10.0 yrs}}{\text{value start}} = \frac{1.97a_0}{a_0} = 1.97$$

33.   For a machine having an initial book value of $100,000 and a depreciation rate of 40%, the first year's depreciation is 40% of $100,000 or $40,000, and the new book value is $100,000 - $40,000 = $60,000. Thus the book values for each year form a GP
$$\$100,000, \$60,000, \$36,000, \cdots$$
Find the book value after 5 years.

The given GP 100,000, 60,000, 36,000, $\cdots$ can be written as
$$100,000, 100,000(0.6), 100,000(0.6)^2, \cdots$$
where $a = 100,000$, $r = 0.6$. Letting $n = 6$ and substituting into $a_n = ar^{n-1}$, we obtain,
$$a = 100,000(0.6)^{6-1} = 100,000(0.6)^5 = \$7776$$

Exercise 4: Infinite Geometric Progressions

1. $\lim\limits_{b \to 0} (b - c + 5) = 0 - c + 5 = 5 - c$

5.   144, 72, 36, 18, $\cdots$. In this GP $r = \frac{1}{2}$ and $a = 144$. Substituting into $S = \frac{a}{(1-r)}$ we obtain,
$$S = \frac{144}{1-\frac{1}{2}} = 288$$

9. $0.57\overline{57}\cdots = 0.57 + 0.0057 + 0.000057 + \cdots$

This forms an infinite GP with $a = 0.57 = \frac{57}{100}$ and $r = 0.01 = \frac{1}{100}$. Substituting into $S = \frac{a}{1-r}$

$$S = \frac{\frac{57}{100}}{1-\frac{1}{100}} = \frac{\frac{57}{100}}{\frac{99}{100}} = \frac{57}{99} = \frac{19}{33}$$

Thus, $0.57\overline{57}\cdots = \frac{19}{33}$

13. Each swing of a certain pendulum is 78% as long as the one before. If the first swing is 10 in., find the entire distance traveled by the pendulum before it comes to rest.

Substituting $a = 10$, $r = 0.78$ into $S = \frac{a}{1-r}$

$$S = \frac{10}{1-0.78} = \frac{10}{0.22} = 45.5 \text{ in.}$$

Exercise 5: The Binomial Theorem

1. $6! = 6 \times 5 \times 4 \times 3 \times 2 \times 1 = 720$

5. $\frac{7!}{3!4!} = \frac{7 \times 6 \times 5 \times 4!}{3 \times 2 \times 1 \times 4!} = 7 \times 5 = 35$

9. $(3a - 2b)^4 = (3a)^4 + 4(3a)^3(-2b) + 6(3a)^2(-2b)^2 + 4(3a)(-2b)^3 + (2b)^4$
$$= 81a^4 - 216a^3b + 216a^2b^2 - 96ab^3 + 16b^4$$

13. $\left(\frac{a}{b} - \frac{b}{a}\right)^6 = \left(\frac{a}{b}\right)^6 + 6\left(\frac{a}{b}\right)^5\left(-\frac{b}{a}\right) + 15\left(\frac{a}{b}\right)^4\left(-\frac{b}{a}\right)^2 + 20\left(\frac{a}{b}\right)^3\left(-\frac{b}{a}\right)^3$
$$+ 15\left(\frac{a}{b}\right)^2\left(\frac{b}{a}\right)^4 + 6\left(\frac{a}{b}\right)\left(-\frac{b}{a}\right)^5 + \left(-\frac{b}{a}\right)^6$$
$$= \left(\frac{a}{b}\right)^6 - 6\left(\frac{a}{b}\right)^4 + 15\left(\frac{a}{b}\right)^2 - 20 + 15\left(\frac{b}{a}\right)^2 - 6\left(\frac{b}{a}\right)^4 + \left(\frac{b}{a}\right)^6$$

17. $(x^2 + y^3)^8 = (x^2)^8 + 8(x^2)^7 y^3 + \left[\frac{8(7)}{2}\right](x^2)^6(y^3)^2 + \left[\frac{8(7)(6)}{3(2)}\right](x^2)^5(y^3)^3 + \cdots$
$$= x^{16} + 8x^{14}y^3 + 28x^{12}y^6 + 56x^{10}y^9 + \cdots$$

21. $(2a^2 + a + 3)^3 = [2a^2 + (a + 3)]^3$
$$= (2a^2)^3 + 3(2a^2)^2(a + 3) + 3(2a^2)(a + 3)^2 + (a + 3)^3$$
$$= 8a^6 + 12a^4(a + 3) + 6a^2(a^2 + 6a + 9) + a^3 + 3a^2(3) + 3a(3)^2 + 27$$
$$= 8a^6 + 12a^5 + 36a^4 + 6a^4 + 36a^3 + 54a^2 + a^3 + 9a^2 + 27a + 27$$
$$= 8a^6 + 12a^5 + 42a^4 + 37a^3 + 63a^2 + 27a + 27$$

25. Eleventh term of $(2 - x)^{16}$

In the expansion for $(a + b)^n$ the rth term is given by
$$\frac{n!}{(r-1)!(n-r+1)!}a^{n-r+1}b^{r-1}$$

Thus the eleventh term of $(2 - x)^{16}$ is
$$\frac{16!}{(11-1)!(16-11+1)!}(2)^{16-11+1}(-x)^{11-1} = \frac{16!}{10!6!}2^6(-x)^{10}$$
$$= \frac{16 \times 15 \times 14 \times 13 \times 12 \times 11 \times 10!}{10! \times 6 \times 5 \times 4 \times 3 \times 2 \times 1}(64)x^{10} = 512,512x^{10}$$

29. Ninth term of $(x^2 + 1)^{15}$. The ninth term is
$$\frac{15!}{(9-1)!(15-9+1)!} - (x^2)^{15-9+1}(1)^{9-1} = 6435x^{14}$$

33. $(1+a)^{-3} = 1 + (-3)a + \frac{-3(-4)}{2}a^2 + \frac{-3(-4)(-5)}{6}a^3 + \cdots$

   $\qquad = 1 - 3a + 6a^2 - 10a^3 \cdots$

Chapter 25: Review Problems

1. Find the sum of 7 terms of the AP $\ -4, -1, -2, \cdots$
   $n = 7,\ a = -4,\ d = 3$, so
   $$S_7 = \left(\frac{7}{2}\right)[2(-4) + (7-1)(3)] = \left(\frac{7}{2}\right)[-8 + 18] = 35$$

5. Insert 4 harmonic means between 2 and 12. To insert harmonic means between two terms of a harmonic progression, we take the reciprocals of the given terms, insert arithmetic means between those terms, and take reciprocals again. Taking reciprocals, our AP is
   $$\frac{1}{2}, \underline{\qquad}, \underline{\qquad}, \underline{\qquad}, \underline{\qquad}, \frac{1}{12}$$
   In this AP, $a = \frac{1}{2}$, $n = 6$, $a_6 = \frac{1}{12}$. We find the common difference, d, from the equation $a_n = a + (n-1)d$.
   $\frac{1}{12} = \frac{1}{2} + (6-1)d$
   $1 = 6 + 12(5)d$
   $60d = -5$
   $d = -\frac{1}{12}$
   Filling in the missing terms, we obtain,
   $$\frac{1}{2}, \frac{5}{12}, \frac{4}{12}, \frac{3}{12}, \frac{2}{12}, \frac{1}{12}$$
   Taking reciprocals again, we get
   $$2, 2\tfrac{2}{5}, 3, 4, 6, 12$$

9. $(a-2)^5 = a^5 + 5a^4(-2) + \frac{5\cdot4}{2}a^3(-2)^2 + \frac{5\cdot4\cdot3}{3\cdot2}a^2(-2)^3 + \frac{5\cdot4\cdot3\cdot2}{4\cdot3\cdot2}a(-2)^4 + (-2)^5$

   $\qquad = a^5 - 10a^4 + 40a^3 - 80a^2 + 80a - 32$

13. $\lim\limits_{b\to0}(2b - x + 9) = 2(0) - x + 9 = -x + 9 = 9 - x$

17. $1, -\frac{2}{5}, \frac{4}{25}, \cdots$ This is a GP with $a = 1$ and $r = -\frac{2}{5}$.
   $$S = \frac{1}{1-(-\frac{2}{5})} = \frac{1}{\frac{7}{5}} = \frac{5}{7}$$

21. Find the sixth term of a GP with first term 5 and common ratio 3. Using $a_n = ar^{n-1}$ with $a = 5$, $r = 3$, $n = 6$,
   $$a_6 = 5(3)^{6-1} = 5(3)^5 = 1215$$

25. Insert 2 geometric means between 8 and 125.
   We first find the common ratio. We will use $a_n = ar^{n-1}$, where $n = 4$, $a_4 = 125$, $a = 8$
   $$125 = 8(r)^{4-1} \geq r^3 = \frac{125}{8} \geq r = \frac{5}{2}$$
   Thus 8, 20, 50, 125 is the GP and 20, 50 are the required two geometric means.

27. $\frac{4!5!}{2!} = \frac{24(120)}{2} = 1440$

31. $\left(\frac{2x}{y^2} - y\sqrt{x}\right)^7 = \left(\frac{2x}{y^2}\right)^7 + 7\left(\frac{2x}{y^2}\right)^6(-y\sqrt{x}) + 21\left(\frac{2x}{y^2}\right)^5(-y\sqrt{x})^2 + 35\left(\frac{2x}{y^2}\right)^4(-y\sqrt{x})^3 + \cdots$

   $\qquad = \frac{128x^7}{y^{14}} - \frac{448x^{13/2}}{y^{11}} + \frac{672x^6}{y^8} - \frac{560x^{11/2}}{y^5}\cdots$

35. Find the eighth term of $(3a - b)^{11}$. $r = 8$, $n = 11$, $n - r + 1 = 4$.
   $$a_8 = \frac{11!}{7!4!}(3a)^4(-b)^7 = 330(-81)a^4b^7 = -26,730a^4b^7$$

CHAPTER 26: INTRODUCTION TO STATISTICS AND PROBABILITY

Exercise 1: Definitions and Terminology

1. The number of people in each county who voted for Jones.   Discrete

5. The models of Ford cars sold each day.   Categorical

9.

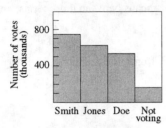

Candidate

Exercise 2: Frequency Distributions

1. (a) $\boxed{\text{Range } = 172 - 111 = 61}$

Class Midpt	Class	Limits	Abs. Freq.	Rel. Freq.	Cumulative Fr. Abs.	Cumulative Fr. Rel.
			1(b)	1(c)	10(a)	10(b)
113	110.5	115.5	3	7.5%	3	7.5%
118	115.5	120.5	4	10.0%	7	17.5%
123	120.5	125.5	1	2.5%	8	20.0%
128	125.5	130.5	3	7.5%	11	27.5%
133	130.5	135.5	0	0.0%	11	27.5%
138	135.5	140.5	2	5.0%	13	32.5%
143	140.5	145.5	2	5.0%	15	37.5%
148	145.5	150.5	6	15.0%	21	52.5%
153	150.5	155.5	7	17.5%	28	70.0%
158	155.5	160.5	2	5.0%	30	75.0%
163	160.5	165.5	5	12.5%	35	87.5%
168	165.5	170.5	3	7.5%	38	95.0%
173	170.5	175.5	2	5.0%	40	100.0%

5.

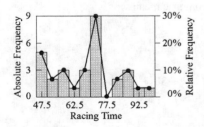

9.

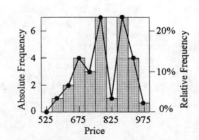

9. See Problem 6 above.

13.

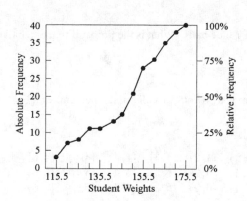

17.

3									
4	8.5	8.9	9.6	8.4	9.4				
5	4.8	9.3	9.3	9.3	0.3				
6	1.4	9.3	6.3	9.3					
7	2.5	1.2	1.4	4.5	3.6	1.4	4.9	2.7	2.8
8	8.2	4.6	9.4	5.7	3.8				
9	9.2	2.4							
10									

Exercise 3: Numerical Description of Data

1. Find the mean of the following set of grades.

$$85 \quad 74 \quad 69 \quad 59 \quad 60 \quad 96 \quad 84 \quad 48 \quad 89 \quad 76 \quad 96 \quad 68 \quad 98 \quad 79 \quad 76$$

To find the mean we add the fifteen pieces of data and divide by 15,

$$\overline{x} = \frac{85+74+69+59+60+96+84+48+89+76+96+68+98+79+76}{15} = 77$$

5. Find the mean of the prices in Exercise 2, Problem 3.

Adding the thirty data values and dividing by 30,
$$\overline{x} = 23873 \div 30 = \$796$$

9. midrange $= \frac{199+116}{2} = 157.5$

13. Find the mode of the times in Exercise 2, Problem 2.

The time 59.3 min occurs 3 times, and is the mode.

17. Find the median of the weights in Exercise 2, Problem 1.
Rewriting the data in order of magnitude gives,

$$111 \quad 114 \quad 114 \quad ... \quad 149 \quad 149 \quad ... \quad 168 \quad 172 \quad 172$$

The median is then 149 lb.

21. Give the 5-number summary for the weights in Problem 2.
From Problem 16,

       the minimum value $= 116$
       the median $= 158$
       the maximum value $= 199$
The lower hinge $= 142$
The upper hinge $= 164$
The 5-number summary is: 116, 142, 158, 164, 199

25. Find the range of the weights in Problem 2.
       range $= 199 - 116 = 83$

29. Find the sample variance and standard deviation of the weights in Problem 2.
From Prob. 2, $\overline{x} = 154.625$. Then,
sample variance: $s^2 = \frac{4881.875}{7} = 697.4$ sample standard deviation: $s = \sqrt{697.4} = 26.4$

Exercise 4: Introduction to Probability

1. We draw a ball from a bag that contains 8 green balls and 7 blue balls. What is the probability that a ball drawn at random will be green?

$$P(G) = \frac{m}{n} = \frac{8}{8+7} = \frac{8}{15}$$

5. If we throw two dice, what is the probability that their sum is 9? Hint: List all possible outcomes, $(1,1)$, $(1,2)$, etc. and count those that have a sum of 9.

The possible outcomes of throwing two dice are as follows,

$(1,1)$ $(1,2)$ $(1,3)$ $(1,4)$ $(1,5)$ $(1,6)$ $\qquad$ $(4,1)$ $(4,2)$ $(4,3)$ $(4,4)$ $(4,5)$ $(4,6)$
$(2,1)$ $(2,2)$ $(2,3)$ $(2,4)$ $(2,5)$ $(2,6)$ $\qquad$ $(5,1)$ $(5,2)$ $(5,3)$ $(5,4)$ $(5,5)$ $(5,6)$
$(3,1)$ $(3,2)$ $(3,3)$ $(3,4)$ $(3,5)$ $(3,6)$ $\qquad$ $(6,1)$ $(6,2)$ $(6,3)$ $(6,4)$ $(6,5)$ $(6,6)$

Since there are four successful ways out of a total of thirty-six possible ways we obtain,

$$P(A) = \frac{4}{36} = \frac{1}{9}$$

9. At a certain school, 55% of the students have brown hair, 15% have blue eyes, and 7% have both brown hair and blue eyes. What is the probability that a student chosen at random will have either brown hair or blue eyes, or both?

The probability that either of two events will occur,

$$P(A + B) = P(A) + P(B) - P(A, B) = 0.55 + 0.15 - 0.07 = 0.63$$

13. For the relative frequency histogram for the population of students' grades (Fig. 26-3), find the probability that a student chosen at random will have gotten a grade between 68 and 74.

$$\frac{4}{30} = 0.133$$

17. $n = 7$, $p = 0.8$, $x = 6$, $q = 0.2$

$$P(6) = \frac{7!}{(7-6)!6!}(0.8)^6(0.2)^{7-6} = \frac{7}{1}(0.8)^6(0.2) = 0.367$$

21. What is the probability of tossing 7 heads in 10 tosses of a fair coin?

$$n = 10, \ x = 7, \ p = 0.5, \ q = 0.5$$
$$P(7) = \frac{10!}{(10-7)!7!}(0.5)^7(0.5)^{10-7} = 0.117$$

Exercise 5: The Normal Curve

1. Find the area under the normal curve between the mean and 1.5 standard deviations, Fig. 26-17.

Referring to Table 26-6, we find that the area under the normal curve within 1.5 standard deviations of the mean is,

$$\text{area} = 0.4332$$

5.   The distribution of the weights of 1000 students at Tech College has a mean of 163 lb with a standard deviation of 18 lb. Assuming that the weights are normally distributed, predict the number of students who have weights between 130 lb and 170 lb.

First we compute the number of students who weigh less than 130.

$$z = \frac{x-\bar{x}}{s} = \frac{130-163}{18} = -1.83$$

From Table 26-6, with $z = 1.8$, we read an area of 0.4641.

Repeating the calculations for students who weigh more than 170, we obtain,

$$z = \frac{x-\bar{x}}{s} = \frac{170-163}{18} = 0.39$$

From Table 26-6, with $z = 0.4$, we find an area of 0.1554. Adding the two areas gives

$$0.4641 + 0.1554 = 0.6195$$

or 61.9% of the students between the given weights. Since there are 1000 students, we predict

$$0.6195(1000) = 620 \text{ students}$$

will be between the given weights.

9.   For the test of Problem 8, predict the number of failing grades (60 or less). First we find the area to the right of z,

$$z = \frac{(60-82.6)}{7.4} = -3.1$$

From Table 26-6 with $z = -3.1$, we find an area of 0.4990. The area to the right of $z = -3.1$ is

$$0.5000 - 0.4990 = 0.001$$

We predict the number of students with grades below 60,

$$500 \times 0.001 = 1 \text{ student}$$

Exercise 6: Standard Errors

1.   The heights of 49 randomly chosen students at Tech College were measured. Their mean $\bar{x}$ was found to be 69.47 in. and their standard deviation s was 2.35 in. Estimate the mean μ of the entire population of students at Tech College with a confidence level of 68%.

Computing the standard error with $s = 2.35$ and $n = 49$,

$$SE_{\bar{x}} = \frac{s}{\sqrt{n}} = \frac{2.35}{\sqrt{49}} = 0.34$$

We predict that there is 68% chance that u will fall within the range

$$\mu = 69.47 \pm 0.34 \text{ inches}$$

5.   A single sample of size 32 drawn from a population is found to have a mean of 164.0 and a standard deviation s of 16.31. Give the population mean with a confidence level of 68%.

Computing the standard error with $s = 16.31$ and $n = 32$,

$$SE_{\bar{x}} = 16.31 \div \sqrt{32} = 2.88$$

We predict that there is a 68% chance that μ will fall within the range

$$\mu = 164.0 \pm 2.88$$

9.   Find the 68% confidence interval for drawing a heart from a deck of cards, for 200 draws from the deck, replacing the card each time before the next draw.

$$p = \frac{13}{52} = 0.250 \text{ for a heart on one draw}$$
$$n = 200 \text{ draws}$$

$$SE_p = \sqrt{\frac{0.25(1-0.25)}{200}} = 0.031$$

Thus the 68% confidence interval is 0.250 ± 0.031.

Exercise 7: Process Control

1. Find the values for the central line, and the upper and lower control limits.

$$\text{central line} = \bar{p} = \frac{\sum \text{defectives}}{nk} = \frac{554}{1000(20)} = 0.0277$$

$$SE_p = \sqrt{\frac{\bar{p}(1-\bar{p})}{n}} = \sqrt{\frac{0.0277(1-0.0277)}{1000}} = 0.00519$$

$$\text{limits} = \bar{p} \pm 3SE_p = 0.0277 \pm 3(0.00519)$$

$$UCL = 0.0277 + 3(0.00519) = 0.0433$$

$$LCL = 0.0277 - 3(0.00519) = 0.0121$$

3. Find the values for the central line, and the upper and lower control limits.

$$\text{central line} = \bar{p} = \frac{2990}{500(20)} = 0.2990$$

$$SE_p = \sqrt{\frac{0.2990(1-0.2990)}{500}} = 0.02047$$

$$UCL = 0.2990 + 3(0.02047) = 0.3604$$

$$LCL = 0.2990 - 3(0.02047) = 0.2376$$

5. Find the values for the central line, and the upper and lower control limits for the mean.

Day	Mean X	Range R
1	192.7	17.8
2	192.7	23.0
3	202.2	32.1
4	199.2	32.9
5	209.1	25.4
6	188.4	23.0
7	204.4	25.7
8	196.5	34.2
9	192.1	34.3
10	201.5	27.9
11	194.5	12.2
12	209.4	19.8
13	199.8	17.7
14	206.6	20.9
15	197.8	33.6
16	195.5	36.6
17	199.2	37.1
18	203.9	13.5
19	210.4	17.1
20	191.6	16.6
21	189.9	18.5
Sums	4177.4	519.9

$$\text{central line} = \overline{\overline{X}} = \frac{\sum \overline{X}}{21} = \frac{4177.4}{21} = 198.9$$

$$\overline{R} = \frac{\sum R}{21} = \frac{519.9}{21} = 24.76$$

$$\text{control limits for mean} = \overline{\overline{X}} \pm A_2 \overline{R} = 198.9 \pm 0.577(24.76)$$

Thus $UCL = 213.2$

 $LCL = 184.6$

9. Find the values for the central line, and the upper and lower control limits for the mean.

Day	Mean X	Range R
1	15.58	8.4
2	14.78	6.9
3	14.50	3.2
4	16.18	5.3
5	15.12	4.4
6	16.84	4.1
7	14.82	8.2
8	14.96	6.5
9	14.80	4.8
10	17.56	7.5
11	16.38	6.8
12	13.56	3.3
13	14.50	7.3
14	14.14	8.6
15	13.06	3.5
16	17.04	4.1
17	14.62	7.7
18	12.64	2.6
19	14.66	7.5
20	13.02	5.7
21	15.16	8.3
Sums	313.92	124.7

central line $= \overline{\overline{X}} = \frac{\sum \overline{X}}{21} = \frac{313.92}{21} = 14.95$

$\overline{R} = \frac{124.7}{21} = 5.938$

control limits for mean $= \overline{\overline{X}} \pm A_2 \overline{R} = 14.95 \pm (0.577)(5.938)$

Thus,   UCL $= 18.4$

LCL $= 11.5$

central line $= 15.0$

Exercise 8: Regression

1. Find the correlation coefficient.

x	y	xy	x^2	y^2
−8	−6.238	49.904	64.000	38.913
−6.66	−3.709	24.702	44.356	13.757
−5.33	−0.712	3.795	28.409	0.507
−4	1.887	−7.548	16.000	3.561
−2.66	4.628	−12.310	7.076	21.418
−1.33	7.416	−9.863	1.769	54.997
0	10.2	0.000	0.000	104.040
1.33	12.93	17.197	1.769	167.185
2.66	15.70	41.762	7.076	246.490
4	18.47	73.880	16.000	341.141
5.33	21.32	113.636	28.409	454.542
6.66	23.94	159.440	44.356	573.124
8	26.70	213.600	64.000	712.890
9.33	29.61	276.261	87.049	876.752
10.6	32.35	342.910	112.360	1046.523
12	35.22	422.640	144.000	1240.448
Sums: 31.93	229.712	1710.006	666.629	5896.288

$n = 16$

$$r = \frac{16(1710.006) - (31.93)(229.712)}{\sqrt{16(666.629) - (31.93)^2}\sqrt{16(5896.288) - (229.712)^2}}$$

$r = 0.9999759 = 1.00$ (rounded)

5. Find the least squares line for the data from Problem 2.

$$\text{Slope } m = \frac{16(-7610.177) - (-143.780)(457.997)}{16(2020.976) - (-143.780)^2} = -4.79$$

$$y \text{ intercept } b = \frac{(2020.976)(457.997) - (-143.78)(-7610.177)}{16(2020.976) - (-143.780)^2} = -14.5$$

Thus, $y = -4.79x - 14.5$

Chapter 26: Review Problems

1. The life of certain radios is continuous.

5. Make a bar graph for the data of Problem 4.

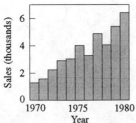

9. At a certain factory, 72% of the workers have brown hair, 6% are left-handed, and 3% have both brown hair and are left-handed. What is the probability that a worker chosen at random will have either brown hair or be left-handed, or both?

$$P(A + B) = P(A) + P(B) - P(A, B)$$
$$= 0.72 + 0.06 - 0.03 = 0.75$$

13. Find the area in the tail of the normal curve to the left of $z = 0.4$.

From Table 26-6, with $z = 0.4$, we read an area of 0.1554. The area to the right of $z = 0.4$ is thus,
$$0.5000 - 0.1554 = 0.3446$$

17.

21. Find the mean. Adding the forty data points and dividing by 40,
$$\overline{x} = \frac{6006}{40} = 150$$

25. Find the standard deviation.
$$\text{standard deviation} = s = \sqrt{489} = 22.1.$$

29. Predict the population standard deviation, with a 95% confidence interval.
$$\sigma = 22.1 \pm 2(2.47) = 22.1 \pm 4.94$$

33. Determine the quartiles and give the quartile range of the data;

167   245   327   486   524   639   797   853   974   1136   1162   1183

First we arrange the data in four groups containing three pieces of data each.

167   245   327   486   524   639   797   853   974   1136   1162   1183

The quartiles are
$$Q_1 = \frac{(327+486)}{2} = 407$$
$$Q_2 = \text{the median} = \frac{(639+797)}{2} = 718$$
$$Q_3 = \frac{(974+1136)}{2} = 1055$$

Quartile range $= 1055 - 407 = 648$

37. Find the correlation coefficient and the least-squares-fit for the following data:

x	y	xy	x^2	y^2
5	6.882	34.410	25.000	47.362
11.2	−7.623	−85.378	125.440	58.110
17.4	−22.75	−395.850	302.760	517.562
23.6	−36.09	−851.724	556.960	1302.488
29.8	−51.13	−1523.674	888.040	2614.277
36.0	−64.24	−2312.640	1296.000	4126.778
42.2	−79.44	−3352.368	1780.840	6310.714
48.4	−94.04	−4551.536	2342.560	8843.522
54.6	−107.8	−5885.880	2981.160	11620.840
60.8	−122.8	−7466.240	3696.640	15079.840
67.0	−138.6	−9286.200	4489.000	19209.960
73.2	−151.0	−11053.200	5358.240	22801.000
79.4	−165.3	−13124.820	6304.360	27324.090
85.6	−177.6	−15202.560	7327.360	31541.760
91.8	−193.9	−17800.020	8427.240	37597.210
98	−208.9	−20472.200	9604.000	43639.210
824.0	−1614.331	−113329.880	55505.600	232634.723

$n = 16$

$$r = \frac{16(-113329.880) - (824)(-1614.331)}{\sqrt{16(55505.6) - (824)^2}\sqrt{16(232634.723) - (-1614.33)^2}}$$

$r = -0.9999 = -1.00$ (rounded)

Thus, correlation coefficient $= -1.00$

slope $m = \dfrac{16(-113329.88) - (824)(-1614.331)}{16(55505.6) - (824)^2} = -2.31$

y intercept $b = \dfrac{(55505.6)(-1614.331) - (824)(-113329.88)}{16(55505.6) - (824)^2} = 18.1$

Thus, the least-squares-fit is:
$$y = -2.31x + 18.1$$

CHAPTER 27: DERIVATIVES OF ALGEBRAIC FUNCTIONS

Exercise 1: Limits

1. $\lim\limits_{x\to 2} (x^2 + 2x - 7) = 1$

5. $\lim\limits_{x\to 5} \frac{x^2-25}{x-5} = \lim\limits_{x\to 5} \frac{(x+5)(x-5)}{x-5}$

$\qquad = \lim\limits_{x\to 5} (x+5) = 5 + 5 = 10$

9. $\lim\limits_{x\to 0} (4x^2 - 5x - 8) = -8$

13. $\lim\limits_{x\to 0} \left(\frac{1}{2+x} - \frac{1}{2}\right) \cdot \frac{1}{x} = \lim\limits_{x\to 0} \left[\frac{2-2-x}{2(2+x)}\right] \cdot \frac{1}{x}$

$\qquad = \lim\limits_{x\to 0} \left[-\frac{1}{2(2+x)}\right] = -\frac{1}{4}$

17. $\lim\limits_{x\to\infty} \frac{x^2+x-3}{5x^2+10} = \lim\limits_{x\to\infty} \frac{1+\frac{1}{x}-\frac{3}{x^2}}{5+\frac{10}{x^2}} = \frac{1}{5}$

21. $\lim\limits_{x\to 0} \frac{\sin x}{\tan x} = \lim\limits_{x\to 0} \cos x = 1$

25. $\lim\limits_{x\to 0^+} \frac{7}{x} = +\infty$

29. $\lim\limits_{x\to 2^-} \frac{5+x}{x-2} = -\infty$

33. $\lim\limits_{d\to 0} \frac{(x+d)^2-x^2}{x^2(x+d)} = 0$

37. $\lim\limits_{d\to 0} 3x + d - \frac{1}{(x+d+2)(x-2)} = 3x - \frac{1}{x^2-4}$

41. $\lim\limits_{d\to 0} \frac{(x+d)^2-2(x+d)-x^2+2x}{d} = \lim\limits_{d\to 0} \frac{2dx+d^2-2d}{d}$

$\qquad = \lim\limits_{d\to 0} 2x + d - 2 = 2x - 2$

Exercise 2: The Derivative

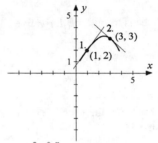

1. $\frac{2-0.5}{1-0} = 1.5$

5. $y = 2x + 5$
    $y + \triangle y = 2(x + \triangle x) + 5$
    $\triangle y = 2x + 2\triangle x + 5 - 2x - 5 = 2\triangle x$
    $\frac{\triangle y}{\triangle x} = 2 \qquad\qquad \frac{dy}{dx} = 2$

9. $y = x^3$
    $y + \triangle y = (x + x)^3 = x^3 + 3x^2\triangle x + 3x\triangle x^2 + \triangle x^3$
    $\triangle y = 3x^2\triangle x + 3x\triangle x^2 + \triangle x^3$
    $\frac{\triangle y}{\triangle x} = 3x^2 + 3x\triangle x + \triangle x^2$
    $\frac{dy}{dx} = 3x^2$

13. $y = \sqrt{3-x}$
    $y + \triangle y = \sqrt{3 - x - \triangle x},$
    $\triangle y = \sqrt{3-x-\triangle x} - \sqrt{3-x} \cdot \frac{\sqrt{3-x-\triangle x}+\sqrt{3-x}}{\sqrt{3-x-\triangle x}+\sqrt{3-x}} = \frac{(3-x-\triangle x)-(3-x)}{\sqrt{3-x-\triangle x}+\sqrt{3-x}} = -\frac{\triangle x}{\sqrt{3-x-\triangle x}+\sqrt{3-x}}$
    $\frac{\triangle y}{\triangle x} = -\frac{1}{\sqrt{3-x-\triangle x}+\sqrt{3-x}}$
    $\frac{dy}{dx} = -\frac{1}{2\sqrt{3-x}}$

17. $y = x + \frac{1}{x}$; $y + \triangle y = x + \triangle x + \frac{1}{(x+\triangle x)}$; $\triangle y = x + \triangle x + \frac{1}{x+\triangle x} - x - \frac{1}{x} = \triangle x + \frac{x - x - \triangle x}{(x+\triangle x)\,x}$

$\frac{\triangle y}{\triangle x} = 1 - \frac{1}{(x+\triangle x)x}$; $\frac{dy}{dx} = 1 - \frac{1}{x^2}$ when $x = 2$, $\frac{dy}{dx} = 1 - \frac{1}{4} = \frac{3}{4}$

21. $y = 2x^2 - 6$; $y + \triangle y = 2(x + \triangle x)^2 - 6 = 2(x^2 + 2x\triangle x + \triangle x^2) - 6$

$\triangle y = 2x^2 - 4x\triangle x + 2\triangle x^2 - 6 - 2x^2 + 6$; $\frac{\triangle y}{\triangle x} = 4x + 2\triangle x$

$\frac{dy}{dx} = 4x$ at $x = 3$, $\frac{dy}{dx} = 4(3) = 12$

25. In Problem 23 $y'(x) = 4x$, therefore, $y'(-1) = 4(-1) = -4$

29. $\frac{d}{dx}$ means to find the derivative

$\frac{d}{dx} = \lim_{\triangle x \to 0} \frac{f(x+\triangle x)-f(x)}{\triangle x} = \lim_{\triangle x \to 0} \frac{3(x+\triangle x)+2-3x-2}{\triangle x} = \lim_{\triangle x \to 0} \frac{3x+3\triangle x+2-3x-2}{\triangle x} = \lim_{\triangle x \to 0} 3$

Therefore, $\frac{d}{dx}(3x + 2) = 3$

33. $D(3x + 2)$ means to find the derivative. This was done in Problem 29. Thus, $D(3x + 2) = 3$.

Exercise 3: Rules for Derivative

1. $y = a^2$

Here a^2 is a constant, therefore, $\frac{dy}{dx} = 0$.

5. $y = 3x^2$

$\frac{dy}{dx} = 3 \cdot 2x^{2-1} = 6x$

9. $y = \frac{1}{x} = x^{-1}$

$\frac{dy}{dx} = -1x^{-1-1} = -1x^{-2} = \frac{-1}{x^2}$

13. $y = 7.5x^{1/3}$

$\frac{dy}{dx} = (7.5) \cdot \frac{1}{3}x^{(1/3)-1} = 2.5x^{-2/3}$

17. $y = -17\sqrt{x^3} = -17x^{3/2}$

$\frac{dy}{dx} = (-17) \cdot \frac{3}{2}x^{(3/2)-1} = -\frac{51}{2}x^{1/2} = \frac{-51\sqrt{x}}{2}$

21. $y = 3x - x^3$

$\frac{dy}{dx} = 3 - 3x^2$

25. $y = ax + b$

$\frac{dy}{dx} = a \cdot 1x^{1-1} + 0 = a$

29. $y = 2x^{3/4} + 4x^{-1/4}$

$\frac{dy}{dx} = (2) \cdot \frac{3}{4}x^{3/4-1} + 4\left(-\frac{1}{4}\right)x^{-1/4-1}$

$= \frac{3}{2}x^{-1/4} - x^{-5/4}$

33. $y = \frac{x+4}{x} = \frac{x}{x} + \frac{4}{x} = 1 + 4x^{-1}$

$\frac{dy}{dx} = 0 - 4x^{-2} = -4x^{-2}$

37. $y'(-3) = 6(-3)^2 = 6 \cdot 9 = 54$

41. $\frac{d}{dx}(3x^5 + 2x) = \frac{d}{dx}3x^5 + \frac{d}{dx}2x = 15x^4 + 2$

45. $D(3x^2 + 2x) = D(3x^2) + D(2x) = 6x + 2$

49. $y = x^3 - 5$

$y' = 3x^2$

$y'(1) = 3$

53. $v = 5t^2 - 3t + 4$

$\frac{dv}{dt} = 10t - 3$

57. $y = \sqrt{5w^3} = \sqrt{5}w^{3/2}$

$\frac{dy}{dw} = \sqrt{5} \cdot \frac{3}{2}w^{1/2} = \frac{3\sqrt{5}}{2}\sqrt{w} = \frac{3\sqrt{5w}}{2}$

Exercise 4: Derivative of a Function Raised to a Power

1. $y = (2x + 1)^5$

$y' = 5(2x + 1)^4(2) = 10(2x + 1)^4$

5. $y = (2 - 5x)^{3/5}$

$y = \frac{3}{5}(2 - 5x)^{-2/5}(-5) = -\frac{3}{(2-5x)^{2/5}}$

9. $y = \frac{3}{x^2+2} = 3(x^2+2)^{-1}$

$y' = -3(x^2+2)^{-2}(2x) = \frac{-6x}{(x^2+2)^2}$

13. $y = \sqrt{1-3x^2} = (1-3x^2)^{1/2}$

$y' = \frac{1}{2}(1-3x^2)^{-1/2}(-6x) = \frac{-3x}{\sqrt{1-3x^2}}$

17. $y = \sqrt[3]{4-9x} = (4-9x)^{1/3}$

$y' = \frac{1}{3}(4-9x)^{-2/3}(-9) = -\frac{3}{(4-9x)^{2/3}}$

21. $\frac{d}{dx}(3x^5+2x)^2$

Letting $u = 3x^5+2x$, where $\frac{du}{dx} = 15x^4+2$ we get, $y' = 2(3x^5+2x)(15x^4+2)$

25. $v = (5t^2-3t+4)^2$

Letting $u = 5t^2-3t+4$, where $\frac{du}{dt} = 10t-3$, we obtain, $\frac{dv}{dt} = 2(5t^2-3t+4)(10t-3)$

29. $y = (4.82x^2-8.25x)^3$

Letting $u = 4.82x^2-8.25x$, where $\frac{du}{dx} = 9.64x-8.25$, we get $y' = 3(4.82x^2-8.25x)^2(9.64x-8.25)$

$y'(3.77) = 3[4.82(3.77)^2-8.25(3.77)]^2 \cdot [(9.64)(3.77)-8.25] = 118,000$

Exercise 5: Derivatives of Products and Quotients

1. $y = x(x^2-3)$

$\frac{dy}{dx} = x\frac{d}{dx}(x^2-3) + (x^2-3)\frac{d}{dx}x = x\cdot 2x + (x^2-3)\cdot 1 = 2x^2+x^2-3 = 3x^2-3$

5. $y = x(x^2-2)^2$

$y' = x(2)(x^2-2)(2x) + (x^2-2)^2 = 4x^2(x^2-2) + (x^2-2)^2 = 4x^4-8x^2+x^4-4x^2+4 = 5x^4-12x^2+4$

9. $y = x\sqrt{1+2x}$

$y' = x\left(\frac{1}{2}\right)(1+2x)^{-1/2}(2) + \sqrt{1+2x} = \frac{x}{\sqrt{1+2x}} + \sqrt{1+2x}$

13. $y = x\sqrt{a+bx}$

$y' = x\left(\frac{1}{2}\right)(a+bx)^{-1/2}(b) + \sqrt{a+bx} = \frac{bx}{2\sqrt{a+bx}} + \sqrt{a+bx}$

17. $v = (5t^2-3t)(t+4)^2$

$\frac{dv}{dt} = (5t^2-3t)\frac{d}{dt}(t+4)^2 + (t+4)^2\frac{d}{dt}(5t^2-3t) = (5t^2-3t)\cdot 2(t+4) + (t+4)^2(10t-3)$

$= (t+4)[(5t^2-3t)\cdot 2 + (t+4)(10t-3)] = (t+4)(10t^2-6t+10t^2+37t-12)$

$= (t+4)(20t^2+31t-12)$

21. $\frac{d}{dx}(2x^5+5x)^2(x-3) = (2x^5+5x)^2\cdot 1 + (x-3)(2)(2x^5+5x)(10x^4+5)$

$= (2x^5+5x)^2 + (2x-6)(2x^5+5x)(10x^4+5)$

25. $y = (x-1)^{1/2}(x+1)^{3/2}(x+2)$

$y' = (x-1)^{1/2}(x+1)^{3/2} + (x-1)^{1/2}\left(\frac{3}{2}\right)(x+1)^{1/2}(x+2) + \left(\frac{1}{2}\right)(x-1)^{-1/2}(x+1)^{3/2}(x+2)$

$= (x-1)^{1/2}(x+1)^{3/2} + \frac{3}{2}(x^2-1)^{1/2}(x+2) + \frac{(x+1)^{3/2}(x+2)}{2\sqrt{x-1}}$

29. $y = \frac{x^2}{4-x^2}$

$y' = \frac{(4-x^2)(2x)-x^2(-2x)}{(4-x^2)^2} = \frac{8x-2x^3+2x^3}{(4-x^2)^2} = \frac{8x}{(4-x^2)^2}$

33. $y = \frac{2x^2-1}{(x-1)^2}$

$y' = \frac{(x-1)^2(4x)-(2x^2-1)(2)(x-1)}{(x-1)^4} = \frac{(x-1)(4x)-(2x^2-1)(2)}{(x-1)^3} = \frac{4x^2-4x-4x^2+2}{(x-1)^3} = \frac{2-4x}{(x-1)^3}$

37. $w = \frac{z}{\sqrt{z^2-a^2}} = \frac{z}{(z^2-a^2)^{1/2}}$

$\frac{dw}{dz} = \frac{\sqrt{z^2-a^2}-\frac{z^2}{\sqrt{z^2-a^2}}}{z^2-a^2} = \frac{\frac{z^2-a^2-z^2}{\sqrt{z^2-a^2}}}{z^2-a^2} = \frac{-a^2}{(z^2-a^2)^{3/2}}$

41. $y = x(8-x^2)^{1/2}$

$y' = x\left(\frac{1}{2}\right)(8-x^2)^{-1/2}(-2x) + (8-x^2)^{1/2} = \frac{-x^2}{\sqrt{8-x^2}} + \sqrt{8-x^2}$

$y'(2) = -\frac{4}{\sqrt{4}} + \sqrt{4} = -2 + 2 = 0$

Exercise 6: Derivatives of Implicit Relations

1. $y = 2u^3$

$\frac{dy}{dw} = 2(3)u^2\frac{du}{dw} = 6u^2\frac{du}{dw}$

5. $\frac{d}{dx}(x^3y^2) = x^3\frac{d}{dx}y^2 + y^2\frac{d}{dx}x^3$

$= x^3 \cdot 2y\frac{dy}{dx} + y^2 \cdot 3x^2$

$= 2x^3y\frac{dy}{dx} + 3x^2y^2$

9. $x = y^2 - 7y$

$\frac{dx}{dy} = 2y - 7$

13. $5x - 2y = 7$

$5 - 2y' = 0$

$y' = \frac{5}{2}$

17. $y^2 = 4ax$

$2yy' = 4a$

$y' = \frac{4a}{2y} = \frac{2a}{y}$

21. $y + y^3 = x + x^3$

$y' + 3y^2y' = 1 + 3x^2$

$y' = \frac{1+3x^2}{1+3y^2}$

25. $x^2 + y^2 = 25$

$2x + 2yy' = 0$

$y' = -\frac{x}{y}$

$y(2) = \sqrt{25 - 2^2} = \sqrt{21}$

$y'(2) = -\frac{2}{\sqrt{21}} = -0.436$

29. $y = x^3$

$\frac{dy}{dx} = 3x^2$

$dy = 3x^2dx$

33. $y = (x+1)^2(2x+3)^3$

$\frac{dy}{dx} = 2(x+1)(2x+3)^3$

$+ (x+1)^2(3)(2x+3)^2(2)$

$= 2(x+1)(2x+3)^2[2x+3+3x+3]$

$= 2(x+1)(2x+3)^2(5x+6)$

$dy = 2(x+1)(2x+3)^2(5x+6)dx$

35. $y = \dfrac{\sqrt{x-4}}{3-2x} = \dfrac{(x-4)^{1/2}}{(3-2x)}$

$\dfrac{dy}{dx} = \dfrac{(3-2x)\left(\frac{1}{2}\right)(x-4)^{-1/2} - (x-4)^{1/2}(-2)}{(3-2x)^2} = \dfrac{\frac{1}{2}(x-4)^{-1/2}[3-2x+4(x-4)]}{(3-2x)^2} = \dfrac{3-2x+4x-16}{2(x-4)^{1/2}(3-2x)^2} = \dfrac{2x-13}{2\sqrt{x-4}(3-2x)^2}$

$dy = \dfrac{2x-13}{2\sqrt{x-4}(3-2x)^2}\,dx$

39. $2\sqrt{x} + 3\sqrt{y} = 4; \quad 2x^{1/2} + 3y^{1/2} = 4; \quad x^{-1/2} + \frac{3}{2}y^{-1/2}\dfrac{dy}{dx} = 0$

$\frac{3}{2}y^{-1/2}\dfrac{dy}{dx} = -x^{-1/2}; \quad \dfrac{dy}{dx} = -x^{-1/2}\left(\frac{2}{3}y^{1/2}\right) = -\dfrac{2y^{1/2}}{3x^{1/2}}; \quad dy = -\dfrac{2\sqrt{y}}{3\sqrt{x}}\,dx$

Exercise 7: Higher-Order Derivatives

1. $y = 3x^4 - x^3 + 5x; \quad y' = 12x^3 - 3x^2 + 5; \quad y'' = 36x^2 - 6x$

5. $y = \sqrt{5-4x^2} = (5-4x^2)^{1/2}$

$y' = \frac{1}{2}(5-4x^2)^{-1/2}(-8x) = -4x(5-4x^2)^{-1/2}$

$y'' = -4\left[x\cdot\left(-\frac{1}{2}\right)(5-4x^2)^{-3/2}(-8x) + (5-4x^2)^{-1/2}\right] = -4\left[4x^2(5-4x^2)^{-3/2} + (5-4x^2)^{-1/2}\right]$

$= -4(5-4x^2)^{-3/2}[4x^2 + (5-4x^2)] = -4(5-4x^2)^{-3/2}(5) = \dfrac{-20}{(5-4x^2)^{3/2}}$

9. $y = x(9+x^2)^{1/2}$

$y' = x\left(\frac{1}{2}\right)(9+x^2)^{-1/2}(2x) + (9+x^2)^{1/2} = x^2(9+x^2)^{-1/2} + (9+x^2)^{1/2}$

$y'' = x^2\left(-\frac{1}{2}\right)(9+x^2)^{-3/2}(2x) + (9+x^2)^{-1/2}(2x) + \left(\frac{1}{2}\right)(9+x^2)^{-1/2}(2x)$

$= -x^3(9+x^2)^{-3/2} + 2x(9+x^2)^{-1/2} + x(9+x^2)^{-1/2}$

$y''(4) = -64(25)^{-3/2} + 8(25)^{-1/2} + 4(25)^{-1/2} = -\dfrac{64}{125} + \dfrac{8}{5} + \dfrac{4}{5} \cong 1.888$

Chapter 27 Review Problems

1. $y = \sqrt{\dfrac{x^2-1}{x^2+1}}$

$y' = \dfrac{\sqrt{x^2+1}(x^2-1)^{1/2}(x) - \sqrt{x^2-1}(x^2+1)^{-1/2}(x)}{x^2+1} = \dfrac{\dfrac{x\sqrt{x^2+1}}{\sqrt{x^2-1}} - \dfrac{x\sqrt{x^2-1}}{\sqrt{x^2+1}}}{x^2+1} = \dfrac{x(x^2+1) - x(x^2-1)}{(x^2+1)\sqrt{x^2-1}\sqrt{x^2+1}}$

$= \dfrac{x^3+x-x^3+x}{(x^2+1)\sqrt{x^4-1}} = \dfrac{2x}{(x^2+1)\sqrt{x^4-1}}$

5. $f(x) = \sqrt{4x^2+9}; \quad f'(x)\frac{1}{2}(4x^2+9)^{-1/2}(8x); \quad f'(2) = \dfrac{4(2)}{\sqrt{16+9}} = \dfrac{8}{5}$

9. $f(x) = \sqrt{25-3x}$

$f'(x) = \frac{1}{2}(25-3x)^{-1/2}(-3)$

$f''(x) = -\frac{3}{2}\left(-\frac{1}{2}\right)(25-3x)^{-3/2}(-3) = -\dfrac{9}{4(25-3x)^{3/2}}$

$f''(3) = -\dfrac{9}{4(25-9)^{3/2}} = -\dfrac{9}{4(64)} \cong -0.03516$

13. $y = (25 - x^2)^{-1/2}$

$y' = -\frac{1}{2}(25 - x^2)^{3/2}(-2x)$

$y'(3) = \frac{3}{(25-9)^{3/2}} = \frac{3}{64}$

17. $\frac{d}{dx}(3x + 2) = \frac{d}{dx}3x + \frac{d}{dx}x^2 = 3 + 0 = 3$

21. $z = 9 - 8w + w^2$

$\frac{dz}{dx} = \frac{d}{dx}9 - \frac{d}{dx}8w + \frac{d}{dx}w^2 = -8\frac{dw}{dx} + 2w\frac{dw}{dx}$

25. $s = 58.3t^3 - 63.8t$

$\frac{ds}{dt} = (3)(58.3)t^2 - 63.8 = 174.9t^2 - 63.8$

29. $y = 16x^3 + 4x^2 - x - 4$

$y' = 48x^2 + 8x - 1$

33. $y = \frac{x^2+3x}{x-1}$

$y' = \frac{(x-1)(2x+3)-(x^2+3x)(1)}{(x-1)^2} = \frac{2x^2-2x+3x-3-x^2-3x}{(x-1)^2} = \frac{x^2-2x-3}{(x-1)^2}$

37. $y = 3x^4$

$\frac{dy}{dx} = 12x^3$

$dy = 12x^3 dx$

CHAPTER 28: GRAPHICAL APPLICATIONS OF THE DERIVATIVE

Exercise 1: Tangents and Normals

1. $y = x^2 + 2$, $y(1) = 3$
   $y' = 2x$
   $y'(1) = 2$
   Tangent: $\frac{y-3}{x-1} = 2$
   $y - 3 = 2x - 2$
   $2x - y + 1 = 0$
   Normal: $\frac{y-3}{x-1} = -\frac{1}{2}$
   $2y - 6 = -x + 1$
   $x + 2y - 7 = 0$

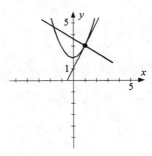

5. $x^2 + y^2 = 25$
   $2x + 2yy' = 0$
   $y' = -\frac{x}{y}$, $y'(3,4) = -\frac{3}{4}$
   Tangent: $\frac{y-4}{x-3} = -\frac{3}{4}$
   $4y - 16 = -3x + 9$
   $3x + 4y = 25$
   Normal: $\frac{y-4}{x-3} = \frac{4}{3}$
   $3y - 12 = 4x - 12$
   $4x - 3y = 0$

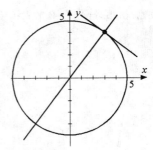

9. $2x + y = 12$, $y = -2x + 12$
   slope $= -2$
   $y = 2x^3 - 6x^2 - 2x + 1$
   $y' = 6x^2 - 12x - 2 = -2$
   $6x^2 - 12x = 0$
   $x^2 - 2x = 0$
   $x(x - 2) = 0$
   $x = 0$
   $y(0) = 1$
   $x = 2$
   $y(2) = 16 - 24 - 4 + 1 = -11$
   so $\frac{y-1}{x-0} = -2$
   $y - 1 = -2x$
   $2x + y = 1$
   and $\frac{y+11}{x-2} = -2$
   $y + 11 = -2x + 4$
   $2x + y = -7$

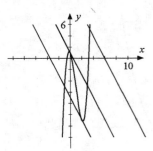

13. $y = -2x$
    $y = x^2(1 - x) = x^2 - x^3$ intersect at $(0,0)$,
    $(2, -4)$ and $(-1, 2)$
    $y' = 2x - 3x^2$
    $y'(0) = 0$
    $y'(2) = 4 - 12 = -8$
    $y'(-1) = -2 - 3 = -5$
    at $(0,0)$, $\tan\phi = \frac{0-(-2)}{1+0} = 2$
    $\phi = 63.4°$
    at $(2, -4)$, $\tan\phi = \frac{-2+8}{1+16} = \frac{6}{17}$
    $\phi = 19.4°$
    at $(-1, 2)$, $\tan\phi = \frac{-2+5}{1+10} = \frac{3}{11}$
    $\phi = 15.3°$

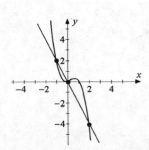

Exercise 2: Maximum, Minimum, and Inflection Points

NOTE: To save space, the 2nd derivative test for maximum and minimum is not shown for many of the following examples, but all points have been checked either by that test or by graphing.

1. $y = 3x + 5$
    $y' = 3$, rising for all x

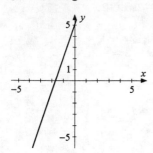

5. $y = 4x^2 - x$
    $y' = 8x - 1$
    $y'(-2) = -16 - 1 = -17$
    (decreasing)

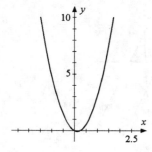

9. $y = -2x^3 - 2(x+2)^{1/2}$
    $y' = -6x^2 - (x+2)^{-1/2}$
    $y'' = -12x + \frac{1}{2}(x+2)^{-3/2}$
    $y''\left(\frac{1}{4}\right) = -3 + \frac{1}{2}\left(\frac{9}{4}\right)^{-3/2}$
    $\qquad = -3 + \frac{1}{2}\left(\frac{2}{3}\right)^3 = -3 + \frac{1}{2}\left(\frac{8}{27}\right)$
    $\qquad = -\frac{81}{27} + \frac{4}{27} = -\frac{77}{27}$ (downward)

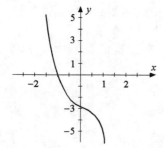

13. $y = 6x - x^2 + 4$
    $y' = 6 - 2x = 0$
    $x = 3$
    $y(3) = 18 - 9 + 4 = 13$
    $(3, 13)$ max
    $y'' = -2$
    Thus no points of inflection

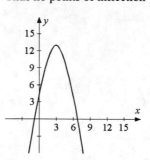

17. $y = 2x^2 - x^4$
    $y' = 4x - 4x^3 = 0$
    $x(1 - x^2) = 0$
    $x = 0$
    $y(0) = 0$
    $(0, 0)$ min
    $x = \pm 1$
    $y(1) = 1, y(-1) = 1$
    $(1, 1)$ max and $(-1, 1)$ max
    $y'' = 4 - 12x^2 = 0$
    $12x^2 = 4$
    $x = \pm \sqrt{\frac{1}{3}} = \pm 0.57735$
    $y\left(\pm \sqrt{\frac{1}{3}}\right) = 0.55556$
    Thus, PI at $(0.577, 0.556)$ and $(-0.577, 0.556)$

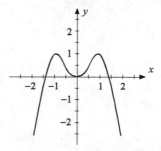

21. $y = 3x^4 - 4x^3 - 12x^2$
$y' = 12x^3 - 12x^2 - 24x = 0$
$x(x^2 - x - 2) = 0$
$x(x - 2)(x + 1) = 0, \ x = 0, \ 2, \ \text{or} \ -1$
at $x = 0, \ y(0) = 0$
$(0, 0)$ max
at $x = 2, y(2) = 48 - 32 - 48 = -32$
$(2, \ -32)$ min
at $x = -1, y(-1) = 3 + 4 - 12 = -5$
$(-1, -5)$ min
$y'' = 36x^2 - 24x - 24$
$12(3x^2 - 2x - 2) = 0$
Using the quadratic formula,
$x = 1.215$ and -0.5486
$y(1.215) = -18.35$
$y(-0.5486) = -2.6793$
Thus PI at $(1.22, -18.4)$ and $(-0.549, -2.68)$

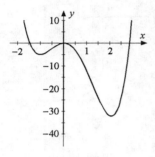

25. $y = (x - 1)^4(x + 2)^3$
$y' = (x - 1)^4(3)(x + 2)^2 + (x + 2)^3(4)(x - 1)^3 = 0$
$(x - 1)^3(x + 2)^2\{(x - 1)(3) + (x + 2)(4)\} = 0$
$(x - 1)^3(x + 2)^2(3x - 3 + 4x + 8) = 0$
$(x - 1)^3(x + 2)^2(7x + 5) = 0$

$x = 1, \ x = -2, \ x = -\frac{5}{7} \simeq -0.714$

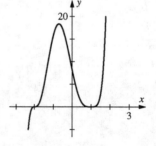

$y(1) = 0, \ y(-2) = 0; \ y\left(-\frac{5}{7}\right) = \left(-\frac{12}{7}\right)^4\left(\frac{9}{7}\right)^3 \simeq 18.4$
$(1, 0)$ min, $(-2, 0)$ PI, $(-0.714, 18.4)$ max
$y' = (x - 1)^3(x + 2)^2(7x + 5)$
$y'' = (x - 1)^3(x + 2)^2(7) + (x - 1)^3(7x + 5)(2)(x + 2) + (x + 2)^2(7x + 5)(3)(x - 1)^2$
$\qquad = (x - 1)^2(x + 2)\big[(x - 1)(x + 2)(7) + (x - 1)(7x + 5)(2) + (x + 2)(7x + 5)(3)\big]$
$\qquad = (x - 1)^2(x + 2)(42x^2 + 60x + 6) = (x - 1)^2(x + 2)(6)(7x^2 + 10x + 1) = 0$
$x = +1 \qquad y(1) = 0$ This is a minimum point. See graph.
$x = -2 \qquad y(-2) = 0$
and using the quadratic formula, $\quad x = -0.10819 \qquad\qquad y(-0.10819) = 10.2$
$\qquad\qquad\qquad\qquad\qquad\qquad x = -1.3204 \qquad\qquad\ y(-1.3204) = 9.10$
Thus PI at $(-2, 0)$, $(-0.108, 10.2)$ and $(-1.32, 9.10)$. NOTE: As we see with $(1, 0)$ above, the second derivative test is not always accurate. $(1, 0)$ is a minimum point, not an inflection point.

29. Using implicit differentiation to find the 1st derivative of $x^2 - x - 2y^2 + 36 = 0$, and setting equal to zero, we obtain,

$2x - 1 - 4yy' = 0$

solving for y': $y' = \frac{2x-1}{4y} = 0$

The only way that this can be zero is for $2x - 1 = 0$, or $x = \frac{1}{2}$. Substituting $x = \frac{1}{2}$ into the original equation and solving for y yields, $y = \pm 4.23$. Therefore, $(0.500, 4.23)$ is a minimum point and $(0.500, -4.23)$ is a maximum point, as verified by the graph.

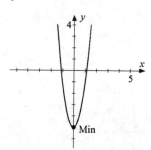

Exercise 3: Sketching, Verifying, and Interpreting Graphs

1. $y = 4x^2 - 5$

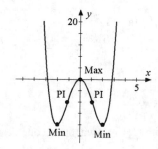

5. $y = x^4 - 8x^2$

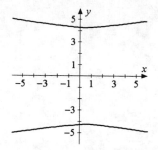

9. $y = 5x - x^5$

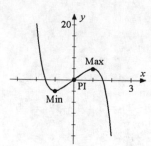

13. $y = x^3 - 6x^2 + 9x + 3$

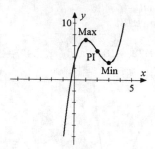

17. $y = \frac{x}{\sqrt{x^2+1}}$

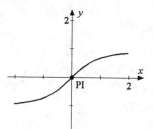

21. $y = x^3 + 4x^2 - 5$

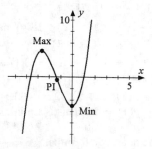

25. $y = 5x^2 - 2x$, y axis, $y = 4$ in the second quadrant

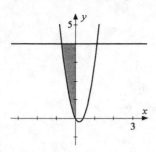

Exercise 4: Approximate Solution of Equations by Newton's Method

1. $f(x) = x^3 + 3x^2 - 40$
   $f'(x) = 3x^2 + 6x$
   guess $x = 2.5$
   correction $\frac{f(2.5)}{f'(2.5)} = \frac{-5.625}{33.75} = -0.1667$
   guess $x = 2.5 - (-0.1667) = 2.667$
   correction $\frac{f(2.67)}{f'(2.67)} = \frac{0.4209}{37.41} = 0.01125$
   guess $x = 2.67 - 0.01125 = 2.66$ to 2 decimal places so $x = 2.66$ to 2 decimal places

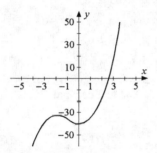

5. $f(x) = x^3 + 2x - 8$
   $f'(x) = 3x^2 - 2$
   guess $x = 1.65$
   correction $= \frac{f(1.65)}{f'(1.65)} = -0.03$
   guess $x = 1.65 + 0.03 = 1.68$
   correction $= \frac{f(1.68)}{f'(1.68)} = 0.02$
   guess $x = 1.68 - 0.02 = 1.66$
   correction $= \frac{f(1.66)}{f'(1.66)} = -0.01$
   so $x = 1.66 + 0.01 = 1.67$ to 2 places

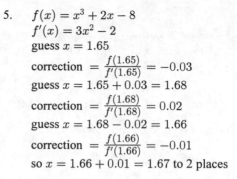

9. $V = x(16 - 2x)^2 = 282$
   $f(x) = x(16 - 2x)^2 - 282$
   $f'(x) = x(2)(16 - 2x)(-2) + (16 - 2x)^2$
   $\qquad = 12x^2 - 128x + 256$

   a. 1st root guess $x = 2.0$
      correction $\frac{f(2.0)}{f'(2.0)} = \frac{6}{48} = 0.125$
      guess $x = 2.0 - 0.125 = 1.875 = 1.88$
      correction $= \frac{f(1.88)}{f'(1.88)} = \frac{-0.3429}{57.77}$
      $\qquad = -0.005936$
      so $x = 1.88 - (-0.005936) = 1.89$ to 2 decimal places

   b. 2nd root guess $x = 3.5$
      correction $\frac{f(3.5)}{f'(3.5)} = \frac{1.5}{-45} = -0.0333$
      guess $x = 3.5 - (-0.0333) = 3.5333$
      correction $\frac{f(3.53)}{f'(3.53)} = \frac{0.1303}{-46.31} = -0.0028$
      guess $x = 3.53 - (-0.0028) = 3.53$ to 2 decimal places. Thus $x = 1.89$ in. or 3.53 in.

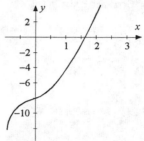

13.
```
10    ' NEWTON
20    '
30    ' This program uses Newton's
40    ' method to find the root of
50    ' an equation.
60    '
70    ' Enter the degree of accuracy
80    ' on line 150, the equation on
90    ' line 160, the derivative of
100   ' the equation on line 170 and
110   ' the appropriate value of the
120   ' root on line 180.
130   '
140   PRINT "Correction", "X"
150   A = .001
160   DEF FNA(X)  = X^4 + 8 * X - 12
170   DEF FND(X) = 4 * X^3 + 8
180   X = 1
190   H = FNA(X) / FND(X)
200   X = X - H
210   PRINT H, X
220   IF ABS(H) > A THEN GOTO 190
230   END
```

Chapter 28 Review Problems

1. $f(x) = x^3 - 4x + 2$
     $f'(x) = 3x^2 - 4$

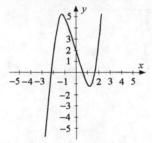

1st root: guess $x = -2$

correction $= \dfrac{f(-2)}{f'(-2)} = 0.25$

guess $x = -2 - 0.25 = -2.25$

correction $= \dfrac{f(-2.25)}{f'(-2.25)} = -0.035$

guess $x = -2.25 + 0.035 = -2.215$

correction $= \dfrac{f(-2.215)}{f'(-2.215)} = -0.00068$

guess $x = -2.215 - (-0.00068)$
$$= -2.214$$

so $x = -2.21$ to 2 decimal places

2nd root: guess $x = 0.5$

correction $= \dfrac{f(0.5)}{f'(0.5)} = -0.04$

guess $x = 0.5 + 0.04 = 0.54$

correction $= \dfrac{f(0.54)}{f'(0.54)} = -0.0007$

guess $x = 0.54 - (-0.0007) = 0.5407$

so $x = 0.54$ to 2 decimal places

3rd root: guess $x = 1.7$

correction $= \dfrac{f(1.7)}{f'(1.7)} = 0.024$

guess $x = 1.7 - 0.024 = 1.676$

correction $= \dfrac{f(1.676)}{f'(1.676)} = 0.0007$

guess $x = 1.676 - 0.0007 = 1.6753$

so $x = 1.68$ to 2 decimal places

So roots are -2.21, 0.54 and 1.68 to 2 places

5. $y = \frac{1+2x}{3-x}$

$y' = \frac{(3-x)(2)-(1+2x)(-1)}{(3-x)^2}$

$y'(2) = \frac{2+5}{1} = 7$

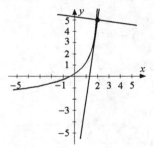

tangent: $\frac{y-5}{x-2} = 7$

$7x - 14 = y - 5$

$7x - y - 9 = 0$

normal: $\frac{y-5}{x-2} = -\frac{1}{7}$

$x - 2 = -7y + 35$

$x + 7y - 37 = 0$

9. $y = 2x^{1/2}$

$y' = x^{-1/2} = 0$

$x = 0$

rising for $x > 0$, never falls

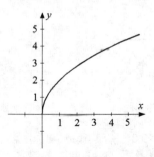

13. Calculate y' to find slope of the tangent line at $x = 2$.

$y' = 9x^2 - 2 \Rightarrow y'(2) = 36 - 2 = 34$

Using the point-slope formula at $x = 2$, $y = 24$,

we obtain, $\frac{y-24}{x-2} = 34 \Rightarrow 34x - y = 44$ (tangent line). Normal line has negative reciprocal slope, or $-\frac{1}{34}$. Again using the point-slope formula, we obtain, $\frac{y-24}{x-2} = -\frac{1}{34} \Rightarrow x + 34y = 818$ (normal line).

CHAPTER 29: MORE APPLICATIONS OF THE DERIVATIVE

Exercise 1: Rate of Change

1. $P = \frac{KT}{V}$, $25.5 = \frac{KT}{146}$
   $KT = 3723$
   $\frac{dP}{dV} = -\frac{KT}{V^2} = -\frac{3723}{(146)^2} = -0.175$ (lb/in.^{2})/in.^{3}

5. $P = 0.324\sqrt{L}$
   $\frac{dP}{dL} = \frac{0.162}{\sqrt{L}}$
   $\left.\frac{dP}{dL}\right|_{L=9.00} = 0.0540$ s/in.

9. $q = 3.48t^2 - 1.64t$ coulomb
   $i = \frac{dq}{dt} = 6.96t - 1.64$ A

13. $q = 22.4t + 41.6t^3$
    $i = \frac{dq}{dt} = 22.4 + 125t^2$ A
    $i(2.50) = 22.4 + 125(2.50)^2 = 804$ A

17. $i = 8.22 + 5.83t^3$
    $v = L\frac{di}{dt} = 8.75(17.49t^2)$
    $v(25.0) = 8.75(17.49)(25.0)^2 = 95,600$ V
    $\quad = 95.6$ kV

Exercise 2: Motion of a Point

1. $v(t) = 32 - 16t$; $v(2) = 0$
   $a(t) = -16$; $a(2) = -16$

5. $v(t) = 120 - 32t$; $v(4) = -8$
   $a(t) = -32$; $a(4) = -32$

9. $v(t) = 12t$, $v(5) = 60$ ft/s

13. $s = 16t^2 - 64t + 64$
    $v(t) = 32t - 64$
    $a(t) = 32$
    Rest point ($v = 0$) occurs at $t = 2$, at which time
    $s = 0$ and $a = 32$ units/s^2

17. $R^2 = 32.3^2 + (-27.3)^2$
    $R = 42.3$ cm/s
    $\tan\theta = -\frac{27.3}{32.3}$
    $\quad \theta = -40.2°$
    $\alpha = 360 - 40.2 = 320°$

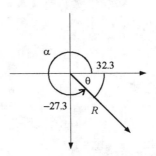

21. $\theta = 44.8t^3 + 29.3t^2 + 81.5$ rad
    $\omega = \frac{d\theta}{dt} = 134.4t^2 + 58.6t$ rad/s
    $\omega(4.25) = 134.4(4.25)^2 + 58.6(4.25)$
    $\quad = 2680$ rad/s (rounded)

Exercise 3: Related Rates

1.

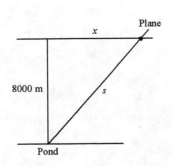

$\frac{dx}{dt} = 100$ m/sec
$s^2 = x^2 + (8000)^2$
$2s\frac{ds}{dt} = 2x\frac{dx}{dt}$
At $t = 1$ min, $x = 100 \cdot 60 = 6000$ m
and $s = 10,000$ m (3-4-5 triangle)
$2(10,000)\frac{ds}{dt} = 2(6000)(100)$
$\frac{ds}{dt} = 60$ m/s

5. $\frac{dx}{dt} = 4.00 \text{ m/s}$
$s^2 = x^2 + 40.0^2$
$2s\frac{ds}{dt} = 2x\frac{dx}{dt}$
When 50.0 m of cable are out,
$x = \sqrt{(50.0)^2 - (40.0)^2} = 30.0 \text{ m}$
$50.0\frac{ds}{dt} = 30.0(4) = 120$
$\frac{ds}{dt} = 2.40 \text{ m/s}$

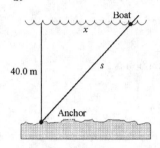

9. $\frac{dx}{dt} = 1.00 \text{ ft/s}$
$s^2 = x^2 + 54.0^2$
$2s\frac{ds}{dt} = 2x\frac{dx}{dt} + 0$
$\frac{ds}{dt} = \frac{x}{s}\frac{dx}{dt}$
When $x = 80.0$,
$s = \sqrt{(80.0)^2 + (54.0)^2} = 96.52 \text{ in.}$
and $\frac{ds}{dt} = \frac{80.0}{96.52}(1.00) = 0.83 \text{ ft/s}$

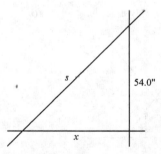

13. x = jogger's distance from starting point
y = boat's distance from starting point
$s^2 = x^2 + y^2 + (30.0)^2$
$2s\frac{ds}{dt} = 2x\frac{dx}{dt} + 2y\frac{dy}{dt} + 0$
$\frac{dx}{dt} = 5.00 \text{ ft/sec}, \frac{dy}{dt} = 10.0 \text{ ft/sec}$ At $t = 3.00$ s
$x = 3.00 \bullet 5.00 = 15.0 \text{ ft}$, and
$y = 3.00 \bullet 10.0 = 30.0 \text{ ft}$
and $s = \sqrt{(15.0)^2 + (30.0)^2 + (30.0)^2} = 45.0 \text{ ft}$
So, $45.0\frac{ds}{dt} = 15.0(5.00) + 30.0(10.0) = 375$
$\frac{ds}{dt} = 8.33 \text{ ft/s}$

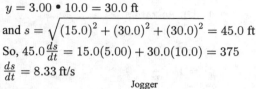

17. $A = (10.0 + x)^2$
$\frac{dA}{dt} = 2(10.0 + x)\frac{dx}{dt}$
$\frac{dx}{dt} = 0.00500 \text{ in./sec}$
At $t = 20.0$ sec, $x = 20.0(0.00500)$
$= 0.100 \text{ in.}$
and $\frac{dA}{dt} = 2(10.1)(0.00500)$
$= 0.101 \text{ in.}^2/\text{s}$

21. We want $\frac{dh}{dt}$ when $\frac{dV}{dt} = 0$, $h = 20.0$, $d = 10.0$
We know   1. $V = \pi r^2 h$

           2. $\frac{dd}{dt} = 2\frac{dr}{dt} = 1.00 \text{ in./min}$
Taking the derivative of $V = \pi r^2 h$ with respect to
time t, and setting $\frac{dV}{dt} = 0$, $r = 5.0$; $h = 20.0$;
we obtain,
$\frac{dV}{dt} = \pi\left(r^2\frac{dh}{dt} + h \bullet 2r\frac{dr}{dt}\right) = 0$
$\left(5.0^2\frac{dh}{dt} + 20.0 \bullet 5.0 \bullet 1.00 \text{ in./min}\right) = 0$
$25\frac{dh}{dt} = -100 \text{ in./min} \Rightarrow \frac{dh}{dt} = -4.00 \text{ in./min}$

25. $A = \frac{1}{2}(2h)(h) = h^2$
$V = 10.0h^2$
$\frac{dV}{dt} = 20.0h\frac{dh}{dt} = 8.00 \text{ ft}^3/\text{min}$
When $h = 2.00$ ft, $(20.0)(2.00)\frac{dh}{dt} = 8.00$
$\frac{dh}{dt} = \frac{1}{5} \text{ ft/min} = 0.200 \text{ ft/min}$

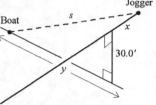

29. $y = \frac{Px^2}{3EI}$

$\frac{dy}{dt} = \frac{3Px^2}{3EI}\frac{dx}{dt} = \frac{Px^2}{EI}\frac{dx}{dt}$

Substituting $x = 75.0$ and $\frac{dx}{dt} = 25.0$,

$\frac{dy}{dt} = \frac{(165)(75.0)^2}{(1,320,000)(10.9)}(25.0) = 1.61$ in./s

Exercise 4: Optimization

1. $s = x + \frac{1}{2x^2}$

$\frac{ds}{dx} = 1 - \frac{1}{x^3} = 0$

$x^3 = 1$

$x = 1$

5. Cost $= \frac{1}{2}y + y + 2x$

$C = \frac{3}{2}y + 2x$

$xy = 432$

$y = \frac{432}{x}, \frac{dy}{dx} = -\frac{432}{x^2}$

$\frac{dC}{dx} = \frac{3}{2} \cdot \frac{dy}{dx} + 2 = \frac{3}{2}\left(-\frac{432}{x^2}\right) + 2$

Setting $\frac{dC}{dx} = 0 = -(3)(432) + 4x^2$

$x^2 = \frac{3(432)}{4} = 324$

$x = 18$ m

$y = \frac{432}{18} = 24$ m

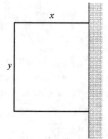

9. Let r = radius of cylinder and h = altitude of cylinder

Equation: (1) $V = \pi r^2 h$

Surface area $= 2\pi rh + \pi r^2 = 100$

Solving for h: $h = \frac{100 - \pi r^2}{2\pi r}$

Substituting into equation (1), we obtain,

$V = 50r - \frac{\pi r^3}{2}$

Derivative: $V'(r) = 50 - \frac{3}{2}\pi r^2$

Critical Points: $50 - \frac{3}{2}\pi r^2 = 0 \Rightarrow$

$r = \frac{10}{\sqrt{3\pi}} = 3.257$

Testing critical points:

$V'(3) = +$
$V'(4) = -$ $\Rightarrow r = 3.257,$

$h = \frac{100 - \pi(3.26)^2}{2\pi(3.26)} = 3.257$

thus, $d = 2r = 6.51$ cm, $h = 3.26$ cm yields maximum volume

13. $h = 2x_0$ and $3x_1 + h = 30$

$x_0 = \frac{h}{2}$ and $x_1 = \frac{30 - h}{3}$

$x_1 - x_0 = \frac{30 - h}{3} - \frac{h}{2} = \frac{60 - 2h - 3h}{6} = \frac{60 - 5h}{6}$

$A = bh = (x_1 - x_0)h = \left(\frac{60 - 5h}{6}\right)h$

$\qquad = 10h - \frac{5}{6}h^2$

$\frac{dA}{dh} = 10 - \frac{5}{3}h$

Setting $\frac{dA}{dh} = 0$, we get $h = \frac{10}{5/3} = 6$

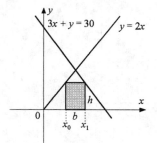

17. $y = \frac{x^2}{2}$

Let the distance from a point $\left(x, \frac{x^2}{2}\right)$ on the curve to the point $(4, 1)$ be $S(x)$.

$S^2 = (x - 4)^2 + \left(\frac{x^2}{2} - 1\right)^2$

$2S\frac{dS}{dx} = 2(x - 4) + 2\left(\frac{x^2}{2} - 1\right)(x)$

Setting $\frac{dS}{dx} = 0$, we have $2x - 8 + x^3 - 2x = 0$

$x^3 = 8; x = 2$

$y = \frac{2^2}{2} = 2; (2, 2)$

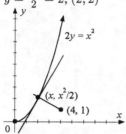

21. $\frac{x^2}{100} + \frac{y^2}{49} = 1$

$y = \sqrt{49\left(1 - \frac{x^2}{100}\right)} = \left(\frac{7}{10}\right)\left(\sqrt{100 - x^2}\right)$

width $= 2x$

height $= 2\left(\frac{7}{10}\right)\left(\sqrt{100 - x^2}\right)$

$A = 2x\left(\frac{7}{5}\right)\left(\sqrt{100 - x^2}\right)$

$\frac{dA}{dx} = \frac{14}{5}x\left(\frac{1}{2}\right)(100 - x^2)^{-1/2}(-2x)$

$\qquad + (100 - x^2)^{1/2}\left(\frac{14}{5}\right)$

Setting $\frac{dA}{dx} = 0$,

$-\frac{14}{5}x^2(100 - x^2)^{-1/2} + \frac{14}{5}(100 - x^2)^{1/2} = 0$

$x^2 = 100 - x^2$

$2x^2 = 100$

$x = \sqrt{50} = 7.07$

width $= 14.1$ units

height $= 9.90$ units

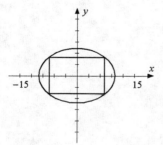

25. $AO = BO = 9.00; AC = h; BC = r$

$\overline{OC}^2 + \overline{BC}^2 = \overline{OB}^2; (h - 9)^2 + r^2 = 9^2$

$r^2 = 81 - (h^2 - 18h + 81); r^2 = 18h - h^2$

$V = \frac{1}{3}\pi r^2 h = \frac{1}{3}\pi(18h - h^2)h = 6\pi h^2 - \frac{1}{3}\pi h^3$

$\frac{dV}{dh} = 12\pi h - \pi h^2$

Setting $\frac{dV}{dh} = 0$, $\pi h^2 = 12\pi h; h = 0, h = 12$ ft

29.

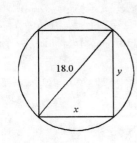

$$\sqrt{18.0^2 - x^2} = y$$

$$\text{Strength, } S = kxy^2 = kx\left(\sqrt{18^2 - x^2}\right)^2$$

$$S = (18^2 x - x^3)k$$

$$\frac{dS}{dx} = (18^2 - 3x^2)k$$

$$\text{Setting } \frac{dS}{dx} = 0,\ 3x^2 = 18^2$$

$$x^2 = 6(18) = 108$$

$$x = \sqrt{108} = 10.4 \text{ in.}$$

$$y = \sqrt{18^2 - x^2} = 14.7 \text{ in.}$$

33. $P = 30i - 2i^2;\ \frac{dP}{di} = 30 - 4i$

Setting $\frac{dP}{di} = 0,\ 30 = 4i;\ i = 7.5$ A

37. $E = \frac{x - \mu x^2}{x + \mu} = (x - \mu x^2)(x + \mu)^{-1}$

$\frac{dE}{dx} = (x - \mu x^2)(-1)(x + \mu)^{-2} + (1 - 2\mu x)(x + \mu)^{-1} = -\frac{x - \mu x^2}{(x+\mu)^2} + \frac{1 - 2\mu x}{(x+\mu)}$

Setting $\frac{dE}{dx} = 0$,

$(x - \mu x^2)(x + \mu) = (x + \mu)^2(1 - 2\mu x)$

$x - \mu x^2 = x + \mu - 2\mu x^2 - 2\mu^2 x$

$\mu x^2 + 2\mu^2 x - \mu = 0$

$x^2 + 2\mu x - 1 = 0$

For $\mu = 0.45,\ x^2 + 0.9x - 1 = 0$

$x = \frac{-0.9 \pm \sqrt{0.81 + 4}}{2} = \frac{-0.9 \pm 2.19}{2} = 0.65$

Chapter 29 Review Problems

1. $x = \frac{160 \text{ ft}}{\text{s}}(12 \min)\left(\frac{60 \text{ s}}{1 \min}\right) = 115,200 \text{ ft}$

$y = 120(24)(60) = 172,800 \text{ ft}$

$s = \sqrt{(115,200)^2 + (172,800)^2} = 207,700 \text{ ft}$

$s^2 = x^2 + y^2$

$2s\frac{ds}{dt} = 2x\frac{dx}{dt} + 2y\frac{dy}{dt}$

$207,700\frac{ds}{dt} = 115,200(160) + 172,800(120)$

$\frac{ds}{dt} = 189 \frac{\text{ft}}{\text{s}}$

5. $r = \dfrac{D}{\sqrt{3}}$

$$V = \tfrac{1}{3}\pi r^2 D = \tfrac{1}{3}\pi \left(\dfrac{D}{\sqrt{3}}\right)^2 D = \dfrac{\pi}{9}D^3$$

$$\dfrac{dV}{dt} = \dfrac{\pi}{3}D^2 \dfrac{dD}{dt} = -5.00\ \text{cm}^3/\text{min}$$

$$A = \pi r\sqrt{r^2 + D^2}$$

$$= \pi\left(\dfrac{D}{\sqrt{3}}\right)\sqrt{\left(\dfrac{D^2}{3}\right) + D^2}$$

$$= \pi\left(\dfrac{D}{\sqrt{3}}\right)(D)\left(\sqrt{\dfrac{4}{3}}\right)$$

$$= \left(\dfrac{2}{3}\right)\pi D^2$$

$$\dfrac{dA}{dt} = \dfrac{4}{3}\pi D \dfrac{dD}{dt}$$

$$\dfrac{\left(\dfrac{dA}{dt}\right)}{\left(\dfrac{dV}{dt}\right)} = \dfrac{\left(\dfrac{4}{3}\right)\pi D\left(\dfrac{dD}{dt}\right)}{\left(\dfrac{\pi}{3}\right)D^2\left(\dfrac{dD}{dt}\right)} = \dfrac{4}{D}$$

$$\dfrac{dA}{dt} = \dfrac{4}{D}\cdot\dfrac{dV}{dt} = \dfrac{4}{D}(-5.00)$$

When $D = 6.00$ cm,

$$\dfrac{dA}{dt} = \dfrac{4}{6.00}(-5.00) = -3.33\ \text{cm}^2/\text{min}$$

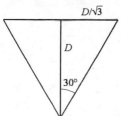

9. (1) $h^2 + (2r)^2 = 12^2$

$$2h + 2(2r)\left(2\dfrac{dr}{dh}\right) = 0$$

$$\dfrac{dr}{dh} = -\dfrac{h}{4r}$$

$$V = \pi r^2 h$$

$$\dfrac{dV}{dh} = \pi r^2(1) + h2\pi r \dfrac{dr}{dh}$$

$$= \pi r^2 + h2\pi r\left(-\dfrac{h}{4r}\right)$$

$$= \pi r^2 - \dfrac{\pi h^2}{2}$$

Setting $\dfrac{dV}{dh} = 0$, $\pi r^2 = \dfrac{\pi h^2}{2}$

$$r = \dfrac{h}{\sqrt{2}}$$

Substituting in (1), $h^2 + \left(2\dfrac{h}{\sqrt{2}}\right)^2 = 12^2$

$$h^2 + 2h^2 = 12^2;\ h^2 = \dfrac{12^2}{3} = 48$$

$$h = \sqrt{48} = 6.93$$

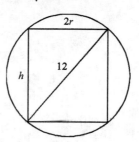

13. $s = -t^3 + 3t^2 + 24t + 28$

$$\dfrac{ds}{dt} = -3t^2 + 6t + 24$$

Setting $\dfrac{ds}{dt} = 0$, $t^2 - 2t - 8 = 0$

$$(t - 4)(t + 2) = 0$$

$$t = 4, -2$$

$$t = 4,\ s = 108$$

$$t = -2,\ s = 0$$

17. $P = x^3(10 - x)^2 = x^3(100 - 20x + x^2)$
$$= 100x^3 - 20x^4 + x^5$$
$\frac{dP}{dx} = 300x^2 - 80x^3 + 5x^4$

Setting $\frac{dP}{dx} = 0$, $300 - 80x + 5x^2 = 0$
and $x^2 = 0$
$x^2 - 16x + 60 = 0$; $(x - 6)(x - 10) = 0$
$x = 6, 10$
$x = 0$ and $x = 10$ are minima
$x = 6$ is a max, so $10 - x = 4$.
The numbers are then 4 and 6.

21. $\theta = 18.5t^2 + 12.8t + 14.8$

$\omega = \frac{d\theta}{dt} = 37.0t + 12.8$

$\alpha = \frac{d\omega}{dt} = 37.0$
$\omega(3.50) = 37.0(3.50) + 12.8 = 142$ rad/s
$\alpha(3.50) = 37.0$ rad/s^2

CHAPTER 30: INTEGRATION

Exercise 1: The Indefinite Integral

1. $\int dx = x + C$

5. $\int \frac{dx}{x^2} = \int x^{-2} \, dx = -x^{-1} + C = -\frac{1}{x} + C$

9. $\int 3x^3 \, dx = 3\int x^3 \, dx = \frac{3}{4}x^4 + C$

13. $\int \frac{dx}{\sqrt{x}} = \int x^{-1/2} \, dx = \frac{x^{1/2}}{1/2} + C = 2\sqrt{x} + C$

17. $\int \frac{7}{2}x^{5/2} \, dx = \frac{7}{2}\int x^{5/2} \, dx = \frac{7}{2} \cdot x^{7/2} \cdot \frac{2}{7} + C$
$\qquad = x^{7/2} + C$

21. $\int \frac{4}{3}x^{1/3} \, dx = \frac{4}{3} \cdot \frac{x^{4/3}}{(4/3)} + C = x^{4/3} + C$

25. $\int 4s^{1/2} \, ds = \frac{4s^{3/2}}{3/2} + C$
$\qquad = \frac{8}{3}s^{3/2} + C$

29. $\int \frac{4x^2 - 2\sqrt{x}}{x} \, dx = \int \left(4x - 2x^{-1/2}\right) dx$
$\qquad = \frac{4x^2}{2} - \frac{2x^{1/2}}{(1/2)} + C = 2x^2 - 4x^{1/2} + C$

33. $\frac{dy}{dx} = 4x^2$
$dy = 4x^2 dx$
$y = \int 4x^2 \, dx = \frac{4x^3}{3} + C$

37. $s = \int \frac{1}{2}t^{-2/3} \, dt = \frac{1}{2} \cdot \frac{t^{1/3}}{(1/3)} + C = \frac{3}{2}t^{1/3} + C$

Exercise 2: Rules for Finding Integrals

1. $\int (x^4 + 1)^3 4x^3 \, dx = \frac{(x^4+1)^4}{4} + C$

5. $\int \frac{dx}{(1-x)^2} = -\int (1-x)^{-2}(-dx)$
$\qquad = -\frac{(1-x)^{-1}}{-1} + C = \frac{1}{1-x} + C$

9. $\int \frac{y^2 \, dy}{\sqrt{1-y^3}} = -\frac{1}{3}\int (1-y^3)^{-1/2}(-3y^2 \, dy)$
$\qquad = \frac{-1/3(1-y^3)^{1/2}}{1/2} + C$
$\qquad = -\frac{2}{3}(1-y^3)^{1/2} + C$

13. $\int \frac{3}{x} \, dx = 3 \ln |x| + C$

17. $\int \frac{x+1}{x} \, dx = \int \left(1 + \frac{1}{x}\right) dx = x + \ln |x| + C$

21. $\frac{dy}{dx} = (3x^2 + x)(2x^3 + x^2)^3$
$dy = \frac{1}{2}(2x^3 + x^2)^3(6x^2 + 2x)dx$
$y = \frac{1}{8}(2x^3 + x^2)^4 + C$

25. $\frac{dy}{dx} = \frac{7}{x}$
$dy = \frac{7}{x}dx \qquad \text{let } u = x$
$y = \int \frac{7}{x}dx = 7\int \frac{dx}{x} = 7 \ln |x| + C$

29. $\int \frac{dt}{4-9t^2} = -\frac{1}{3}\int \frac{3 \, dt}{9t^2-4}$
no. 57: $u = 3t$, $a = 2$, $du = 3 \, dt$
$\qquad = -\frac{1}{3}\left(\frac{1}{4}\ln \left|\frac{3t-2}{3t+2}\right| + C_1\right)$
$\qquad = -\frac{1}{12}\ln \left|\frac{3t-2}{3t+2}\right| + C$

33. $\int \frac{dx}{x^2+9}$
no. 56: $u = x$, $a = 3$, $du = dx$
$\qquad = \frac{1}{3}\text{Tan}^{-1}\frac{x}{3} + C$

37. $\int \frac{dx}{9-4x^2} = -\frac{1}{2}\int \frac{2 \, dx}{4x^2-9}$
no. 57: $u = 2x$, $a = 3$, $du = 2 \, dx$
$\qquad = -\frac{1}{2} \cdot \frac{1}{2 \cdot 3}\ln \left|\frac{2x-3}{2x+3}\right| + C$
$\qquad = -\frac{1}{12}\ln \left|\frac{2x-3}{2x+3}\right| + C$

41. $\int \sqrt{\frac{x^2}{4} - 1} \, dx = \frac{1}{2}\int \sqrt{x^2 - 4} \, dx$
no. 66: $u = x$, $a = 2$
$\qquad = \frac{1}{2}\left(\frac{x}{2}\sqrt{x^2 - 4}\right.$
$\qquad \left. - \frac{4}{2}\ln \left|x + \sqrt{x^2 - 4}\right|\right) + C$
$\qquad = \frac{x}{4}\sqrt{x^2 - 4} - \ln \left|x + \sqrt{x^2 - 4}\right| + C$

Exercise 3: Constant of Integration

1. $\frac{dy}{dx} = 3$

   $y = \int 3\, dx = 3x + C$

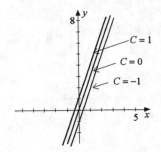

5. $y = \int x^2\, dx = \frac{x^3}{3} + C$

   at $(1,1)$, $C = 1 - \frac{1}{3} = \frac{2}{3}$

   so $y = \frac{x^3}{3} + \frac{2}{3}$ or $x^3 - 3y + 2 = 0$

9. $y' = \frac{7}{2}$ at $(3,0)$

   $y'' = x$

   $y' = \int x\, dx = \frac{x^2}{2} + C$

   $\frac{7}{2} = \frac{9}{2} + C$

   $C = -1$

   $y' = \frac{x^2}{2} - 1$

   $y = \int \left(\frac{x^2}{2} - 1\right) dx$

   $y = \frac{x^3}{6} - x + C$

   $0 = \frac{27}{6} - 3 + C$

   $C = \frac{6}{2} - \frac{9}{2} = -\frac{3}{2}$

   $y = \frac{x^3}{6} - x - \frac{3}{2}$

   $6y = x^3 - 6x - 9$

Exercise 4: The Definite Integral

1. $\int_1^2 x\, dx = \frac{x^2}{2}\Big|_1^2 = \frac{4}{2} - \frac{1}{2} = \frac{3}{2} = 1.50$

5. $\int_0^4 (x^2 + 2x)dx = \frac{x^3}{3} + \frac{2x^2}{2}\Big|_0^4 = \frac{64}{3} + 16 = 37.3$

9. $\int_1^{10} \frac{dx}{x} = \ln|x|\Big|_1^{10} = \ln 10 - \ln 1 = \ln 10 = 2.30$

13. $\int_0^1 \frac{dx}{\sqrt{3-2x}} = -\frac{1}{2}\int_0^1 \frac{-2dx}{\sqrt{3-2x}} = -\frac{1}{2}\left[2\sqrt{3-2x}\right]_0^1$

   $= -\frac{1}{2}\left[2\sqrt{1} - 2\sqrt{3}\right] = 0.732$

Exercise 5: Approximate Area Under a Curve

1. Guess: Area = about $6\frac{1}{2}$ rectangles (1 by 5)

   $= 6\frac{1}{2}(1)(5) = 32\frac{1}{2}$ sq. units

   By graphing calculator: Area = 33 sq. units

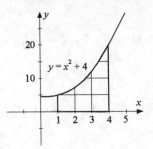

5. Guess: Area = about 2 triangles

   $= 2\left(\frac{1}{2}\right)(1.5)(4) = 6$ sq. units

   By graphing calculator: Area $\simeq 5.33$ sq. units

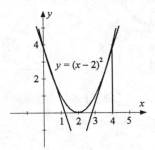

9. Guess: Area = 1 triangle = $\left(\frac{1}{2}\right)(8)(0.5)$

   $= 2.00$ sq. units

   By graphing calculator: Area = 1.61 sq. units

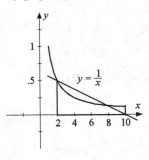

13. $\sum_{n=1}^{7} 3n = 3(1) + 3(2) + 3(3) + 3(4) + 3(5)$

   $+ 3(6) + 3(7)$

   $= 84$

17. $y = x^2 + 1$, for $x = 0$ to 8. Computing y at the midpoint of each panel, we get

x	1	3	5	7
y	2	10	26	50

Then $A \simeq 2(2) + 2(10) + 2(26) + 2(50)$
$= 176$ sq. units.
$\left(A = 178\frac{2}{3} \text{sq. units by integration.}\right)$

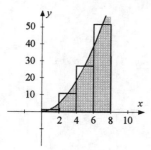

21.
```
10   '    PANELS
20   '
30   '    COMPUTES AREA UNDER A CURVE
40   '    BY MIDPOINT METHOD
50   '
60   DEF FNA(X) = X^2 + 1
70   INPUT "LOWER LIMIT"; A
80   INPUT "UPPER LIMIT"; B
90   INPUT "NUMBER OF PANELS"; N
100  W = (B-A)/N
110  FOR X=A+W/2 TO B-W/2 STEP W
120  Y = FNA(X)
130  AREA = AREA + Y*W
140  NEXT X
150  PRINT "PANEL WIDTH IS"; W
160  PRINT "AREA IS"; AREA
```

Exercise 6: Exact Area Under a Curve

1. $A = \int_0^{10} 2x \, dx = x^2 \Big]_0^{10} = 10^2 = 100$

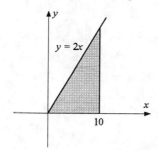

5. $A = \int_0^4 x^3 \, dx = \frac{x^4}{4} \Big]_0^4 = \frac{4^4}{4} = 64$

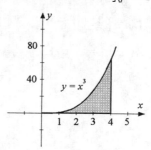

9. $y = x^2 + x + 1$, $x = 2$ to 3
$A = \int_2^3 (x^2 + x + 1) dx = \frac{x^3}{3} + \frac{x^2}{2} + x \Big]_2^3$
$= 9 + \frac{9}{2} + 3 - \left(\frac{8}{3} + 2 + 2\right) = 9.83$

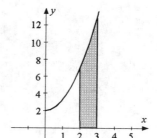

Chapter 30: Review Problems

1. $\int \frac{dx}{\sqrt[3]{x}} = \int x^{-1/3} \, dx = \frac{3x^{2/3}}{2} + C$

5. $\int \sqrt{4x} \, dx = 2\int x^{1/2} \, dx = 2(2)\frac{x^{3/2}}{3} + C$
$= \frac{4x\sqrt{x}}{3} + C$

9. $\int \frac{dx}{x+5} = \ln |x + 5| + C$

13. $\int_0^a \left(\sqrt{a} - \sqrt{x}\right)^2 dx = \int_0^a \left(a - 2\sqrt{a}\,\sqrt{x} + x\right) dx$
$= ax - \frac{4\sqrt{a}}{3}x^{3/2} + \frac{x^2}{2} \Big]_0^a$
$= a^2 - \frac{4a^2}{3} + \frac{a^2}{2} = \frac{a^2}{6}$

17. $\int_1^4 x\sqrt{1 + 5x} \, dx$
no. 52: $u = x$, $a = 1$, $b = 5$
$= \frac{2(15x-2)}{15(5)^2}(1 + 5x)^{3/2} \Big]_1^4$
$= 29.7684 - 1.0190 = 28.75$

21. $\int \frac{dx}{9+4x^2}$

     no. 56: $u = x, \ a = 3, \ b = 2$

$$= \tfrac{1}{6} \ \text{Tan}^{-1} \left(\tfrac{2x}{3} \right) + C$$

$$= \tfrac{1}{6} \ \text{Arctan} \left(\tfrac{2x}{3} \right) + C$$

25. $\frac{dN}{dt} = 0.5N$

       $\frac{dN}{0.5N} = dt$

       $\int \frac{dN}{0.5N} = \int dt$

       $2\ln N = t + C$

       $\ln N = \frac{t}{2} + \frac{C}{2}$

       $N = e^{\{(t/2)+(C/2)\}} = e^{C/2} e^{t/2} = k e^{t/2}$

       Since $N = 100$ when $t = 0$,

       $100 = k e^0 = k$

       So $N = 100 e^{t/2}$

29. $A = \int_0^2 \sqrt{8} \, x^{1/2} \, dx = \sqrt{8} \cdot \frac{2}{3} x^{3/2} \Big]_0^2 = \frac{16}{3}$

CHAPTER 31: APPLICATIONS OF THE INTEGRAL

Exercise 1: Applications to Motion

1. $a = \frac{dv}{dt} = 4.00 - t^2$

   $v = \int (4 - t^2)dt = 4t - \frac{t^3}{3} + C$
   Since $v = 2.00$ when $t = 3.00$,
   $2 = 4(3) - \frac{27}{3} + C$
   $2 = 12 - 9 + C$
   $C = -1$
   so $v = 4t - \frac{1}{3}t^3 - 1$

5. $a = \frac{dv}{dt} = -32$

   $v = \int -32 \, dt = -32t + C_1$
   Since $v = 20$ when $t = 0$,
   $C_1 = 20$
   $v = \frac{ds}{dt} = -32t + 20$
   $s = \int (-32t + 20)dt = \frac{-32t^2}{2} + 20t + C_2$
   since $s = 0$ when $t = 0$,
   $C_2 = 0$
   So $s = 20t - 16t^2$

9. $a_x = \frac{dv_x}{dt} = 3t \Rightarrow v_x = \int 3t \, dt$

   $\qquad v_x = \frac{3}{2}t^2 + C_1$
   At $t = 0$,
   $v_x = 6.00$, $C_1 = 6.00$, $v_x(t) = \frac{3}{2}t^2 + 6.00$
   $a_y = \frac{dv_y}{dt} = 2t^2 \Rightarrow v_y = \int 2t^2 \, dt$
   $\qquad \Rightarrow v_y = \frac{2}{3}t^3 + C_2$
   At $t = 0$,
   $v_y = 2.00 \Rightarrow C_2 = 2 \Rightarrow v_y(t) = \frac{2}{3}t^3 + 2$
   We now evaluate $v_x(t)$ and $v_y(t)$ at $t = 15.0$
   $v_x(15.0) = \frac{3 \cdot 15^2}{2} + 6 = 344$ cm/s and
   $v_y(15.0) = \frac{2 \cdot 15^3}{3} + 2 = 2250$ cm/s

13. $\omega = \int 7.24 \, dt = 7.24t + C$
    At $t = 0$,
    $\omega = 1.25$, $C = 1.25$, $\omega = 7.24t + 1.25$
    When $t = 2.00$,
    $\omega(t) = \omega(2.00) = 7.24(2.00) + 1.25 = 15.7$ rad/s

Exercise 2: Applications to Electric Circuits

1. $q = \int i \, dt = \int (2t + 3)dt = t^2 + 3t + C$
   Since $q = 8.13$ coulombs when $t = 0$, we obtain,
   $C = 8.13$.
   Therefore, $q = t^2 + 3t + 8.13$ and at $t = 1.00$ s we
   obtain, $q(1) = 1^2 + 3 \cdot 1 + 8.13 = 12.1$ coulombs

5. $V = \frac{1}{C}\int i \, dt = \frac{1}{15.2}\int t\sqrt{5 + t^2} \, dt$

   $\qquad = \frac{1}{2(15.2)}\int \sqrt{5 + t^2}(2t)dt$
   $\qquad = \frac{1}{2(1.52)}(5 + t^2)^{3/2}\left(\frac{2}{3}\right) + C_1$
   $\qquad = 0.0219(5 + t^2)^{3/2} + C_1$
   At $t = 0$, $V = 2$, so,
   $2 = 0.0219(5 + t^2)^{3/2} + C_1 \Rightarrow C_1 = 1.755$
   Therefore, $V = 0.0129(5 + t^2)^{3/2} + 1.755$
   At $t = 1.75$,
   $V = 0.0129(5 + 1.75^2)^{3/2} + 1.755 = 2.26$ volts

9. $i = \frac{1}{L}\int V \, dt = \frac{1}{15.0}\int \left(28.5 + \sqrt{6}\,t^{1/2}\right)dt$

   $\qquad = \frac{1}{15.0}\left(28.5t + \frac{2}{3}\sqrt{6}\,t^{3/2}\right) + C$
   $\qquad = 1.9t + 0.1088t^{3/2} + C$
   At $t = 0$, $i = 15.0$, therefore, $C = 15.0$. Then,
   $i(2.50) = 1.9(2.50) + (0.1088)\left(2.5^{3/2}\right) + 15.0$
   $\qquad = 20.2$ A

Exercise 3: Finding Areas by Means of the Definite Integral

1. Using a horizontal strip of width dy and length
   $x_2 - x_1$, we obtain, $dA = (x_2 - x_1)dy$,
   where the right curve $y = x^2 + 2$
   solved for x becomes $x_2 = \sqrt{y - 2}$ and the
   left curve or y-axis is $x_1 = 0$. Integrating gives us
   $A = \int_3^5 \left(\sqrt{y - 2} - 0\right)dy = \frac{2}{3}(y - 2)^{3/2}\Big|_3^5$
   $\qquad = \frac{2}{3} \cdot 3^{3/2} - \frac{2}{3} = 2.797$ square units

5. Using a horizontal strip, we obtain
$dA = (x_2 - x_1)dy$
The right curve is $x_2 = 16 - y^2$, while the left curve (y-axis) is $x_1 = 0$.
Thus, $dA = (16 - y^2 - 0)dy$
Integrating from $y = 0$ to $y = 4$ yields,
$$A = \int_0^4 (16 - y^2)dy = 16y - \frac{y^3}{3}\Big|_0^4 = 64 - \frac{64}{3}$$
$$= \frac{128}{3} = 42\frac{2}{3} \text{ square units}$$

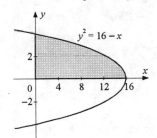

9. $y^2 = x,\ x = 4$
$$A = 2\int_0^4 x^{1/2}\ dx = \frac{4}{3}x^{3/2}\Big]_0^4 = \frac{4}{3}\left(4^{3/2}\right)$$
$$= 10\frac{2}{3}\ \text{square units}$$

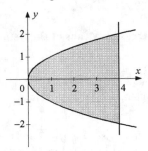

13. The graph $10y = x^2 - 80$ crosses the x-axis when $y = 0$ or $x = \sqrt{80} = 8.9$. Thus, as we can see from the graph our required area lies entirely below the x-axis. Using a vertical strip, we obtain, $dA = (y_2 - y_1)dx$ where the upper curve is the x-axis or $y_2 = 0$, and the lower curve is $y_1 = \frac{1}{10}x^2 - 8$. Integrating from $x = 1$ to $x = 6$ yields:
$$A = \int_1^6 \left[0 - \left(\frac{1}{10}x^2 - 8\right)\right]dx = \int_1^6 \left(8 - \frac{1}{10}x^2\right)dx$$
$$= 8x - \frac{1}{30}x^3\Big|_1^6$$
$$= 48 - \frac{216}{30} - \left(8 - \frac{1}{30}\right)$$
$$= 32.83 \text{ square units (rounded)}$$

17. $y_1 = 3x_1,\ x_1 = \frac{y_1}{3}$
$y_2 = 15 - 3x_2,\ x_2 = \frac{15 - y_1}{3} = 5 - \frac{y_1}{3}$
$$A = \int_0^{15/2}(x_2 - x_1)dy = \int_0^{15/2}\left[\left(5 - \frac{y}{3}\right) - \frac{y}{3}\right]dy$$
$$= \int_0^{15/2}\left(5 - \frac{2y}{3}\right)dy = 5y - \frac{y^2}{3}\Big]_0^{15/2}$$
$$= 5\left(\frac{15}{2}\right) - \frac{15^2}{2^2(3)} = 18.75 \text{ square units}$$

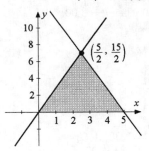

21. $y - 4 = \frac{4-1}{8-(-1)} \bullet (x - 8)$
$y - 4 = \frac{1}{3}(x - 8)$
$y = \frac{x+4}{3}$
$$A = \int_{-1}^8 \left[\frac{x+4}{3} - x^{2/3}\right]dx$$
$$= \int_{-1}^8 \left[-x^{2/3} + \frac{x}{3} + \frac{4}{3}\right]dx$$
$$= -\frac{3}{5}x^{5/3} + \frac{x^2}{6} + \frac{4}{3}x\Big]_{-1}^8$$
$$= 2.70 \text{ square units}$$

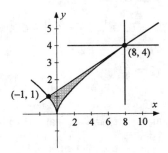

25. $A = \int_0^a b\ dx = bx\Big]_0^a = ab - 0 = ab$

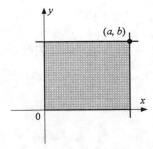

29.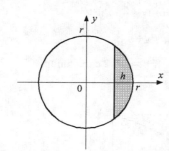

$$A = 2\int_{r-h}^{r} \sqrt{r^2 - x^2}\, dx = 2\left[\frac{x}{2}\sqrt{r^2-x^2} + \frac{r^2}{2}\,\text{Arcsin}\,\frac{x}{r}\right]_{r-h}^{r}$$

(Table of Integrals #69)

$$= 2\left(\frac{r}{2}\cdot 0 + \frac{r^2}{2}\,\text{Arcsin}\,1\right) - 2\left(\frac{r-h}{2}\sqrt{r^2-(r-h)^2} + \frac{r^2}{2}\,\text{Arcsin}\,\frac{r-h}{r}\right)$$

$$= r^2\cdot\frac{\pi}{2} - (r-h)\sqrt{2rh-h^2} - r^2\,\text{Arcsin}\,\frac{r-h}{r}$$

$$= r^2\left[\frac{\pi}{2} - \text{Arcsin}\,\frac{r-h}{r}\right] - (r-h)\sqrt{2rh-h^2}$$

$$= r^2\,\text{Arccos}\,\frac{r-h}{r} - (r-h)\sqrt{2rh-h^2}$$

33. Volume = cross sectional area (A) × length

$$y = kx^2 + 5$$
$$15 = k(10)^2 + 5$$
$$k = 0.1$$
$$A = 2(3\bullet 15) + 2\int_0^{10}(0.1x^2 + 5)dx$$
$$= 90 + 2\left[\frac{0.1x^3}{3} + 5x\right]_0^{10}$$
$$= 90 + 2\left(\frac{100}{3} + 50\right) = 256\frac{2}{3}$$
$$V = \left(256\frac{2}{3}\right)(50) = 12,800\,\text{cm}^3$$

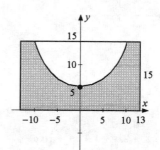

37. For the circle of radius 80, $x_1^2 + y^2 = 80^2$ from which $x_1 = \sqrt{80^2 - y^2}$. Since we want only the semicircle for which x is positive, we drop the minus sign. For the circle of radius 60, $(x_2 - 120)^2 + y^2 = 60^2$ from which $x_2 = 120 - \sqrt{60^2 - y^2}$. Here we have dropped the plus on the radical, so as to obtain only the left portion of the circle.

$$A = 2\int_0^{28}(x_1 - x_2)dy = 2\left[\int_0^{28}\sqrt{80^2 - y^2}\,dy + \int_0^{28}\sqrt{60^2 - y^2}\,dy - \int_0^{28}120\,dy\right]$$

Using #69, Table of Integrals:

$$\int_0^{28}\sqrt{80^2 - y^2}\,dy = \left[\frac{y}{2}\sqrt{80^2 - y^2} + \frac{80^2}{2}\,\text{Arcsin}\,\frac{y}{80}\right]_0^{28} = \frac{28}{2}\sqrt{80^2 - 28^2} + \frac{80^2}{2}\,\text{Arcsin}\,\frac{28}{80} - (0+0) = 2193.4$$

$$\int_0^{28}\sqrt{60^2 - y^2}\,dy = \left[\frac{y}{2}\sqrt{60^2 - y^2} + \frac{60^2}{2}\,\text{Arcsin}\,\frac{y}{60}\right]_0^{28} = 1616.9$$

$$\int_0^{28}120\,dy = 120(28) = 3360$$

$$A = 2[2193.4 + 1616.9 - 3360] = 900\,\text{mm}^2 = 9.00\,\text{cm}^2$$

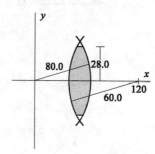

Exercise 4: Volumes by Integration

1. $dV = \pi y^2 \ dx = \pi x^6 \ dx$

     $V = \pi \int_0^2 x^6 \ dx = \pi \frac{x^7}{7} \Big|_0^2 = \frac{2^7 \pi}{7} = 57.4$ cubic units

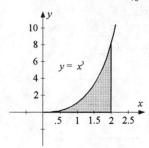

5. $y^2 (2 - x) = x^3$

     $y^2 = \frac{x^3}{2-x} = -x^2 - 2x - 4 + \frac{8}{2-x}$

     $dV = \pi y^2 \ dx$

     $V = \pi \int_0^1 \left(-x^2 - 2x - 4 + \frac{8}{2-x} \right) dx$

         $= \pi \left(-\frac{x^3}{3} - x^2 - 4x - 8 \ln |2 - x| \right) \Big|_0^1$

         $= \pi \left\{ \left(-\frac{1}{3} - 1 - 4 - 0 \right) - (0 - 8 \ln 2) \right\}$

         $= \pi \left(-\frac{16}{3} + 8 \ln 2 \right) = 0.666$ cubic units

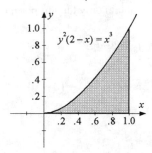

9. $y = x^3$, the y-axis, and $y = 8$.

     $V = \pi \int x^2 \ dy = \pi \int_0^8 y^{2/3} \ dy$

         $= \frac{3\pi y^{5/3}}{5} \Big|_0^8 = \frac{3\pi}{5} (8)^{5/3}$

         $= 60.3$ cubic units

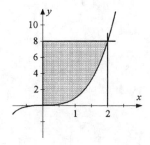

13. $dV = \pi x^2 \ dy = \pi \frac{y^4}{16} \ dy$

     $V = \pi \int_0^4 \frac{y^4}{16} \ dy = \frac{\pi}{16} \cdot \frac{y^5}{5} \Big|_0^4$

         $= \frac{64\pi}{5} = 40.2$ cubic units

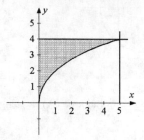

17. $dV = \pi (y - 4)^2 dx = \pi (6x - 2x^2)^2 dx$

         $= \pi (36x^2 - 24x^3 + 4x^4) dx$

     $V = 4\pi \int_0^3 (9x^2 - 6x^3 + x^4) dx$

         $= 4\pi \left[\frac{9x^3}{3} - \frac{6x^4}{4} + \frac{x^5}{5} \right]_0^3$

         $= 102$ cubic units

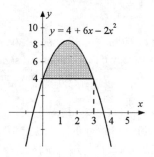

21. $dV = 2\pi x \cdot y \cdot dx$

         $= 2\pi x \sqrt{100 \left(1 + \frac{x^2}{1225} \right)} \ dx$

         $= \frac{20\pi x}{35} \sqrt{1225 + x^2} \ dx$

     $V = \int_{10.0}^{100} \frac{10\pi}{35} (1225 + x^2)^{1/2} \cdot 2x \ dx$

         $= \frac{10\pi}{35} \left(\frac{2}{3} (1225 + x^2)^{3/2} \right)_{10.0}^{100}$

         $= 682,800$ cm^3 $= 0.683$ m^3

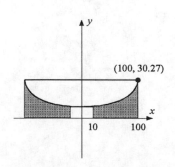

Chapter 31: Review Problems

1. $4x = 5 - x$

   $5x = 5$

   $x = 1$

   $V = \int_0^1 \pi(4x)dx + \int_1^5 \pi(5-x)dx$

   $\quad = \pi\{2x^2\}_0^1 - \pi\left(\dfrac{(5-x)^2}{2}\right)_1^5$

   $\quad = \pi\{2 - (-8)\} = 10\pi$

   $\quad = 31.4$ cubic units

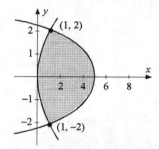

5. $A = 2\int_0^4 x^{3/2} \, dx = \dfrac{4}{5}\left[x^{5/2}\right]_0^4$

   $\quad = \dfrac{128}{5} = 25.6$ square units

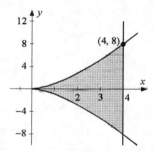

9. $\alpha' = 7.25t^2 \, \text{rad/s}^2$

   $\omega = 7.25 \int t^2 \, dt = \left(\dfrac{7.25}{3}\right)t^3 + \omega_0$

   $\quad = 2.417t^3 \ \text{rad/s}$

   $\theta = 2.417 \int t^3 \, dt = \dfrac{2.417t^4}{4} + \theta_0 = 0.604t^4$

   $\omega(20.0) = 2.417(20.0)^3 = 19,300 \, \text{rad/s}$

   $\theta(20.0) = (0.604)(20.0)^4$

   $\quad\quad = 96,600 \, \text{rad} = 15,400 \, \text{rev}$

13. $y^2 = 9\left(1 - \dfrac{x^2}{16}\right) = \dfrac{9}{16}(16 - x^2)$

    $y = \dfrac{3}{4}\sqrt{16 - x^2}$

    $V = 2\int_0^4 \pi y^2 \, dx = 2\pi\int_0^4 \dfrac{9}{16}(16 - x^2)dx$

    $\quad = \dfrac{9}{8}\pi\left\{16x - \dfrac{x^3}{3}\right\}_0^4$

    $\quad = \dfrac{9}{2}\pi\left(16 - \dfrac{16}{3}\right) = 48\pi$

    $\quad = 151$ cubic units

17. $A = 4\int_0^4 \sqrt[3]{1 - \dfrac{x^2}{16}} \, dx = 3\int_0^4 \sqrt{16 - x^2} \, dx$

    $\quad = 3\left[\dfrac{x}{2}\sqrt{16 - x^2} + \dfrac{16}{2}\text{Arcsin}\dfrac{x}{4}\right]_0^4$

    $\quad = 3\{(0 + 8\,\text{Arcsin}\,1) - (0 + 0)\}$

    $\quad = 3\left(\dfrac{8\pi}{2}\right) = 12\pi$ square units

CHAPTER 32: MORE APPLICATIONS OF THE INTEGRAL

Exercise 1: Length of Arc

1.

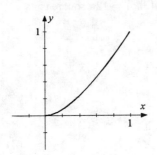

$y^2 = x^3$

$y = x^{3/2}, \quad y' = \frac{3}{2}x^{1/2}$

$s = \int_0^{5/9} \sqrt{1 + \left(\frac{3}{2}x^{1/2}\right)^2}\, dx = \int_0^{5/9} \sqrt{1 + \frac{9x}{4}}\, dx = \frac{1}{2}\int_0^{5/9} \sqrt{4 + 9x}\, dx$

$\quad = \frac{1}{18}\int_0^{5/9}(4 + 9x)^{1/2} \cdot 9\, dx = \frac{1}{18} \cdot \frac{2}{3}(4 + 9x)^{3/2}\Big|_0^{5/9}$

$\quad = \frac{1}{27}(27 - 8) = 0.704$

5.

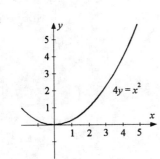

$4y = x^2$

$y = \frac{x^2}{4}, y' = \frac{x}{2}$

$s = \int_0^4 \sqrt{1 + \left(\frac{x}{2}\right)^2}\, dx = \frac{1}{2}\int_0^4 \sqrt{4 + x^2}\, dx \quad \text{(use rule no. 66)}$

$\quad = \frac{1}{2}\left(\frac{x}{2}\sqrt{4 + x^2} + \frac{4}{2}\ln\left|x + \sqrt{4 + x^2}\right|\right)_0^4 = 5.92$

9.

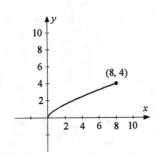

(8, 4)

$y^3 = x^2$

$x = y^{3/2}$

$x' = \frac{3}{2}y^{1/2}$

$(x')^2 = \frac{9}{4}y$

$s = \int_0^4 \sqrt{1 + \frac{9}{4}y}\, dy = \frac{4}{9}\int_0^4 \sqrt{1 + \frac{9}{4}y} \cdot \frac{9}{4}dy$

$\quad = \frac{4}{9} \cdot \frac{2}{3}\left(1 + \frac{9}{4}y\right)^{3/2}\Big|_0^4 = \frac{8}{27}\left(10^{3/2} - 1\right) = 9.07$

13.

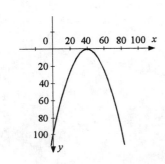

$y = 0.0625x^2 - 5x + 100$

$s = 2\int_0^{40} \sqrt{1 + y'^2}\, dx \quad \text{(use rule 66)}$

$\quad = 2\int_0^{40} \sqrt{1 + (0.125x - 5)^2}\, dx$

$\quad = \frac{2}{0.125}\int_0^{40} \sqrt{1 + (0.125x - 5)^2}\,(0.125)dx$

$\quad = 16\left(\frac{0.125x - 5}{2}\sqrt{1 + (0.125x - 5)^2}\right.$

$\qquad \left. + \frac{1}{2}\ln\left|(0.125x - 5) + \sqrt{1 + (0.125x - 5)^2}\right|\right)_0^{40} = 223 \text{ ft}$

Exercise 2: Area of Surface of Revolution

1. First we find $\frac{dy}{dx}$: $\frac{dy}{dx} = \frac{x^2}{3} \Rightarrow \left(\frac{dy}{dx}\right)^2 = \frac{x^4}{9}$

$$S = 2\pi \int_a^b y \sqrt{1 + \left(\frac{dy}{dx}\right)^2}\, dx = 2\pi \int_0^3 \frac{x^3}{9} \sqrt{1 + \frac{x^4}{9}}\, dx = \frac{2\pi}{9} \int_0^3 \frac{\sqrt{9+x^4}}{3}\, (x^3\, dx)$$

$$= \frac{2\pi}{27} \cdot \frac{1}{4} \int_0^3 (9 + x^4)^{1/2} (4x^3\, dx) = \frac{\pi}{54}(9 + x^4)^{3/2} \cdot \frac{2}{3}\Big|_0^3 = 32.1$$

5. $y = \sqrt{4 - x}$

$y' = \frac{1}{2}(4 - x)^{-1/2}(-1) = -\frac{1}{2\sqrt{4-x}}$

$\sqrt{1 + y'^2} = \sqrt{1 + \frac{1}{16-4x}} = \sqrt{\frac{17-4x}{16-4x}}$

$s = \int_0^4 2\pi \sqrt{4-x} \sqrt{\frac{17-4x}{16-4x}}\, dx = 2\pi \int_0^4 \sqrt{\frac{17-4x}{4}}\, dx$

$= -\frac{1}{4}\pi \int_0^4 \sqrt{17 - 4x}\,(-4\, dx) = -\frac{\pi}{4} \cdot \frac{2}{3}\left\{(17 - 4x)^{3/2}\right\}_0^4 = 36.2$

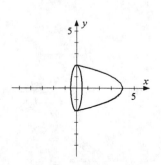

9. $y = 3x^2, \qquad \frac{dy}{dx} = 6x, \qquad \left(\frac{dy}{dx}\right)^2 = 36x^2$

Substituting into the formula for surface area and integrating between $x = 0$ and $x = 5$, we obtain,

$$S = 2\pi \int_a^b x \sqrt{1 + \left(\frac{dy}{dx}\right)^2}\, dx = 2\pi \int_0^5 x\sqrt{1 + 36x^2}\, dx = \frac{2\pi}{72} \int_0^5 (1 + 36x^2)^{1/2}(72x\, dx)$$

$$= \frac{2\pi}{108}(1 + 36x^2)^{3/2}\Big|_0^5 = 1570$$

13. $x^2 + y^2 = r^2$

$y = \sqrt{r^2 - x^2}$

$y' = \frac{1}{2}(r^2 - x^2)^{-1/2}(-2x) = -\frac{x}{\sqrt{r^2-x^2}}$

$\sqrt{1 + y'^2} = \sqrt{1 + \frac{x^2}{r^2-x^2}} = \sqrt{\frac{r^2}{r^2-x^2}} = \frac{r}{\sqrt{r^2-x^2}}$

$s = 2\int_0^r 2\pi \sqrt{r^2 - x^2}\left(\frac{r}{\sqrt{r^2-x^2}}\right) dx = 4\pi \int_0^r r\, dx = 4\pi\{rx\}_0^r = 4\pi r^2$

Exercise 3: Centroids

1. $0 + 4 + 3 + (-2) = 4\overline{x}; \quad \overline{x} = \frac{5}{4}$

$0 + 2 + (-5) + (-3) = 4\overline{y}; \quad \overline{y} = -\frac{6}{4} = -\frac{3}{2} : \left(\frac{5}{4}, -\frac{3}{2}\right)$

5.
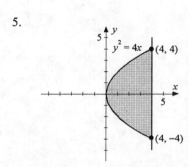

$A = 2\int_0^4 2x^{1/2} dx = 4 \cdot \frac{2}{3}\{x^{3/2}\}_0^4 = \frac{64}{3}$

$A \cdot \overline{x} = \int_0^4 x\{2x^{1/2} - (-2x^{1/2})\} dx = \int_0^4 4x^{3/2} dx$

$= 4 \cdot \frac{2}{5}\{x^{5/2}\}_0^4 = \frac{256}{5}$

$\overline{x} = \frac{(256/5)}{(64/3)} = \frac{3 \cdot 4}{5} = 2.40$

$\overline{y} = 0$

9.

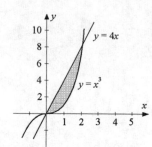

$$A\bar{x} = \int_0^2 x(4x - x^3)dx = \int_0^2 (4x^2 - x^4)dx = \frac{4}{3}x^3 - \frac{1}{5}x^5\Big|_0^2 = 4.267$$

$$A = \int_0^2 (4x - x^3)dx = \left\{2x^2 - \frac{x^4}{4}\right\}_0^2 = 4$$

$$\bar{x} = \frac{4.267}{4} = 1.067$$

13.

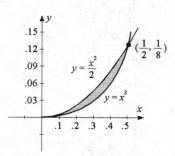

$$y = x^3, \quad 2y = x^2$$

$$x^3 = \frac{x^2}{2}$$

$$x = \frac{1}{2}, 0$$

$$A = \int_0^{1/2}\left(\frac{x^2}{2} - x^3\right)dx = \frac{x^3}{6} - \frac{x^4}{4}\Big|_0^{1/2} = 0.005208$$

$$A \cdot \bar{x} = \int_0^{1/2} x\left(\frac{x^2}{2} - x^3\right)dx = \frac{x^4}{8} - \frac{x^5}{5}\Big|_0^{1/2} = 0.001562$$

$$\bar{x} = \frac{0.001562}{0.005208} = 0.2999 = \frac{3}{10}$$

17.

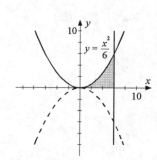

$$V = \int_0^6 \pi\left(\frac{x^2}{6}\right)^2 dx = \frac{\pi}{36}\int_0^6 x^4 dx = \frac{\pi}{36\cdot 5}\{x^5\}_0^6 = \frac{\pi}{5} \cdot 6^3$$

$$\bar{x} \cdot V = \int_0^6 x \cdot \pi\left(\frac{x^2}{6}\right)^2 dx = \frac{\pi}{36}\int_0^6 x^5 dx = \frac{\pi}{36\cdot 6}\{x^6\}_0^6 = \pi \cdot 6^3$$

$$\bar{x} = \frac{\pi\cdot 6^3}{(\pi/5)6^3} = 5$$

21.

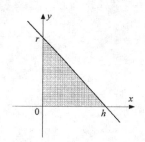

$$V = \frac{1}{3}\pi r^2 h$$

$$\bar{x} \cdot V = \int_0^h x \cdot \pi\left(r - \frac{r}{h}x\right)^2 dx = \pi r^2 \int_0^h\left(x - 2\frac{x^2}{h} + \frac{x^3}{h^2}\right)dx$$

$$= \pi r^2\left(\frac{x^2}{2} - \frac{2x^3}{3h} + \frac{x^4}{4h^2}\right)_0^h = \pi r^2\left(\frac{h^2}{2} - \frac{2h^2}{3} + \frac{h^2}{4}\right) = \frac{\pi r^2 h^2}{12}$$

$$\bar{x} = \frac{\pi r^2 h^2/12}{\pi r^2 h/3} = \frac{h}{4}$$

25. $y = 18 + kx^2$

$90 = 18 + k(82)^2$

$k = \frac{72}{(82)^2} = 0.01071$

$V = \int_0^{82} 2\pi x(18 + kx^2)dx + \pi(95^2 - 82^2)(90.0)$

$= 2\pi \int_0^{82}(18x + kx^3)dx + 650,592$

$= 2\pi\left\{9x^2 + \frac{kx^4}{4}\right\}_0^{82} + 650,592 = 1,791,440$

$\bar{y} \cdot V = \int_0^{82}\left(\frac{18+kx^2}{2}\right)2\pi x(18 + kx^2)dx + \frac{90}{2}(650592) = \pi\int_0^{82}(18^2 \cdot x + 36kx^3 + k^2x^5)dx + 29,276,640$

$= \pi\left\{9 \cdot 18x^2 + 9kx^4 + \frac{k^2}{6}x^6\right\}_0^{82} + 29,276,640 = \pi \cdot 82^2\left\{9 \cdot 18 + 9k(82^2) + \frac{k^2}{6}82^4\right\} + 29,276,640$

$= 35,371,479 + 29,276,640 = 64,648,119$

$\bar{y} = \frac{64,648,119}{1,791,440} = 36.1$ mm

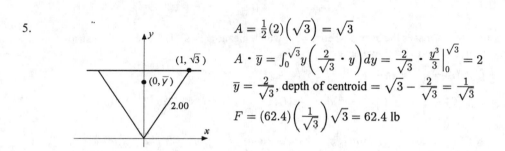

Exercise 4: Fluid Pressure

1. $F = (62.4)(50.0)(\pi)(3.00)^2 = 88,200$ lb

5. $A = \frac{1}{2}(2)\left(\sqrt{3}\right) = \sqrt{3}$

$A \cdot \bar{y} = \int_0^{\sqrt{3}} y\left(\frac{2}{\sqrt{3}} \cdot y\right)dy = \frac{2}{\sqrt{3}} \cdot \frac{y^3}{3}\Big|_0^{\sqrt{3}} = 2$

$\bar{y} = \frac{2}{\sqrt{3}}$, depth of centroid $= \sqrt{3} - \frac{2}{\sqrt{3}} = \frac{1}{\sqrt{3}}$

$F = (62.4)\left(\frac{1}{\sqrt{3}}\right)\sqrt{3} = 62.4$ lb

Exercise 5: Work

1. $F = k \cdot x$

$50.0 = k \cdot 1$

$k = 50.0$

$W = \int_2^4 F dx = 50.0\int_2^4 x \, dx = 25.0x^2\Big|_2^4 = 25.0(16 - 4) = 25.0 \cdot 12$ in. $\cdot$ lb $= 25.0$ ft $\cdot$ lb

5. $V = \int_2^6 \pi x^2 dy = \int_2^6 \pi(36 - y^2)dy = \pi\left\{36 - \frac{y^3}{3}\right\}_2^6 = \pi\left\{\left(216 - \frac{216}{3}\right) - \left(72 - \frac{8}{3}\right)\right\} = 74.67\pi = 234.6$

$\bar{y} \cdot V = \int_2^6 y \cdot \pi x^2 dx = \pi\int_2^6 y(36 - y^2)dy = \pi\left\{18y^2 - \frac{y^4}{4}\right\}_2^6$

$= \pi\left\{y^2\left(18 - \frac{y^2}{4}\right)\right\}_2^6 = \pi\{36(18 - 9) - 4(18 - 1)\} = 256\pi$

$\bar{y} = \frac{256\pi}{74.67\pi} = 3.43$ ft

work $= V(62.4)(3.43) = 50,200$ ft $\cdot$ lb

9.

$$p_1 v_1 = p_2 v_2$$

$$v_2 = \frac{p_1}{p_2}(v_1) = \frac{15.0}{80.0}(200) = 37.5 \text{ ft}^3$$

from Problem 8,

$$W = -c\int_{200}^{37.5} \frac{dv}{v} = -c \ln v \Big|_{200}^{37.5} = -432,000(\ln 37.5 - \ln 200) = 723,000 \text{ ft} \cdot \text{lb}$$

13. $\quad W = \int_0^{20.0}\{200 + (500 - x)\}dx = 700x - \frac{x^2}{2}\Big|_0^{20.0} = 13,800 \text{ ft} \cdot \text{lb}$

Exercise 6: Moment of Inertia

1.

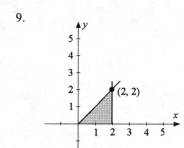

$$I_x = \tfrac{1}{3}\int_0^1 y^3 dx = \tfrac{1}{3}\int_0^1 x^3 dx = \tfrac{1}{3}\cdot\tfrac{x^4}{4}\Big|_0^1 = \tfrac{1}{12}$$

5. The parabola $y = 4 - x^2$ crosses the x axis when $y = 0$, which means $x = 2$. Using a vertical strip and integrating from $x = 0$ to $x = 2$, we obtain,

$$I_x = \tfrac{1}{3}\int_a^b y^2 dA = \tfrac{1}{3}\int_0^2 y^2(4 - x^2)dx = \tfrac{1}{3}\int_0^2 (4 - x^2)^3 dx$$

$$= \tfrac{1}{3}\int_0^2 (64 - 48x^2 + 12x^4 - x^6)dx = \tfrac{1}{3}\left(64x - 16x^3 + \tfrac{12}{5}x^5 - \tfrac{1}{7}x^7\right)\Big|_0^2 = 19.5$$

9.

$$I_x = m\int_0^2 y^2 \cdot 2\pi y(2 - x)dy = m\int_0^2 y^2 \cdot 2\pi y(2 - y)dy$$

$$= 2\pi m\int_0^2 (2y^3 - y^4)dy = 2\pi m\left(\tfrac{y^4}{2} - \tfrac{y^5}{5}\right)\Big|_0^2 = 10.1m$$

13.

$$M = \tfrac{1}{3}\pi r^2 h \cdot m$$

$$I_x = \int_0^r y^2 2\pi y(h - x)m\,dy$$

since $\frac{r - y}{h - x} = \frac{r}{h}$, $h - x = \frac{h(r - y)}{r}$

so $I_x = 2\pi m\int_0^r y^3(h - x)dy = \frac{2\pi mh}{r}\int_0^r (ry^3 - y^4)dy = \frac{2\pi mh}{r}\left(\tfrac{r}{4}y^4 - \tfrac{1}{5}y^5\right)\Big|_0^r$

$$= \frac{2\pi mh}{r}\cdot r^4\left(\tfrac{r}{4} - \tfrac{r}{5}\right) = 2\pi mh \cdot r^3\left(\tfrac{r}{20}\right) = \frac{\pi mhr^4}{10}$$

$$\frac{I_x}{M} = \frac{3\pi mhr^4}{10\pi r^2 hm} = \frac{3r^2}{10}$$

$$I_x = \frac{3Mr^2}{10}$$

Chapter 32: Review Problems

1. $\bar{y} = 0, \ A = \frac{1}{2}\pi \cdot 10^2 = 50\pi$

$\bar{x} \cdot A = \int_0^{10.0} x \cdot 2y \, dx = 2\int_0^{10.0} x\sqrt{100 - x^2} \, dx = -\int_0^{10.0}\sqrt{100 - x^2}(-2x \, dx) = -\frac{2}{3}(100 - x^2)^{3/2}\Big|_0^{10.0}$

$= 0 - \left\{-\frac{2}{3}(100)^{3/2}\right\} = \frac{2}{3} \cdot 1000$

$\bar{x} = \frac{(2/3)(1000)}{50\pi} = \frac{2(20)}{3\pi} = 4.24$

5. $V = \frac{1}{3}\pi(4)^2(12) = 64\pi$

$\bar{y} \cdot V = \int_0^{12.0} y \cdot \pi x^2 dy$

$\frac{x}{y} = \frac{4}{12}, \ x = \frac{y}{3}$

$\bar{y} \cdot V = \pi\int_0^{12.0}\frac{y^3}{9}\,dy = \frac{\pi}{36}y^4\Big|_0^{12.0} = \frac{\pi \cdot 12^3}{3}$

$\bar{y} = \frac{\pi \cdot 12^3}{3(\pi)(64)} = 3^2 = 9$

centroid to be raised $12 - 9 = 3$ ft

$W = (64\pi)(80.0)(3) = 48,300$ ft lb

9. $W = \int_{1.00}^{3.00} 5.45x \, dx = 5.45\frac{x^2}{2}\Big|_{1.00}^{3.00} = (5.45)(4) = 21.8$ in. lb

13.

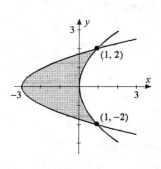

$y^2 = x + 3$

$x = y^2 - 3, \ \frac{dx}{dy} = 2y$

outer surface $= \int_0^2 2\pi y\sqrt{1 + (2y)^2}\,dy = \frac{\pi}{4}\int_0^2\sqrt{1 + 4y^2}\,(8y\,dy)$

$= \frac{\pi}{4} \cdot \frac{2}{3}(1 + 4y^2)^{3/2}\Big|_0^2 = \frac{\pi}{6}(17^{3/2} - 1) = 36.18$

$\frac{y^2}{4} = x, \ \frac{dx}{dy} = \frac{y}{2}$

inner surface $= \int_0^2 2\pi y\sqrt{1 + \left(\frac{y}{2}\right)^2}\,dy = \frac{\pi}{2}\int_0^2\sqrt{4 + y^2}(2y\,dy)$

$= \frac{\pi}{2} \cdot \frac{2}{3}(4 + y^2)^{3/2}\Big|_0^2 = \frac{\pi}{3}(8^{3/2} - 4^{3/2}) = 15.32$

total $s = 51.5$

17.

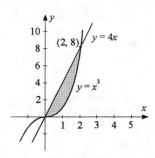

area of conical surface $= \frac{1}{2}(2\pi)(8)\left(\sqrt{2^2 + 8^2}\right) = 8\pi\left(\sqrt{68}\right) = 207.25$

area of parabolic surface:

$y = x^3, \ \frac{dy}{dx} = 3x^2$

$S = 2\pi\int_0^2 x^3\sqrt{1 + (3x^2)^2}\,dx = 2\pi\int_0^2 x^3\sqrt{1 + 9x^4}\,dx$

$= \frac{2\pi}{36}\int_0^2(1 + 9x^4)^{1/2}36x^3\,dx = \frac{\pi}{18}\left(\frac{2}{3}\right)\left[(1 + 9x^4)^{3/2}\right]_0^2$

$= \frac{\pi}{27}(1746 - 1) = 203.04$

total area $= 207.25 + 203.04 = 410.29 = 410$

CHAPTER 33: DERIVATIVES OF TRIGONOMETRIC, LOGARITHMIC, AND EXPONENTIAL FUNCTIONS

Exercise 1: Derivatives of the Sine and Cosine Functions

1. $y = \sin x$
   $y' = \cos x$

5. $y = \sin 3x$
   $y' = (\cos 3x)(3) = 3\cos 3x$

9. $y = 3.75x \cos x$
   $y' = 3.75(\cos x - x\sin x)$

13. $y = \sin 2\theta \cos \theta$
    $y' = \sin 2\theta(-\sin \theta) + \cos \theta(\cos 2\theta)(2)$
    $\quad = 2\cos 2\theta \cos \theta - \sin 2\theta \sin \theta$

17. $y = 1.23 \sin^2 x \cos 3x$
    $y' = 1.23(-3\sin^2 x \sin 3x + 2\sin x \cos x \cos 3x)$

21. $y = \sin x$
    $y' = \cos x$
    $y'' = -\sin x$

25. $f(x) = x \sin \frac{\pi}{2}x; \; f'(x)$
    $\quad = x\cos \frac{\pi}{2}x\left(\frac{\pi}{2}\right) + (1)\sin \frac{\pi}{2}x$
    $f''(x) = \frac{\pi}{2}\left\{x\left(-\sin \frac{\pi}{2}x\right)\left(\frac{\pi}{2}\right) + (1)\cos \frac{\pi}{2}x\right\}$
    $\qquad + \left(\cos \frac{\pi}{2}x\right)\left(\frac{\pi}{2}\right)$
    $f''(1) = \frac{\pi}{2}\left\{1(-1)\left(\frac{\pi}{2}\right) + 0\right\} + 0 = -\frac{\pi^2}{4}$

29. $x = \sin(x + y)$
    $1 = \cos(x + y)(1 + y')$
    $\quad = \cos(x + y) + y'\cos(x + y)$
    $y' = \frac{1 - \cos(x+y)}{\cos(x+y)} = \sec(x + y) - 1$

33. $y = x \sin \frac{x}{2}$
    $y' = x\left(\cos \frac{x}{2}\right)\left(\frac{1}{2}\right) + \sin \frac{x}{2}$
    $y'(2) = 1.381$

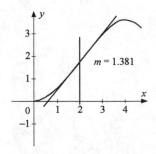

37. $y = 3\sin x - 4\cos x$
    $y' = 3\cos x + 4\sin x$
    $y'' = 4\cos x - 3\sin x \qquad (= -y)$
    $y' = 0$ when $3\cos x = -4\sin x$
    $-\frac{3}{4} = \frac{\sin x}{\cos x} = \tan x$
    $x = -0.65 \text{ rad} = 5.64 \text{ rad and}$
    $x = -0.65 \text{ rad} + \pi = 2.50 \text{ rad}$
    At $x = 2.50$, $y'' = -5$, so $x = 2.50$ is a maxmum
    At $x = 5.64$, $y'' = 5$, so $x = 5.64$ is a minimum
    maximum: $(2.50, 5)$
    minimum: $(5.64, -5)$
    $y'' = 0$ when $4\cos x = 3\sin x$
    $x = 0.927, 4.07$
    inflection points: $(0.927, 0)$, $(4.07, 0)$

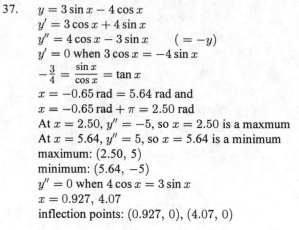

41. $y = 2\sin x - x^2$
    $y' = 2\cos x - 2x$
    First guess, $x = 1.4$
    $h = \frac{y(1.4)}{y'(1.4)} = -0.0044$
    Second guess, $x = 1.4 + 0.0044 = 1.4044$
    $h = \frac{y(1.4044)}{y'(1.4044)} = -0.000014$
    So, $x = 1.404$ to 3 decimal places

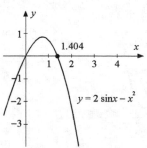

Exercise 2: Derivatives of the Tangent, Cotangent, Secant, and Cosecant

1. $y = \tan 2x$
$y' = 2 \sec^2 2x$

5. $y = 3.25 \tan x^2$
$y' = 3.25(2x)\sec^2 x^2 = 6.50x \sec^2 x^2$

9. $y = x \tan x$
$y' = x(\sec^2 x) + \tan x = x \sec^2 x + \tan x$

13. $w = \sin\theta \tan 2\theta$
$w' = 2\sin\theta \sec^2 2\theta + \tan 2\theta \cos\theta$

17. $y = 5.83 \tan^2 2x$
$\Rightarrow y'(x) = 5.83(2)\tan 2x \cdot 2\sec^2 2x$
$y'(x) = 23.32 \tan 2x \sec^2 2x$
$y'(1) = (23.32)(-2.185)(5.774) = -294$

21. $y = 3\tan x \Rightarrow y' = 3\sec^2 x$
$\Rightarrow y'' = 3 \cdot 2\sec x \cdot \sec x \cdot \tan x$
$y'' = 6\sec^2 x \tan x$

25. $y\tan x = 2,\ y\sec^2 x + y'\tan x = 0$
Solving for y',
$y'\tan x = -y\sec^2 x,$
$y' = \dfrac{-y\sec^2 x}{\tan x} = -y\sec x \csc x$

29. $y = \tan x,\ m(x) = y'(x) = \sec^2 x$
$m(1) = y'(1) = \sec^2(1) = (1.85)^2 = 3.43$
At $x = 1$, $y = \tan(1) = 1.557$
Using the point-slope formula,
$\dfrac{y-1.557}{x-1} = 3.43 \Rightarrow 3.43x - 3.43 = y - 1.56$
$\Rightarrow 3.43x - y = 1.87$

33. $y = 6\sin 4t$
$y' = 6\cos 4t(4) = 24\cos 4t$
$y'' = 24(-\sin 4t)(4) = -96\sin 4t$
At $t = 0.0500$ sec, $v = y' = 23.5$ cm/s
$a = y'' = -19.1$ cm/s^2

37. $l = \dfrac{x+2.67}{\cos\theta}$
$x = 9\cot\theta$
$l = \dfrac{9\cot\theta+2.67}{\cos\theta}$
$\dfrac{dl}{d\theta} = \dfrac{\cos\theta(-9\csc^2\theta)-(9\cot\theta+2.67)(-\sin\theta)}{\cos^2\theta}$
Setting $\dfrac{dl}{d\theta} = 0$,
$(9\cot\theta + 2.67)(\sin\theta) = 9\cos\theta\csc^2\theta$
$9\cos\theta + 2.67\sin\theta = \dfrac{9\cos\theta}{\sin^2\theta}$
$\left(9 - \dfrac{9}{\sin^2\theta}\right)\cos\theta + 2.67\sin\theta = 0$
$9\left(\dfrac{\sin^2\theta-1}{\sin^2\theta}\right)\cos\theta + 2.67\sin\theta = 0$
$9\left(\dfrac{-\cos^2\theta}{\sin^2\theta}\right)\cos\theta + 2.67\sin\theta = 0$
$2.67\sin^3\theta - 9\cos^3\theta = 0$
$2.67\tan^3\theta = 9$
$\tan^3\theta = \dfrac{9}{2.67}$
$\theta = 0.983$ rad
$x = 9\cot 0.983 = 6.00$ ft
$l = \dfrac{6+2.67}{\cos 0.983} = 15.6$ ft

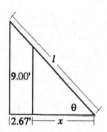

41.

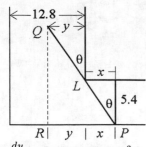

As P moves to the right down the second corridor, with the girder held against the inside corner, θ and x increase, and y first increases to a maximum and then decreases. If the maximuma value of $y =$ the width of the first corridor (12.8'), then L is the greatest length that will fit.

From $\triangle PQR$ we have: $\sin\theta = \frac{(x+y)}{L}$.

$x + y = L\sin\theta$

$y = L\sin\theta - x$

But $x = 5.4\tan\theta$, so $y = L\sin\theta - 5.4\tan\theta$

$\frac{dy}{d\theta} = L\cos\theta - 5.4\sec^2\theta$

Setting $\frac{dy}{d\theta} = 0$, $L\cos\theta = 5.4\sec^2\theta$

$L = 5.4\sec^3\theta$

Substituting in the equation for y,

$y = (5.4\sec^3\theta)\sin\theta - 5.4\tan\theta = 5.4(\sec^2\theta\tan\theta - \tan\theta) = 5.4\tan\theta\tan^2\theta = 5.4\tan^3\theta$

So, when y is a maximum, $y = 5.4\tan^3\theta$

We want the max to be 12.8, so $12.8 = 5.4\tan^3\theta$

$\tan\theta = \sqrt[3]{\frac{12.8}{5.4}} = 1.33$

$\theta = 53.13°$

$L = 5.4\sec^3 53.13° = 25$ ft

Exercise 3: Derivatives of the Inverse Trigonometric Functions

1. $y = x\operatorname{Sin}^{-1}x$

$y' = x\dfrac{1}{\sqrt{1-x^2}} + (1)\operatorname{Sin}^{-1}x = \operatorname{Sin}^{-1}x + \dfrac{x}{\sqrt{1-x^2}}$

5. $y = \operatorname{Sin}^{-1}\left(\dfrac{\sin x - \cos x}{\sqrt{2}}\right)$

$y' = \dfrac{1}{\sqrt{1-(\sin x-\cos x)^2/2}}\left(\dfrac{1}{\sqrt{2}}(\cos x + \sin x)\right) = \dfrac{\cos x+\sin x}{\sqrt{2-\sin^2 x+2\sin x\cos x - \cos^2 x}}$

$= \dfrac{\cos x+\sin x}{\sqrt{2-(\sin^2 x+\cos^2 x)+2\cos x\sin x}} = \dfrac{\cos x+\sin x}{\sqrt{1+\sin 2x}}$

9. $y = \operatorname{Arctan}(1 + 2x)$

$y' = \dfrac{1}{1+(1+2x)^2}(2) = \dfrac{2}{2+4x+4x^2} = \dfrac{1}{1+2x+2x^2}$

13. $y = \operatorname{Arccsc}2x$

$y' = \dfrac{-1}{2x\sqrt{4x^2-1}}(2) = \dfrac{-1}{x\sqrt{4x^2-1}}$

17. $y = \operatorname{Sec}^{-1}\dfrac{a}{\sqrt{a^2-x^2}}$

$y' = \dfrac{1}{\dfrac{a}{\sqrt{a^2-x^2}}\sqrt{\dfrac{a^2}{a^2-x^2}-1}}\left(-\tfrac{1}{2}a(a^2-x^2)^{-3/2}(-2x)\right) = \dfrac{\sqrt{a^2-x^2}}{a\sqrt{\dfrac{x^2}{a^2-x^2}}}\{ax(a^2-x^2)^{-3/2}\}$

$= \dfrac{a^2-x^2}{ax}(ax)(a^2-x^2)^{-3/2} = \dfrac{1}{\sqrt{a^2-x^2}}$

21. $y = \sqrt{x}\,\text{Arccot}\,\frac{x}{4}$

$y' = \sqrt{x}\left(\frac{-1}{\frac{x^2}{16}+1}\right)\left(\frac14\right) + \frac{1}{2\sqrt{x}}\,\text{Arccot}\,\frac{x}{4}$

$\quad = \dfrac{\text{Arccot}\,\frac{x}{4}}{2\sqrt{x}} - \dfrac{4\sqrt{x}}{x^2+16}$

$y'(4) = \dfrac{\text{Arccot}(1)}{2(2)} - \dfrac{4(2)}{16+16} = \dfrac{\pi/4}{4} - \dfrac{8}{32}$

$\quad = \dfrac{\pi-4}{16} = -0.0537$

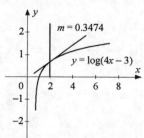

Exercise 4: Derivatives of Logarithmic Functions

1. $y = \log 7x$

$y' = \frac{7}{7x}\log e = \frac1x \log e$

5. $y = \log x\sqrt{5+6x} = \log x + \frac12 \log(5+6x)$

$y' = \frac{1}{x\ln 10} + \frac12\left(\frac{1}{5+6x}\cdot\frac{1}{\ln 10}\right)(6) = \frac{1}{\ln 10}\left(\frac1x + \frac{3}{5+6x}\right) = \frac{1}{\ln 10}\left(\frac{5+9x}{5x+6x^2}\right) = \frac{(5+9x)\log e}{5x+6x^2}$

9. $y = \ln 3x \qquad y' = \frac{1}{3x}(3) = \frac1x$

13. $y = 2.75x \ln 1.02x^3$

$y' = (2.75x)\left(\frac{1}{1.02x^3}\right)(3.06x^2) + (\ln 1.02x^3)(2.75) = 8.25 + 2.75\ln 1.02x^3$

17. $s = \ln\sqrt{t-5} = \ln(t-5)^{1/2} = \frac12\ln(t-5)$

$s' = \frac{1/2}{t-5} = \frac{1}{2t-10}$

21. $y = \sin x \ln \sin x$

$y' = \sin x\left(\frac{\cos x}{\sin x}\right) + \cos x \ln \sin x$

$\quad = \cos x(1 + \ln\sin x)$

25. $x - y = \ln(x+y)$

$1 - \frac{dy}{dx} = \frac{1+(dy/dx)}{x+y}$

$(x+y) - (x+y)\frac{dy}{dx} = 1 + \frac{dy}{dx}$

$x + y - 1 = (x+y+1)\frac{dy}{dx}$

$\frac{dy}{dx} = \frac{x+y-1}{x+y+1}$

29. $y = \frac{\sqrt{a^2-x^2}}{x}$

$\ln y = \frac12\ln(a^2-x^2) - \ln x$

$\frac1y \cdot \frac{dy}{dx} = \frac12\cdot\frac{-2x}{a^2-x^2} - \frac1x = \frac{-2x^2-2(a^2-x^2)}{2x(a^2-x^2)}$

$\frac{dy}{dx} = \frac{a^2 y}{x(x^2-a^2)} = \frac{a^2}{x(x^2-a^2)}\cdot\frac{\sqrt{a^2-x^2}}{x}$

$\quad = -\frac{a^2}{x^2\sqrt{a^2-x^2}}$

33. $y = (\text{Cos}^{-1} x)^x$

$\ln y = x\ln(\text{Cos}^{-1} x)$

$\frac1y \cdot \frac{dy}{dx} = x\left(\frac{1}{\text{Cos}^{-1} x}\right)\left(\frac{-1}{\sqrt{1-x^2}}\right) + \ln(\text{Cos}^{-1} x)$

$\frac{dy}{dx} = y\left\{\ln(\text{Cos}^{-1} x) - \frac{x}{(\text{Cos}^{-1} x)\sqrt{1-x^2}}\right\}$

$\quad = (\text{Arccos}\,x)^x\left[\ln \text{Arccos}\,x - \frac{x}{\sqrt{1-x^2}\,\text{Arccos}\,x}\right]$

37. $y = \log(4x - 3)$

$y' = \frac{4\log e}{4x-3}$

$y'(2) = \frac45 \cdot \frac{1}{\ln 10} = 0.3474$

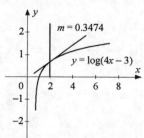

41. $y = x \ln x$
$y' = x\frac{1}{x} + \ln x = 1 + \ln x$
$y'' = 0 + \frac{1}{x} = \frac{1}{x}$
$y' = 0$ when $\ln x = -1$, $x = e^{-1} = 0.3679$
At $x = 0.3679$, $y'' = 2.770$,
so $x = 0.3679$ is a min
$y(0.3679) = -0.3679$
min: $(0.3679, -0.3679)$ or $\left(\frac{1}{e}, -\frac{1}{e}\right)$

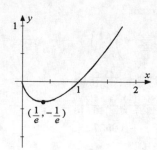

45. $y = x - 10 \log x$
$y' = 1 - \frac{10 \log e}{x} = 1 - \frac{4.343}{x}$
1st guess $x = 1.5$
$h = \frac{y(1.5)}{y'(1.5)} = 0.138$
2nd guess $x = 1.5 - 0.138 = 1.36$
$h = \frac{y(1.36)}{y'(1.36)} = -0.011$
3rd guess $x = 1.36 + 0.011 = 1.37$

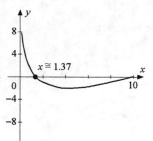

49. $\text{pH} = -\log_{10} C$
$\frac{d\text{pH}}{dC} = -\frac{1}{C \ln 10}$
$\frac{d\text{pH}}{dt} = \frac{d\text{pH}}{dC}\frac{dC}{dt} = -\frac{1}{C \ln 10}\left(\frac{dC}{dt}\right) = -\frac{1}{(2.0\times10^{-5})\ln 10}(-5.5 \times 10^{-5}) = 1.2$ per min
Thus, the pH is increasing at 1.2 per min.

Exercise 5:   Derivatives of the Exponential Function

1. $y = 3^{2x}$
$y' = 3^{2x}(2)\ln 3 = 2(3^{2x})\ln 3$

5. $y = 2^{x^2}$
$y' = 2^{x^2}(2x)\ln 2$

9. $y = e^{e^x}$
$y' = e^{e^x} \cdot e^x = e^{x+e^x}$

13. $y = \frac{2}{e^x} = 2e^{-x}$
$y' = -2e^{-x} = -\frac{2}{e^x}$

17. $y = \frac{e^x}{x}$
$y' = \frac{xe^x - e^x}{x^2} = \frac{e^x(x-1)}{x^2}$

21. $y = (x + e^x)^2$
$y' = 2(x + e^x)(1 + e^x) = 2(x + xe^x + e^x + e^{2x})$

25. $y = \sin^3 e^x$
$y' = 3 \sin^2(e^x)(\cos e^x)(e^x) = 3e^x \sin^2 e^x \cos e^x$

29. $e^x + e^y = 1$
$e^x + e^y \cdot y' = 0$
$y' = -\frac{e^x}{e^y} = -e^{(x-y)}$

33. $f(t) = e^{\sin t} \cos t$
$f'(t) = e^{\sin t}(-\sin t) + e^{\sin t}(\cos t)(\cos t)$
$f''(t) = e^{\sin t}(-\cos t) + e^{\sin t}(\cos t)(-\sin t)$
$\qquad + e^{\sin t}(2 \cos t)(-\sin t)$
$\qquad + e^{\sin t}(\cos t)(\cos^2 t)$
$f''(0) = e^0(-1) + e^0(1)(0) + e^0(2 \cdot 1)(0) + e^0(1)$
$\qquad = -1 + 1 = 0$

37. $y = \ln x e^x = e^x \ln x$
$y' = e^x\left(\ln x + \frac{1}{x}\right)$

41. $y = e^x \sin x$
$y' = e^x(\cos x) + e^x(1)(\sin x) = e^x(\sin x + \cos x)$
$y'' = e^x\{\cos x - \sin x\} + e^x(1)(\sin x + \cos x)$
$\qquad = e^x(\cos x - \sin x + \sin x + \cos x)$
$\qquad = 2e^x \cos x$

45. $y = 5e^{-x} + x - 5$
$y' = -5e^{-x} + 1$
1st guess $x = 5$

$h = \dfrac{y(5)}{y'(5)} = 0.035$

2nd guess $x = 5 - 0.035 = 4.965$

$h = \dfrac{y(4.965)}{y'(4.965)} = -0.0001$

So, $x = 4.97$ to 2 places

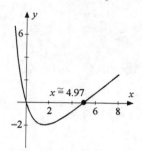

49. $y = 10e^{-x} \sin x$
$y' = 10\{e^{-x} \cos x + e^{-x}(-1)\sin x\}$
$\quad = 10\,e^{-x}(\cos x - \sin x)$

$y' = 0$ when $\cos x = \sin x$, $x = \dfrac{\pi}{4}, \dfrac{5\pi}{4}$

max: $\left(\dfrac{\pi}{4}, 3.224\right)$

min: $\left(\dfrac{5\pi}{4}, -0.139\right)$

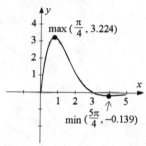

53. $y = \frac{1}{2}(e^x + e^{-x})$

$y' = \frac{1}{2}(e^x - e^{-x})$

$y'(5) = \frac{1}{2}(e^5 - e^{-5}) = 74.2$

57. $N = 10{,}000e^{0.1t}$
$N' = 1000e^{0.1t}$
(a) At $t = 0$, $N' = 1000$ bacteria/h
(b) At $t = 10\overline{0}$ h, $N' = 22 \times 10^6$ bacteria/h

61. $d = 64e^{0.00676(h)}$ lb/ft^3
$d' = (0.00676)(64)e^{0.00676(h)}$
For $h = 1$ mi
$d' = 0.4864e^{0.00676} = 0.0436$ lb/ft^3

Chapter 33 Review Problems

1. $y = \frac{a}{2}\left(e^{x/a} - e^{-x/a}\right)$
$y' = \frac{a}{2}\left(e^{x/a} \cdot \frac{1}{a} - e^{-x/a} \cdot -\frac{1}{a}\right) = \frac{1}{2}\left(e^{x/a} + e^{-x/a}\right)$

5. $y = x \operatorname{Arctan} 4x$
$y' = x\dfrac{1}{16x^2+1} \cdot 4 + \operatorname{Arctan} 4x$
$\quad = \dfrac{4x}{16x^2+1} + \operatorname{Arctan} 4x$

9. $y = x \sin x$
$y' = x \cos x + \sin x$

13. $y = \dfrac{\sin x}{x}$
$y' = \dfrac{x \cos x - \sin x}{x^2}$

17. $y = \ln(x + \sqrt{x^2 + a^2})$
$y' = \dfrac{1 + \frac{1}{2}(x^2+a^2)^{-1/2}(2x)}{x+\sqrt{x^2+a^2}}$
$\quad = \dfrac{\sqrt{x^2+a^2}+x}{\sqrt{x^2+a^2}(x+\sqrt{x^2+a^2})} = \dfrac{1}{\sqrt{x^2+a^2}}$

21. $y = \dfrac{\sin x}{\cos x} = \tan x$
$y' = \sec^2 x$

25. $y = x^2 \operatorname{Arccos} x$
$y' = x^2\dfrac{-1}{\sqrt{1-x^2}} + 2x \operatorname{Arccos} x$
$\quad = 2x \operatorname{Arccos} x - \dfrac{x^2}{\sqrt{1-x^2}}$

29. $y = 2\tan x - \tan^2 x$
$y' = 2\sec^2 x - 2\tan x \sec^2 x = 2\sec^2 x(1 - \tan x)$
$y' = 0$ when $\tan x = 1$; $x = \dfrac{\pi}{4}$

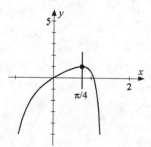

33. $y = \sin x$, $y' = \cos x$
At $x = \dfrac{\pi}{6}$, $y = \sin \dfrac{\pi}{6} = 0.5$, $y' = \cos x = 0.866$
$y - \frac{1}{2} = \dfrac{\sqrt{3}}{2}\left(x - \dfrac{\pi}{6}\right)$

37. $y = 2 \sin \frac{x}{2} - \cos 2x$

$y' = 2\left(\frac{1}{2}\right)\cos \frac{x}{2} + 2 \sin 2x$

1st guess $x = 0.5$

$h = \frac{y(0.5)}{y'(0.5)} = -0.017$

2nd guess $x = 0.5 + 0.017 = 0.517$

$h = \frac{y(0.517)}{y'(0.517)} = -0.00005$

So $x = 0.517$ to 3 places

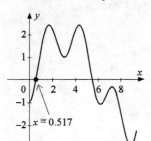

CHAPTER 34: METHODS OF INTEGRATION

Exercise 1: Integrals of Exponential and Logarithmic Functions

1. $\int a^{5x}dx = \frac{1}{5}\int a^{5x}(5dx) = \frac{a^{5x}}{5\ln a} + C$

5. $\int a^{3y}dy = \frac{1}{3}\int a^{3y} \cdot 3dy = \frac{a^{3y}}{3\ln a} + C$

9. $\int xe^{x^2}dx = \frac{1}{2}\int e^{x^2}2x\,dx = \frac{1}{2}e^{x^2} + C$

13. $\int \frac{e^{\sqrt{x}}\,dx}{\sqrt{x}} = \int e^{\sqrt{x}}x^{-1/2}dx = 2\int e^{x^{1/2}}\left(\frac{1}{2}x^{-1/2}dx\right) = 2e^{\sqrt{x}} + C$

17. $\int \frac{e^{\sqrt{x-2}}}{\sqrt{x-2}}dx = 2\int e^{\sqrt{x-2}}\left(\frac{1}{2}\right)(x-2)^{-1/2}dx = 2e^{\sqrt{x-2}} + C$

21. $\int \ln 3x\,dx = \frac{1}{3}\int \ln 3x(3dx) = \frac{1}{3}(3x)(\ln 3x - 1) + C = x\ln 3x - x + C$

25. $\int_2^4 x\log(x^2+1)dx = \int_2^4 x\frac{\ln(x^2+1)}{\ln 10}dx = \frac{1}{2\ln 10}\int_2^4 \ln(x^2+1)(2x\,dx) = \frac{1}{2\ln 10}(x^2+1)(\ln(x^2+1)-1)\Big|_2^4$

$= \frac{(x^2+1)\ln(x^2+1)-x^2-1}{2\ln 10}\Big|_2^4 = 6.106$

29.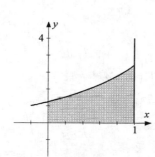

$dV = 2\pi(1-x)\cdot y\cdot dx = 2\pi(1-x)e^x dx$

$V = 2\pi\int_0^1(e^x - xe^x)dx = 2\pi\{e^x - e^x(x-1)\}_0^1$

$= 2\pi\{2e^x - xe^x\}_0^1 = 2\pi\{e^x(2-x)\}_0^1 = 2\pi(e-2) = 4.51$

33.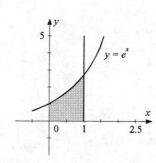

$A = \int_0^1 e^x dx = e^x\Big|_0^1 = e - 1$

$A\cdot\overline{y} = \int_0^1\frac{e^x}{2}\cdot e^x dx = \frac{1}{4}\int_0^1 e^{2x}\cdot 2dx = \frac{1}{4}\{e^{2x}\}_0^1 = \frac{1}{4}(e^2-1)$

$\overline{y} = \frac{(1/4)(e^2-1)}{e-1} = \frac{1}{4}(e+1) = 0.930$

Exercise 2: Integrals of the Trigonometric Functions

1. $\int \sin 3x\,dx = \frac{1}{3}\int \sin 3x(3\,dx) = -\frac{1}{3}\cos 3x + C$

5. $\int \sec 4x\,dx = \frac{1}{4}\int \sec 4x(4\,dx) = \frac{1}{4}\ln|\sec 4x + \tan 4x| + C$

9. $\int x\sin x^2 dx = \frac{1}{2}\int \sin x^2(2x\,dx) = -\frac{1}{2}\cos x^2 + C$

13. $\int \sin(x+1)dx = -\cos(x+1) + C$

17. $\int x\sec(4x^2-3)dx = \frac{1}{8}\int \sec(4x^2-3)(8x\,dx) = \frac{1}{8}\ln\left|\sec(4x^2-3) + \tan(4x^2-3)\right| + C$

21. $\int_0^\pi \sin\phi\,d\phi = -\cos\phi\Big]_0^\pi = -(-1) - 1 = 2$

25. $y = \sin x$

$A = \int_0^\pi \sin x\,dx = -\cos x\Big|_0^\pi = -[(-1) - 1] = 2$

29. $A = \int_0^{\pi/2}\cos x\,dx - \int_{\pi/2}^{3\pi/2}\cos x\,dx$

$= \sin x\Big]_0^{\pi/2} - \sin x\Big]_{\pi/2}^{3\pi/2}$

$= (1 - 0) - (-1 - 1) = 3$

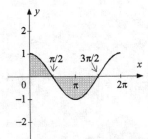

33. $\bar{x} = \frac{\pi}{2}$

$A = \int_0^\pi \sin x\,dx = -\cos x\Big|_0^\pi = 2$

$\bar{y} \cdot A = 2\int_0^{\pi/2}(\sin x)\frac{\sin x}{2}dx = \int_0^{\pi/2}\sin^2 x\,dx$

$= \frac{x}{2} - \frac{\sin 2x}{4}\Big|_0^{\pi/2} = \left(\frac{\pi}{4} - 0\right) - (0 - 0) = \frac{\pi}{4}$

$\bar{y} = \frac{\pi/4}{2} = \frac{\pi}{8}$

$\bar{x} = \frac{\pi}{2},\ \bar{y} = \frac{\pi}{8}$

Exercise 3: Average and Root-Mean-Square Values

1. $y_{avg} = \frac{1}{6}\int_0^6 x^2 dx = \frac{1}{6}\cdot\frac{x^3}{3}\Big|_0^6 = \frac{6^3}{18} = 12$

5. $y_{avg} = \frac{2}{\pi}\int_0^{\pi/2}\sin^2 x\,dx = \frac{2}{\pi}\left(\frac{x}{2} - \frac{\sin 2x}{4}\right)_0^{\pi/2}$

$= \frac{2}{\pi}\left(\frac{\pi}{4} - 0\right) = \frac{1}{2}$

9. $rms^2 = \frac{1}{4-1}\int_1^4 (x + 2x^2)^2 dx$

$= \frac{1}{3}\int_1^4 (x^2 + 4x^3 + 4x^4)dx$

$= \frac{1}{3}\left(\frac{x^3}{3} + x^4 + \frac{4x^5}{5}\right)\Big|_1^4 = 365$

$rms = \sqrt{365} = 19.1$

Exercise 4: Integration by Parts

1. $u = x,\ du = dx,\ v = -\cos x,\ dv = \sin x\,dx$

$\int x\sin x\,dx = -x\cos x - \int -\cos x\,dx = \int \cos x\,dx - x\cos x = \sin x - x\cos x + C$

5. $u = x,\ du = dx,\ v = \sin x,\ dv = \cos x\,dx$

$\int x\cos x\,dx = x\sin x - \int \sin x\,dx = x\sin x + \cos x + C$

9. $u = \ln(x+1),\ du = \frac{1}{x+1}dx,\ v = 2\sqrt{x+1},\ dv = \frac{dx}{\sqrt{x+1}}$

$\int_1^3 \frac{\ln(x+1)dx}{\sqrt{x+1}} = \{\ln(x+1)\}\left(2\sqrt{x+1}\right)\Big|_1^3 - \int_1^3 2\sqrt{x+1}\cdot\frac{1}{x+1}dx = 2\sqrt{x+1}\ln(x+1)\Big|_1^3 - 2\int_1^3 \frac{dx}{\sqrt{x+1}}$

$= 2\sqrt{x+1}\ln(x+1)\Big|_1^3 - 4\sqrt{x+1}\Big|_1^3 = 4\ln 4 - 2\sqrt{2}\ln 2 - 8 + 4\sqrt{2} = 1.24$

13. $u = x,\ du = dx,\ v = \frac{1}{2}e^{2x},\ dv = e^{2x}dx$

$\int_0^4 xe^{2x}dx = x\cdot\frac{1}{2}e^{2x}\Big|_0^4 - \int_0^4 \frac{1}{2}e^{2x}dx = \frac{1}{2}xe^{2x} - \frac{1}{4}e^{2x}\Big|_0^4 = \frac{1}{4}e^{2x}(2x-1)\Big|_0^4 = \frac{1}{4}e^8\cdot 7 - \frac{1}{4}e^0(-1)$

$= 5217$

17. $u = x, \; du = dx, \; v = -\frac{1}{2} \cdot \frac{(1-x^2)^{-1/2}}{-1/2} = (1-x^2)^{-1/2} = \frac{1}{\sqrt{1-x^2}}, \; dv = \frac{x\,dx}{(1-x^2)^{3/2}}$

$\int \frac{x^2 dx}{(1-x^2)^{3/2}} = \frac{x}{\sqrt{1-x^2}} - \int \frac{dx}{\sqrt{1-x^2}} = \frac{x}{\sqrt{1-x^2}} - \text{Arcsin}\,x + C$

21. $u = x^2, \; du = 2x\,dx, \; v = -e^{-x}, \; dv = e^{-x}dx$

$\int x^2 e^{-x} dx = x^2(-e^{-x}) - \int (-e^{-x} \cdot 2x)\,dx = 2\int xe^{-x}dx - x^2 e^{-x}$

$\qquad (u = x, \; du = dx, \; v = -e^{-x}, \; dv = e^{-x}dx)$

$\qquad = 2\{x(-e^{-x}) - \int(-e^{-x})dx\} - x^2 e^{-x} = -2xe^{-x} - 2e^{-x} - x^2 e^{-x} + C$

$\qquad = -x^2 e^{-x} - 2xe^{-x} - 2e^{-x} + C$

25. $u = e^{-x}, \; du = -e^{-x}dx, \; v = -\frac{1}{4}\cos 4x, \; dv = \sin 4x\,dx$

$\int_1^3 e^{-x}\sin 4x\,dx = -\frac{e^{-x}}{4}\cos 4x - \int\left(-\frac{1}{4}\cos 4x\right)(-e^{-x})dx\Big|_1^3 = -\frac{e^{-x}}{4}\cos 4x - \frac{1}{4}\int e^{-x}\cos 4x\,dx\Big|_1^3$

$\qquad \left(u = e^{-x}, \; du = -e^{-x}dx, \; v = \frac{1}{4}\sin 4x, \; dv = \cos 4x\,dx\right)$

$\qquad = -\frac{e^{-x}}{4}\cos 4x - \frac{1}{4}\left(\frac{e^{-x}}{4}\sin 4x - \int\left(\frac{1}{4}\sin 4x\right)(-e^{-x})dx\right)\Big|_1^3 = -\frac{e^{-x}}{4}\cos 4x - \frac{e^{-x}}{16}\sin 4x - \frac{1}{16}\int e^{-x}\sin 4x\,dx\Big|_1^3$

$\left(1+\frac{1}{16}\right)\int_1^3 e^{-x}\sin 4x\,dx = -\frac{e^{-x}}{16}(4\cos 4x + \sin 4x)\Big|_1^3$

$\int_1^3 e^{-x}\sin 4x\,dx = -\frac{e^{-x}}{17}(4\cos 4x + \sin 4x)\Big|_1^3 = -0.0083 - 0.0730 = -0.0813$

Exercise 5: Integrating Rational Fractions

1. $\int \frac{(4x-2)dx}{x^3 - x^2 - 2x} = \int\left(\frac{A}{x} + \frac{B}{x-2} + \frac{C}{x+1}\right)dx$

$A(x^2 - x - 2) + B(x^2 + x) + C(x^2 - 2x) = 4x - 2$

$(A + B + C) = 0, \; (B - A - 2C) = 4, \; -2A = -2$

$A = 1, \; B + C = -1$

$B - 2C = 5$

$3C = -6$

$C = -2, \; B = 1$

$\int \frac{dx}{x} + \int \frac{dx}{x-2} - 2\int \frac{dx}{x+1} = \ln|x| + \ln|x-2| - 2\ln|x+1| + C = \ln\left|\frac{x(x-2)}{(x+1)^2}\right| + C$

5. $\int \frac{x+1}{x^3 + 2x^2 - 3x}dx$

Multiplying both sides by $x(x+3)(x-1)$

$x + 1 = A(x+3)(x-1) + Bx(x-1) + Cx(x+3)$

$\quad$ If $x = 0 \Rightarrow 1 = -3A \Rightarrow A = -\frac{1}{3}$

$\quad$ If $x = -3 \Rightarrow -2 = 12B \Rightarrow B = -\frac{1}{6}$

$\quad$ If $x = 1 \Rightarrow 2 = 4C \Rightarrow C = \frac{1}{2}$

We now integrate as follows:

$\int \frac{x+1}{x^3 + 2x^2 - 3x}dx = -\frac{1}{3}\int \frac{1}{x}dx - \frac{1}{6}\int \frac{1}{x+3}dx + \frac{1}{2}\int \frac{1}{x-1}dx = -\frac{1}{3}\ln|x| - \frac{1}{6}\ln|x+3| + \frac{1}{2}\ln|x-1| + C$

$\qquad = \ln\left|\frac{(x-1)^{1/2}}{x^{1/3}(x+3)^{1/6}}\right| + C$

9. $\int \frac{(2x-5)dx}{(x-2)^3} = \int\left(\frac{A}{x-2} + \frac{B}{(x-2)^2} + \frac{C}{(x-2)^3}\right)dx$

$A(x^2 - 4x + 4) + B(x-2) + C = 2x - 5$

$Ax^2 + (B - 4A)x + (4A - 2B + C) = 2x - 5$

$A = 0, \; B = 2, \; C = -1$

$\int \frac{2dx}{(x-2)^2} - \int \frac{dx}{(x-2)^3} = -\frac{2}{(x-2)} + \frac{1}{2(x-2)^2} + C = \frac{1-4(x-2)}{2(x-2)^2} + C = \frac{9-4x}{2(x-2)^2} + C$

13. $\int_1^2 \frac{dx}{x^4+x^2} = \int \frac{dx}{x^2(x^2+1)} = \int\left(\frac{A}{x} + \frac{B}{x^2} + \frac{Cx+D}{x^2+1}\right)dx$

$Ax(x^2+1) + B(x^2+1) + (Cx+D)x^2 = 1$

$(A+C)x^3 + (B+D)x^2 + Ax + B = 1$

$A = 0,\ B = 1,\ C = 0,\ D = -1$

$\int\left(\frac{1}{x^2} - \frac{1}{x^2+1}\right)dx = -\frac{1}{x} - \text{Arctan}\,x\Big|_1^2 = -\frac{1}{2} - 1.107149 - [-1 - 0.785398] = 0.1782$ (rounded)

17. $\int \frac{x^2\,dx}{1-x^4} = \int \frac{x^2\,dx}{(1+x^2)(1-x^2)}$

$\frac{Ax+B}{1+x^2} + \frac{Cx+D}{1-x^2} = \frac{(Ax+B)(1-x^2)+(Cx+D)(1+x^2)}{1-x^4} = \frac{x^2}{1-x^4}$

$Ax - Ax^3 + B - Bx^2 + Cx + Cx^3 + D + Dx^2 = x^2$

$(C-A)x^3 + (D-B)x^2 + (A+C)x + (B+D) = x^2$

$B + D = 0$

$-B + D = 1$

$2D = 1,\ D = \frac{1}{2},\ B = -\frac{1}{2}$

$A + C = 0$

$-A + C = 0,\ A = 0,\ C = 0$

$\frac{1}{2}\int \frac{dx}{1-x^2} - \frac{1}{2}\int \frac{dx}{1+x^2} = \frac{1}{2}\left(-\frac{1}{2}\ln\left|\frac{x-1}{x+1}\right|\right) - \frac{1}{2}\text{Arctan}\,x + C = \frac{1}{4}\ln\left|\frac{x+1}{x-1}\right| - \frac{1}{2}\text{Arctan}\,x + C$

21. $\int \frac{x^5\,dx}{(x^2+4)^2} = \int\left\{x - \frac{8x^3+16x}{(x^2+4)^2}\right\}dx = \int\left\{x - \left(\frac{Ax+B}{(x^2+4)} + \frac{Cx+D}{(x^2+4)^2}\right)\right\}dx$

$(Ax+B)(x^2+4) + (Cx+D) = 8x^3 + 16x$

$Ax^3 + Bx^2 + (4A+C)x + (4B+D) = 8x^3 + 16x$

$A = 8,\ B = 0$

$32 + C = 16$

$C = -16$

$4 \cdot 0 + D = 0$

$D = 0$

$\int\left\{x - \left(\frac{8x}{x^2+4} - \frac{16x}{(x^2+4)^2}\right)\right\}dx = \int x\,dx - 4\int \frac{2x}{(x^2+4)}\,dx + 8\int \frac{2x}{(x^2+4)^2}\,dx = \frac{x^2}{2} - 4\ln|x^2+4| - \frac{8}{(x^2+4)} + C$

Exercise 6: Integrating by Algebraic Substitution

1. $\int \frac{dx}{1+\sqrt{x}}$

Let $z = x^{1/2}$. Then $x = z^2$ and $dx = 2z\,dz$.

$\int \frac{dx}{1+\sqrt{x}} = \int \frac{2z\,dz}{1+z} = 2\int \frac{z\,dz}{z+1} = 2\int\left(1 - \frac{1}{z+1}\right)dz = 2[z - \ln|z+1|] + C = 2\left[\sqrt{x} - \ln|\sqrt{x}+1|\right] + C$

5. $\int \frac{x\,dx}{\sqrt[3]{1+x}}$ $z = \sqrt[3]{1+x},\ x = z^3 - 1,\ dx = 3z^2\,dz$

$= \int \frac{(z^3-1)3z^2\,dz}{z} = \int(z^3-1)3z\,dz = \int(3z^4 - 3z)dz = \frac{3z^5}{5} - \frac{3z^2}{2} + C = \frac{3}{5}(1+x)^{5/3} - \frac{3}{2}(1+x)^{2/3} + C$

$= \frac{3}{10}(1+x)^{2/3}\{2(1+x) - 5\} + C = \frac{3}{10}(1+x)^{2/3}(2x - 3) + C$

9. $\int \frac{x^2 dx}{(4x+1)^{5/2}}$ $z = \sqrt{4x+1}, \ x = \frac{1}{4}(z^2-1), \ dx = \frac{1}{2}z\,dz$

$= \int \frac{\frac{1}{16}(z^2-1)^2 \cdot \frac{1}{2}z\,dz}{z^5} = \frac{1}{32}\int \frac{z^4 - 2z^2 + 1}{z^4}dz = \frac{1}{32}\int (1 - 2z^{-2} + z^{-4})dz = \frac{1}{32}\left(z + \frac{2}{z} - \frac{1}{3z^3}\right) + C$

$= \frac{1}{32}\left\{(4x+1)^{1/2} + \frac{2}{(4x+1)^{1/2}} - \frac{1}{3(4x+1)^{3/2}}\right\} + C = \frac{1}{32}\left(\frac{3(4x+1)^2 + 6(4x+1) - 1}{3(4x+1)^{3/2}}\right) + C$

$= \frac{1}{32}\left(\frac{48x^2 + 24x + 3 + 24x + 5}{3(4x+1)^{3/2}}\right) + C = \frac{1}{4}\left(\frac{6x^2 + 6x + 1}{3(4x+1)^{3/2}}\right) + C = \frac{6x^2 + 6x + 1}{12(4x+1)^{3/2}} + C$

13. $\int \frac{dx}{x\sqrt{1-x^2}}$ $z = \sqrt{1-x^2}, \ x = \sqrt{1-z^2}, \quad dx = -\frac{z}{\sqrt{1-z^2}}$

$= -\int \frac{\frac{z}{\sqrt{1-z^2}}}{\sqrt{1-z^2}\cdot z}dz = -\int \frac{dz}{1-z^2} = -\frac{1}{2}\int \left(\frac{1}{1-z} + \frac{1}{1+z}\right)dz = -\frac{1}{2}(-\ln|1-z| + \ln|1+z|) + C$

$= \frac{1}{2}\ln\left|\frac{1-z}{1+z}\right| + C = \frac{1}{2}\ln\left|\frac{1-\sqrt{1-x^2}}{1+\sqrt{1-x^2}}\right| + C = \frac{1}{2}\ln\left|\frac{\left(1-\sqrt{1-x^2}\right)^2}{1-(1-x^2)}\right| + C = \frac{1}{2}\ln\left(\frac{1-\sqrt{1-x^2}}{x}\right)^2 + C$

$= \ln\left|\frac{1-\sqrt{1-x^2}}{x}\right| + C$

17. $\int_0^1 \frac{x^{3/2}dx}{x+1}$ $z = x^{1/2}, \ x = z^2, \ dx = 2z\,dz$
$\qquad\qquad\qquad x = 1, \ z = 1$
$\qquad\qquad\qquad x = 0, \ z = 0$

$= \int_0^1 \frac{z^3 \cdot 2z\,dz}{z^2+1} = 2\int_0^1 \left(z^2 - 1 + \frac{1}{z^2+1}\right)dz = 2\left(\frac{z^3}{3} - z + \text{Arctan } z\right)\Big)_0^1 = 0.2375$

Exercise 7: Integrating by Trigonometric Substitution

1. $\int \frac{5dx}{(5-x^2)^{3/2}}$ $x = \sqrt{5}\sin\theta, \ 5 - x^2 = 5 - 5\sin^2\theta = 5\cos^2\theta, \ dx = \sqrt{5}\cos\theta d\theta$

$= \int \frac{5\cdot\sqrt{5}\cos\theta d\theta}{(5\cos^2\theta)^{3/2}} = \int \frac{5^{3/2}\cos\theta d\theta}{5^{3/2}\cos^3\theta} = \int \frac{d\theta}{\cos^2\theta} = \int \sec^2\theta d\theta = \tan\theta + C = \frac{x}{\sqrt{5-x^2}} + C$

5. Let $x = 2\sin\theta$ and $dx = 2\cos\theta d\theta$
$\qquad \sqrt{4-x^2} = \sqrt{4 - 4\sin^2\theta} = 2\sqrt{1-\sin^2\theta} = 2\cos\theta$
$\int \frac{dx}{x^2\sqrt{4-x^2}} = \int \frac{2\cos\theta d\theta}{4\sin^2\theta(2\cos\theta)} = \frac{1}{4}\int \csc^2\theta d\theta = -\frac{1}{4}\cot\theta + C$
Since $\cot\theta = \frac{\sqrt{4-x^2}}{x}$, we have, $\int \frac{dx}{x^2\sqrt{4-x^2}} = \frac{-\sqrt{4-x^2}}{4x} + C$

9. $\int \frac{\sqrt{16-x^2}dx}{x^2}$ Let $x = 4\sin\theta, \ \sqrt{16-x^2} = 4\cos\theta, \ \theta = \text{Arcsin }\frac{x}{4}, \ dx = 4\cos\theta d\theta$
$\int \frac{4\cos\theta \cdot 4\cos\theta d\theta}{16\sin^2\theta} = \int \cot^2\theta d\theta = -\cot\theta - \theta + C = -\frac{\sqrt{16-x^2}}{x} - \text{Arcsin }\frac{x}{4} + C$

13. $\int \frac{dx}{x^2\sqrt{x^2-7}}$ $x = \sqrt{7}\sec\theta, \ \sqrt{x^2-7} = \sqrt{7}\tan\theta, \ dx = \sqrt{7}\sec\theta\tan\theta d\theta$

$= \int \frac{\sqrt{7}\sec\theta\tan\theta d\theta}{7\sec^2\theta\sqrt{7}\tan\theta} = \int \frac{d\theta}{7\sec\theta} = \frac{1}{7}\int \cos\theta d\theta = \frac{1}{7}\sin\theta + C = \frac{1}{7}\cdot\frac{\sqrt{x^2-7}}{x} + C$

Exercise 8: Improper Integrals

1. $\int_1^\infty \frac{dx}{x^3} = \lim_{b\to\infty} \left[\frac{x^{-2}}{-2}\right]_1^b = \lim_{b\to\infty}\left[-\frac{1}{2b^2} + \frac{1}{2}\right] = \frac{1}{2}$

5. $\int_{-\infty}^\infty \frac{dx}{1+x^2} = \lim_{\substack{b\to\infty \\ a\to-\infty}} [\text{Arctan } x]_a^b$

   $= \lim_{\substack{b\to\infty \\ a\to-\infty}} [\text{Arctan } b - \text{Arctan } a]$

   $= \frac{\pi}{2} - \left(\frac{-\pi}{2}\right) = \pi$

9. $\int_0^{\pi/2} \tan x \, dx = \lim_{z\to\pi/2^-} \left[-\ln|\cos x|\right]_0^z$

   Diverges

13. $V = \pi \int_a^b r^2 dh$
   For the curve $y = \frac{1}{x}$ for $x = 1$ to ∞,
   $V = \pi \int_1^\infty \left(\frac{1}{x}\right)^2 dx = \pi \lim_{z\to\infty} \left| -\frac{1}{x} \right|_1^z$
   $= \pi \lim_{z\to\infty} \left| -\frac{1}{z} + 1 \right| = \pi$ cubic units

Exercise 9: Approximate Value of a Definite Integral

NOTE: The trapezoid rule was used for the following solutions.

1. We compute y at intervals of $h = 0.2$, then
   $A = 0.2\left\{\frac{1}{2}(1.5 + 0.583) + 1.29 + 1.13 + 1.01 + 0.913 + 0.833 + 0.767 + 0.711 + 0.662 + 0.620\right\} = 1.79$

5. Let $h = 0.5$, then
   $A = 0.5\left\{\frac{1}{2}(0.5 - 0.333) + 0.471 + 0.447 + 0.426 + 0.408 + 0.392 + 0.378 + 0.365 + 0.354 + 0.343\right\} = 2.00$

9.

x	0	1	2	3	4	5	6	7	8	9	10
y	415	593	577	615	511	552	559	541	612	745	893

   $A = 1\left[\frac{(415+893)}{2} + 593 + 577 + 615 + 511 + 552 + 559 + 541 + 612 + 745\right] = 5960$

13. $A = 30.0\left\{\frac{1}{2}(5.5 + 12.0) + 34.0 + 53.5 + \cdots + 45.2\right\} = 13,\overline{0}00 \text{ ft}^2$

17. Output $\simeq 20\left\{\frac{1}{2}(0 + 0) + 21.5 + \cdots + 64.6\right\} = 86,600 \text{ W} = 86.6 \text{ kW}$

Chapter 34: Review Problems

1. $\int \cot^2 x \csc^4 x \, dx = \int \cot^2 x(\cot^2 x + 1) \cdot \csc^2 x \, dx = -\int (\cot^4 x + \cot^2 x)(-\csc^2 x \, dx) = -\frac{1}{5}\cot^5 x - \frac{1}{3}\cot^3 x + C$

5. $\int x\ln x^2 dx$ use Rule 43: $u = x^2 \quad du = 2x \, dx$
   $\frac{1}{2}\int (\ln x^2)(2x \, dx) = \frac{1}{2}[x^2(\ln x^2 - 1) + C_1]$
   $= \frac{x^2}{2}(\ln x^2 - 1) + \frac{C_1}{2} = \frac{x^2}{2}(\ln x^2 - 1) + C$

9. $\int \frac{(3x-1)dx}{x^2+9} = \int \frac{3x \, dx}{x^2+9} - \int \frac{dx}{x^2+9} = \frac{3}{2}\int \frac{2x \, dx}{x^2+9} - \int \frac{dx}{x^2+9} = \frac{3}{2}\ln|x^2 + 9| - \frac{1}{3}\text{Arctan}\frac{x}{3} + C$

13. $\int \frac{dx}{2x^2-2x+1} = \int \frac{2dx}{(4x^2-4x+1)+1} = \int \frac{2dx}{(2x-1)^2+1} = \text{Tan}^{-1}(2x - 1) + C$

17. $\int \frac{(x+2)dx}{x^2+x+1} = \frac{1}{2}\int \frac{(2x+1)+3}{x^2+x+1}dx = \frac{1}{2}\int \frac{(2x+1)dx}{x^2+x+1} + \frac{3}{2}\int \frac{dx}{x^2+x+1} = \frac{1}{2}\ln|x^2 + x + 1| + \frac{3}{2}\int \frac{dx}{\left(x+\frac{1}{2}\right)^2+\frac{3}{4}}$

   $= \frac{1}{2}\ln|x^2 + x + 1| + \frac{3}{2}\left(\frac{1}{\sqrt{3/4}}\text{Arctan}\frac{x+(1/2)}{\sqrt{3/4}}\right) + C = \frac{1}{2}\ln|x^2 + x + 1| + \sqrt{3}\,\text{Arctan}\left(\frac{2x+1}{\sqrt{3}}\right) + C$

21. $\int_0^4 \frac{9x^2 dx}{(2x+1)(x+2)^2} = \int_0^4 \left(\frac{A}{2x+1} + \frac{B}{x+2} + \frac{C}{(x+2)^2} \right) dx$

$A(x+2)^2 + B(2x+1)(x+2) + C(2x+1) = 9x^2$

$x = -2, \ C \cdot -3 = 36, \ C = -12$

$x = -\frac{1}{2}, \ A\left(\frac{3}{2}\right)^2 = \frac{9}{4}, \ A = 1$

$x = 0, \ 4A + 2B + C = 0$

$4 + 2B - 12 = 0, \ B = 4$

$\int_0^4 \left(\frac{1}{2x+1} + \frac{4}{x+2} - \frac{12}{(x+2)^2} \right) dx = \frac{1}{2}\ln|2x+1| + 4\ln|x+2| + \frac{12}{x+2} \Big|_0^4 = 1.493$

25. $\int \frac{x \, dx}{(1-x)^5}$ $\quad u = x, \ du = dx, \ v = \frac{1}{4(1-x)^4}, \ dv = \frac{dx}{(1-x)^5}$

$= \frac{x}{4(1-x)^4} - \int \frac{dx}{4(1-x)^4} = \frac{x}{4(1-x)^4} + \frac{1}{4} \int \frac{-dx}{(1-x)^4} = \frac{x}{4(1-x)^4} + \frac{1}{4} \cdot -\frac{1}{3(1-x)^3} + C$

$= \frac{3x}{12(1-x)^4} - \frac{1-x}{12(1-x)^4} + C = \frac{4x-1}{12(1-x)^4} + C$

29. $\int x \ln(3x^2 - 2) dx$ $\quad$ use Rule 43: $u = 3x^2 - 2, \ du = 6x \, dx$

$\frac{1}{6} \int \ln(3x^2 - 2)(6x \, dx) = \frac{1}{6}[(3x^2 - 2)(\ln(3x^2 - 2) - 1) + C_1] = \frac{1}{6}(3x^2 - 2)\ln(3x^2 - 2) - \frac{1}{6}(3x^2 - 2) + \frac{C_1}{6}$

$= \frac{1}{6}(3x^2 - 2)\ln(3x^2 - 2) - \frac{x^2}{2} + \frac{1}{3} + \frac{C_1}{6} = \frac{1}{6}(3x^2 - 2)\ln(3x^2 - 2) - \frac{x^2}{2} + C$

33. $\int_0^1 \frac{(3x^2 + 7x) dx}{(x+1)(x+2)(x+3)} = \int_0^1 \left(\frac{A}{x+1} + \frac{B}{x+2} + \frac{C}{x+3} \right) dx$

$A(x+2)(x+3) + B(x+1)(x+3) + C(x+1)(x+2) = 3x^2 + 7x$

$x = -2, \ B(-1)(1) = 12 - 14, \ B = 2$

$x = -3, \ C(-2)(-1) = 27 - 21, \ C = 3$

$x = -1, \ A(1)(2) = 3 - 7, \ A = -2$

$-2\int_0^1 \frac{dx}{x+1} + 2\int \frac{dx}{x+2} + 3\int \frac{dx}{x+3} = -2\ln|x+1| + 2\ln|x+2| + 3\ln|x+3| \Big|_0^1 = 2\ln\left|\frac{x+2}{x+1}\right| + 3\ln|x+3| \Big|_0^1$

$= 0.2877$

37. $\int_0^2 \frac{dx}{(x-2)^{2/3}} = \lim_{b \to 2^-} \left[3(x-2)^{1/3} \right]_0^b = \lim_{b \to 2^-} \left[3(b-2)^{1/3} - 3(0-2)^{1/3} \right] = 3\sqrt[3]{2}$

41. $y_{\text{avg}} = \frac{1}{2\pi} \int_0^{2\pi} \sin^2 x \, dx = \frac{1}{2\pi} \left(\frac{x}{2} - \frac{\sin 2x}{4} \right)_0^{2\pi} = \frac{1}{2\pi} \left(\frac{2\pi}{2} - 0 - (0) \right) = \frac{1}{2}$

CHAPTER 35: DIFFERENTIAL EQUATIONS

Exercise 1: Definitions

1. $\frac{dy}{dx} + 3xy = 5$ a) 1st order   b) 1st degree   c) ordinary

5. $3(y'')^4 - 5y' = 3y$ a) 2nd order   b) 4th degree   c) ordinary

9. $4x - 3y' = 5$
   Multiplying by dx, $\; 4x\,dx - 3\,dy = 5\,dx$
   Integrating, $\; \int 4x\,dx - \int 3\,dy = \int 5\,dx$
   $$2x^2 - 3y = 5x + C_1$$
   Solving for y,
   $$y = \tfrac{2}{3}x^2 - \tfrac{5}{3}x + C \;\text{ where } C = \tfrac{C_1}{-3}$$

13. $y' = \frac{2y}{x}$ $(y = Cx^2)$
    The derivative of $y = Cx^2$ is $y' = 2Cx$.
    Substituting $y = Cx^2$ and $y' = 2Cx$,
    $$2Cx = \frac{2Cx^2}{x}$$
    or $2Cx = 2Cx$.
    Thus $y = Cx^2$ must be a solution.

Exercise 2: Graphical and Numerical Solution of Differential Equations

1. $y' = x$
   Start at $(0, 1)$. Find $y(2)$.

x	y	m
0.00	1.00	0.00
0.10	1.00	0.10
0.20	1.01	0.20
0.30	1.03	0.30
0.40	1.06	0.40
0.50	1.10	0.50
0.60	1.15	0.60
0.70	1.21	0.70
0.80	1.28	0.80
0.90	1.36	0.90
1.00	1.45	1.00
1.10	1.55	1.10
1.20	1.66	1.20
1.30	1.78	1.30
1.40	1.91	1.40
1.50	2.05	1.50
1.60	2.20	1.60
1.70	2.36	1.70
1.80	2.53	1.80
1.90	2.71	1.90
2.00	2.90	2.00

   $y(2) = 3$

Exercise 3: First-Order Differential Equation, Variables Separable

1. $y' = \frac{x}{y}$

    Separating variables: $\frac{dy}{dx} = \frac{x}{y} \Rightarrow y\,dy = x\,dx$

    Integrating: $\int y\,dy = \int x\,dx$

    $$\frac{y^2}{2} = \frac{x^2}{2} + C_1$$

    Solving for y: $y^2 = x^2 + 2C_1$, $y = \pm\sqrt{x^2 + C}$

5. $\frac{dy}{dx} = \frac{x^2}{y^3}$

    Separating variables: $y^3 dy = x^2 dx$

    Integrating: $\int y^3 dy = \int x^2 dx$

    $$\frac{y^4}{4} = \frac{x^3}{3} + C_1$$

    or $\qquad 4x^3 - 3y^4 = C$

9. $(1 + x^2)dy + (y^2 + 1)dx = 0$

    Separating variables: $(1 + x^2)dy = (y^2 + 1)dx$

    $$\frac{dy}{y^2+1} + \frac{dx}{x^2+1} = 0$$

    Integrating: $\text{Arctan } y + \text{Arctan } x = C$

13. $(y^2 + 1)dx = (x^2 + 1)dy$

    Separating variables: $(y^2 + 1)dx = (x^2 + 1)dy$

    $$\frac{dx}{x^2+1} = \frac{dy}{y^2+1}$$

    Integrating: $\text{Arctan } x = \text{Arctan } y + C$

17. $(x - xy^2)dx = -(x^2y + y)dy$

    Separating variables:

    $$(x - xy^2)dx = -(x^2y + y)dy$$
    $$x(1 - y^2)dx = -y(x^2 + 1)dy$$
    $$\frac{x\,dx}{x^2+1} = \frac{-y\,dy}{1-y^2}$$

    Integrating: $\frac{1}{2}\int \frac{2x\,dx}{x^2+1} = \frac{1}{2}\int \frac{-2y\,dy}{1-y^2}$

    $$\frac{1}{2}\ln|x^2 + 1| = \frac{1}{2}\ln|1 - y^2| + C_1$$

    $\ln\left|\frac{x^2+1}{1-y^2}\right| = 2C_1$

    $\left|\frac{x^2+1}{1-y^2}\right| = e^{2C_1}$

    $\frac{x^2+1}{1-y^2} = C$

    $x^2 + 1 = C - Cy^2$

    $x^2 + Cy^2 = C - 1$

21. $e^{x-y}dx + e^{y-x}dy = 0$

    $e^{x+y} \cdot e^{x-y}dx + e^{x+y} \cdot e^{y-x}dy = 0$

    $e^{2x}dx + e^{2y}dy = 0$

    $\frac{1}{2}\int e^{2x}2dx + \frac{1}{2}\int e^{2y}2dy = C_1$

    $\frac{1}{2}e^{2x} + \frac{1}{2}e^{2x} = C_1$

    $e^{2x} + e^{2y} = C$

25. $\cos x \sin y\,dy + \sin x \cos y\,dx = 0$

    $\cos x \cdot \sin y\,dy + \sin x \cdot \cos y\,dx = 0$

    $\frac{\cos x \cdot \sin y}{\cos x \cdot \cos y}dy + \frac{\sin x \cdot \cos y}{\cos x \cdot \cos y}dx = 0$

    $\tan y\,dy + \tan x\,dx = 0$

    $\int \tan y\,dy + \int \tan x\,dx = C_1$

    $-\ln|\cos y| - \ln|\cos x| = C_1$

    $\ln|\cos y \cdot \cos x| = C_2$

    $|\cos x \cos y| = e^{C_2}$

    $\cos x \cos y = C$

29. $y^2 y' = x^2 \qquad (x = 0 \text{ when } y = 1)$

    $y^2\frac{dy}{dx} = x^2$

    $y^2 dy = x^2 dx$

    Integrating: $\frac{y^3}{3} = \frac{x^3}{3} + C$

    To find C let $x = 0$, $y = 1$

    $C = \frac{1}{3}$

    $\frac{y^3}{3} = \frac{x^3}{3} + \frac{1}{3}$

    $y^3 - x^3 = 1$

Exercise 4: Exact First-Order Differential Equations

1. $y\,dx + x\,dy = 7dx$

    Recognizing the left side as the derivative of xy, we integrate both sides,

    $\int(y\,dx + x\,dy) = 7\int dx$

    $xy = 7x + C$

    $xy - 7x = C$

5. Substituting $\frac{dy}{dx}$ for y' and multiplying both sides by dx,

    $$2x\,dy = x\,dx - 2y\,dx$$
    $$2x\,dy + 2y\,dx = x\,dx$$
    $$2(x\,dy + y\,dx) = x\,dx$$

    Recognizing the left side as the derivative of $2xy$ and integrating,

    $$2\int(x\,dy + y\,dx) = \int x\,dx$$
    $$2xy = \frac{x^2}{2} + C_1$$
    $$4xy - x^2 = C$$

9. $3x^2 + 2y + 2xy' = 0$

    Substituting $\frac{dy}{dx}$ for y' and multiplying both sides by dx,

    $$3x^2 dx + 2y\,dx + 2x\,dy = 0$$

    Recognizing $2y\,dx + 2x\,dx$ as the derivative of $2xy$, we integrate each term, obtaining,

    $$3\int x^2 dx + 2\int(y\,dx + x\,dy) = C$$
    $$x^3 + 2xy = C$$

13. $y\,dx - x\,dy = 2y^2 dx$

Dividing both sides by y^2,

$$\frac{y\,dx - x\,dy}{y^2} = 2dx$$

Recognizing the left side as the derivative of $\frac{x}{y}$, we integrate each term,

$$\int \frac{y\,dx - x\,dy}{y^2} = 2\int dx$$

$$\frac{x}{y} = 2x + C$$

17. $3x - 2y^2 - 4xyy' = 0$

Substituting $\frac{dy}{dx}$ for y' and multiplying both sides by dx,

$$3x\,dx - 2y^2 dx - 4xy\,dy = 0$$

Factoring out the minus sign from the last two terms,

$$3x\,dx - (2y^2 dx + 4xy\,dy) = 0$$

Recognizing the two terms in the parentheses as the derivative of $2xy^2$, we integrate

$$3\int x\,dx - \int (2y^2 dx + 4xy\,dy) = C_1$$

$$\frac{3x^2}{2} - 2xy^2 = C_1$$

$$4xy^2 - 3x^2 = C$$

21. $y = (3y^3 + x)\dfrac{dy}{dx}$ ($x = 1$ when $y = 1$)

$$y\,dx = 3y^3 dy + x\,dy$$

$$y\,dx - x\,dy = 3y^3 dy$$

$$\frac{y\,dx - x\,dy}{y^2} = 3y\,dy$$

The left side is the derivative of $\frac{x}{y}$. Integrating

$$\int \frac{y\,dx - x\,dy}{y^2} = 3\int y\,dy \Rightarrow \frac{x}{y} = \frac{3}{2}y^2 + C$$

Letting $x = 1$, and $y = 1$, we solve for C, obtaining $C = -\frac{1}{2}$. Therefore,

$$\frac{x}{y} = \frac{3}{2}y^2 - \frac{1}{2} \text{ or } 3y^3 - y = 2x.$$

Exercise 5: First-Order Homogeneous Differential Equations

1. $(x - y)dx - 2x\,dy = 0$.

Letting $y = vx$, $dy = v\,dx + x\,dv$, and substituting, we get

$$(x - vx)dx - 2x(v\,dx + x\,dv) = 0$$

Separating variables,

$$x\,dx - vx\,dx - 2xv\,dx - 2x^2 dv = 0$$

$$x(1 - 3v)dx - 2x^2 dv = 0$$

$$\frac{dx}{x} - \frac{2dv}{1 - 3v} = 0$$

Integrating: $\int \frac{dx}{x} - \frac{2}{3}\int \frac{-3dv}{1 - 3v} = C_1$

$$\ln x = \frac{2}{3}\ln(1 - 3v) = C_1 = \ln C_2$$

$$\ln x(1 - 3v)^{2/3} = \ln C_2$$

Taking the antilog, $x(1 - 3v)^{2/3} = C_2$

Substituting $v = \frac{y}{x}$: $x\left(1 - 3\frac{y}{x}\right)^{2/3} = C_2$

$$x^{1/3}(x - 3y)^{2/3} = C_2$$

Cubing: $x(x - 3y)^2 = C$

5. $xy^2 dy - (x^3 + y^3)dx = 0$

Letting $y = vx$, $dy = x\,dv + v\,dx$, and substituting, we obtain,

$$xy^2 dy - (x^3 + y^3)dx = 0$$

$$x^3 v^2(x\,dv + v\,dx) - (x^3 + x^3 v^3)dx = 0$$

$$x^4 v^2 dv + x^3 v^3 dx - x^3 dx - x^3 v^3 dx = 0$$

Separating variables: $v^2 dv - \frac{1}{x}dx = 0$

Integrating: $\int v^2 dv - \int \frac{1}{x}dx = C_1$

$$\frac{v^3}{3} - \ln x = C_1$$

Substituting $v = \frac{y}{x}$: $\frac{y^3}{3x^3} - \ln x = C_1$

Solving for y^3 : $y^3 = x^3(3\ln x + C)$

9. $(2x + y)dx = y\,dy$ $(x = 2, y = 1)$

Letting $y = vx$, $dy = x\,dv + v\,dx$, and substituting, we obtain,

$$(2x + y)\,dx = y\,dy$$
$$(2x + vx)dx = vx(x\,dv + v\,dx)$$
$$2x\,dx + vx\,dx = vx^2 dv + v^2 x\,dx$$

Separating variables:

$$2x\,dx + vx\,dx - v^2 x\,dx = vx^2 dv$$
$$x(2 + v - v^2)dx = vx^2 dv$$
$$\frac{1}{x}dx + \frac{v}{v^2 - v - 2}dv = 0$$

The second term needs to be separated into two fractions by the use of partial fractions to make it integrable.

$$\frac{v}{v^2 - v - 2} = \frac{v}{(v-2)(v+1)} = \frac{A}{v-2} + \frac{B}{v+1}$$

Multiplying by LCD: $v = A(v + 1) + B(v - 2)$

When $v = 2$: $2 = 3A$ or $A = \frac{2}{3}$

When $v = -1$: $-1 = -3B$ or $B = \frac{1}{3}$

Integrating: $\int \frac{dx}{x} + \frac{2}{3}\int \frac{dv}{v-2} + \frac{1}{3}\int \frac{dv}{v+1} = C_1$

$\ln x + \frac{2}{3}\ln(v - 2) + \frac{1}{3}\ln(v + 1) = \ln C_2$

Simplifying: $x(v - 2)^{2/3}(v + 1)^{1/3} = C_2$

Cubing: $x^3 (v - 2)^2 (v + 1) = C_3$

Substituting $v = \frac{y}{x}$ and simplifying:

$(y - 2x)^2 (y + x) = C_3$

At $(2,1)$, $C_3 = 27$, so $(y - 2x)^2(x + y) = 27$

Exercise 6: First-Order Linear Differential Equations

1. $y' + \frac{y}{x} = 4$

This first order linear differential equation is in standard form with $P(x) = \frac{1}{x}$ and $Q(x) = 4$. Calculating $\int P(x)dx$, we obtain,

$$\int \frac{1}{x}dx = \ln x \Rightarrow e^{\int P dx} = e^{\ln x} = x$$

Substituting into the equation

$$ye^{\int P dx} = \int e^{\int P dx} Q\,dx$$

gives $yx = \int 4x\,dx = 2x^2 + C$

or $y = 2x + \frac{C}{x}$

5. Putting the first order linear differential equation into standard form, we obtain,

$$\frac{dy}{dx} + x^2 y = x^2$$

where $P(x) = x^2$, and $Q(x) = x^2$

Calculating $\int P(x)dx$:

$$\int x^2 dx = \frac{x^3}{3} \Rightarrow e^{\int P dx} = e^{x^3/3}$$

Substituting into $ye^{\int P dx} = \int e^{\int P dx} Q\,dx$, we obtain,

$$ye^{x^3/3} = \int e^{x^3/3} \cdot x^2 dx = e^{x^3/3} + C$$

Solving for y: $y = 1 + Ce^{-x^3/3}$

9. Going to standard form, we obtain,

$$\frac{dy}{dx} - \frac{2y}{x} = -1$$

where $P(x) = -\frac{2}{x}$ and $Q(x) = -1$

$\int P(x)dx:$ $-\int \frac{2}{x}dx = -2\ln x \Rightarrow e^{-2\ln x} = x^{-2}$

Substituting into $ye^{\int P dx} = \int e^{\int P dx}Q dx$, we obtain,

$$yx^{-2} = \int(-x^{-2})dx = x^{-1} + C$$

Solving for y: $y = x + Cx^2$

13. $xy' + x^2 y + y = 0$. Going to standard form, we obtain,

$$\frac{dy}{dx} + \left(x + \frac{1}{x}\right)y = 0$$

where $P(x) = x + \frac{1}{x}$ and $Q(x) = 0$

$\int \left(x + \frac{1}{x}\right)dx = \frac{x^2}{2} + \ln x \Rightarrow$

$$e^{\int P dx} = e^{x^2/2} \cdot e^{\ln x} = xe^{x^2/2}$$

Substituting into $ye^{\int P dx} = \int e^{\int P dx}Q\,dx$, we obtain,

$$yxe^{x^2/2} = \int e^{x^2/2} \cdot x \cdot 0\,dx = C$$

$$y = \frac{C}{xe^{x^2/2}}$$

17. $y' = 2y + 4e^{2x}$

Going to standard form, we obtain,

$$\frac{dy}{dx} - 2y = 4e^{2x}$$

where $P(x) = -2$, $Q(x) = 4e^{2x}$

$\int P(x)dx = -2\int dx = -2x$, $e^{\int P dx} = e^{-2x}$

Substituting into $y\,e^{\int P dx} = \int e^{\int P dx}Q\,dx$, we obtain,

$$ye^{-2x} = \int e^{-2x} \cdot 4e^{2x}dx = 4\int dx = 4x + C$$

Solving for y: $y = (4x + C)e^{2x}$

21. $y' + y = \sin x$

$\int P dx = x$ I.F. $= e^x$

$ye^x = \int e^x \sin x\,dx = \frac{e^x}{2}(\sin x - \cos x) + C$

$y = \frac{1}{2}(\sin x - \cos x) + Ce^{-x}$

25. $y' + \frac{y}{x} = 3x^2 y^2$

This is a Bernoulli equation with $P = \frac{1}{2}$, $Q = 3x^2$, and $n = 2$. We first divide through by y^2, obtaining,

$$\frac{1}{y^2} \cdot \frac{dy}{dx} + \frac{y^{-1}}{x} = 3x^2$$

We substitute $z = y^{1-2} = y^{-1}$ and

$$\frac{dz}{dx} = -y^{-2}\frac{dy}{dx} \Rightarrow \frac{dy}{dx} = -y^2\frac{dz}{dx}$$

Substituting: $-\frac{dz}{dx} + \frac{z}{x} = 3x^2 \Rightarrow \frac{dz}{x} - \frac{z}{x} = -3x^2$

This is a linear equation in z, with $P = \frac{-1}{x}$,

$Q = -3x^2$

$\int P(x)dx : -\int \frac{dx}{x} = -\ln x \Rightarrow e^{\int P dx} = e^{-\ln x} = \frac{1}{x}$

Substituting into $ze^{\int P dx} = \int e^{\int P dx} Q\, dx$,
we obtain,

$$z \cdot \frac{1}{x} = -\int 3x\, dx = -\frac{3x^2}{2} + C$$

Solving for z, and substituting back, $z = \frac{1}{y}$,
we obtain,

$$z = \frac{1}{y} = -\frac{3x^3}{2} + Cx \Rightarrow xy\left(C - \frac{3x^2}{2}\right) = 1$$

29. $xy' + y = 4x$ ($x = 1$ when $y = 5$)

Going to standard form,

$$\frac{dy}{dx} + \frac{y}{x} = 4, \text{ where } P(x) = \frac{1}{x}, \ Q(x) = 4$$

$\int P(x)dx = \int \frac{1}{x}dx = \ln x, \ e^{\int P dx} = e^{\ln x} = x$

Substituting into $y\,e^{\int P dx} = \int e^{\int P dx} Q\, dx$,
we obtain, $yx = \int 4x\, dx = 2x^2 + C$

Letting $x = 1$ and $y = 5 \Rightarrow C = 3$. Thus:

$$xy - 2x^2 = 3$$

33. $y' = \tan^2 x + y\cot x$ $\left(x = \frac{\pi}{4} \text{ when } y = 2\right)$

Going to standard form,

$$\frac{dy}{dx} - y\cot x = \tan^2 x$$

where $P(x) = -\cot x$, $Q(x) = \tan^2 x$

$\int P(x)dx : -\int \cot x\, dx = -\ln \sin x$

$$e^{\int P dx} = e^{-\ln \sin x} = \csc x$$

Substituting into $ye^{\int P dx} = \int e^{\int P dx} Q\, dx$,
we obtain,

$$y\csc x = \int \csc x \tan^2 x\, dx = \int \frac{\sin^2 x}{\cos^2 x \cdot \sin x}dx$$

$$= -\int \cos^{-2}x(-\sin x)dx = \frac{1}{\sec x} + C$$

Solving for y :

$$y = \frac{1}{\sec x \cdot \csc x} + \frac{C}{\csc x} = \tan x + C\sin x$$

Letting $x = \frac{\pi}{4}$ when $y = 2$ yields $C = \sqrt{2}$.

Therefore, $y = \tan x + \sqrt{2}\sin x$

Exercise 7: Geometric Applications of First-Order Differential Equations

1. Find the equation of the curve which passes through the point $(2, 9)$ and whose slope is

$$y' = x + \frac{1}{x} + \frac{y}{x}.$$

Putting the equation into standard form, we obtain,

$$\frac{dy}{dx} - \frac{y}{x} = x + \frac{1}{x}$$

where $P(x)dx = -\frac{1}{x}$ and $Q(x) = x + \frac{1}{x}$

$\int P(x)dx = -\int \frac{1}{x}dx = -\ln x \Rightarrow$

$$e^{\int P dx} = e^{-\ln x} = \frac{1}{x}$$

Substituting into $y\, e^{\int P dx} = \int e^{\int P dx} Q dx$,
we obtain,

$$y\left(\frac{1}{x}\right) = \int \frac{1}{x}\left(x + \frac{1}{x}\right)dx = \int (1 + x^{-2})dx$$

$$= x - \frac{1}{x} + C$$

Solving for y : $y = x^2 - 1 + Cx$

Substituting $(2, 9)$ for x and y, yields $C = 3$

Therefore: $y = x^2 + 3x - 1$

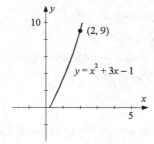

5.　A curve passes through the point $(2, 3)$ and has a slope equal to the sum of the abscissa and ordinate at each point. Find its equation.

The slope equals the sum of the abscissa and ordinate therefore:

$$\frac{dy}{dx} = x + y$$

Putting into standard form for a first order linear D.E.,

$$\frac{dy}{dx} - y = x$$

where $P(x) = -1$ and $Q(x) = x$

$$\int P(x)dx = -\int dx = -x \Rightarrow e^{\int Pdx} = e^{-x}$$

Substituting into $ye^{\int Pdx} = \int e^{\int Pdx} Qdx$, we

obtain,　　$ye^{-x} = \int xe^{-x}dx$

Integrating by parts:　$u = x$　　$dv = e^{-x}dx$

　　　　　　　　　　　$du = dx$　　$v = -e^{-x}$

Therefore,

$$ye^{-x} = -xe^{-x} + \int e^{-x}dx$$
$$= -xe^{-x} - e^{-x} + C$$
$$y = -x - 1 + Ce^{x}$$

At $x = 2$, $y = 3 \Rightarrow C = \frac{6}{e^2}$, therefore,

$$y = 6e^{x-2} - x - 1$$

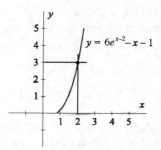

9.　Find the equation of the curve passing through $(0, 1)$ for which the length PC of a normal through P equals the square of the ordinate of P, Fig. 35-10, where

$$PC = y\sqrt{1 + (y')^2}$$

Since PC equals the square of the ordinate of P, we write,

$$PC = y\sqrt{1 + (y')^2} = y^2$$

Solving for $\frac{dy}{dx}$ and separating the variables, we obtain,

$$\frac{dy}{\sqrt{y^2-1}} = dx \Rightarrow \int \frac{dy}{\sqrt{y^2-1}} = \int dx \Rightarrow \ln\left(y + \sqrt{y^2 - 1}\right) = x + C$$

Since the curve passes through $(0, 1)$, we obtain $C = 0$.

$$\ln\left(y + \sqrt{y^2 - 1}\right) = x \Rightarrow y + \sqrt{y^2 - 1} = e^x$$
$$e^x - y = \sqrt{y^2 - 1}$$

Squaring both sides　$e^{2x} - 2e^x y + y^2 = y^2 - 1$

Solving for y,　$y = \frac{e^{2x}+1}{2e^x} = \left(\frac{1}{2}\right)\left(e^x + \frac{1}{e^x}\right) = \frac{e^x + e^{-x}}{2}$

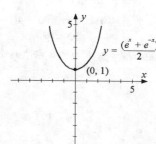

13. $x^2 - y^2 = Cy$

Differentiating and solving for $\frac{dy}{dx}$ to find the slope:
$$2x - 2y\frac{dy}{dx} = C\frac{dy}{dx}, \quad \frac{dy}{dx} = \frac{2x}{2y+C}$$
Solving for C and substituting in last equation, we obtain,
$$C = \frac{x^2-y^2}{y}$$
$$\frac{dy}{dx} = \frac{2x}{2y+\frac{(x^2-y^2)}{y}} = \frac{2xy}{2y^2+x^2-y^2} = \frac{2xy}{x^2+y^2}$$

Taking the negative reciprocal of this slope to obtain the slope of the orthogonal trajectory, setting it equal to $\frac{dy}{dx}$ and solving, we obtain,
$$\frac{dy}{dx} = -\frac{x^2+y^2}{2xy}$$
$$2xy\,dy + (x^2+y^2)dx = 0$$
Letting $y = vx$ and $dy = v\,dx + x\,dv$, we obtain,
$$2x(vx)(v\,dx + x\,dv) + (x^2 + x^2v^2)dx = 0$$
$$2x^2v^2dx + 2x^3v\,dv + x^2dx + x^2v^2dx = 0$$
$$3x^2v^2dx + x^2dx + 2x^3v\,dv = 0$$
$$x^2(3v^2 + 1)dx + 2x^3v\,dv = 0$$
$$\frac{1}{x}dx + \frac{2v\,dv}{3v^2+1} = 0$$
$$\int \frac{1}{x}dx + \frac{1}{3}\int \frac{6v\,dv}{3v^2+1} = 0$$
$$\ln|x| + \frac{1}{3}\ln|3v^2+1| = C_1$$
$$\ln|x^3(3v^2+1)| = 3C_1 = C$$
$$x^3(3v^2+1) = k$$
Substituting $v = \frac{y}{x}$, we obtain, $3xy^2 + x^3 = k$

Exercise 8: Exponential Growth and Decay

1. A quantity grows with time such that its rate of growth $\frac{dy}{dt}$ is proportional to the present amount y. Use this statement to derive the equation for exponential growth, $y = ae^{nt}$.

$\frac{dy}{dt} = ny$ where n is the constant of proportionality. Separating variables and integrating, we obtain,
$$\int \frac{dy}{y} = n\int dt \Rightarrow \ln y = nt + C \Rightarrow e^{nt+C} = y$$
$$y = e^{nt} \cdot e^c \Rightarrow y = ae^{nt}$$
Note: At $t = 0$, $y = a \Rightarrow a$ is the initial amount present.

5. An iron ingot is 1850°F above room temperature. If it cools exponentially at 3.50 percent/min, find its temperature (above room temperature) after 2.50 h.

This is exponential decay which obeys the formula $T = T_0e^{-nt}$, where

T_0 = temp above room temp at $t = 0$
T = temp above room temp at time t
t = time
n = constant

Substituting $T_0 = 1850$, $n = 0.035$, $t = 2.5(60) = 150$ min gives
$$T = 1850e^{-(0.035)(150)}$$
$$= 9.71°F \text{ (above room temperature)}$$

9. A 45.5 lb carton is initially at rest. It is then pulled horizontally by a 19.4 lb force in the direction of motion, and resisted by a frictional force which is equal (in pounds) to four times the carton's velocity (in feet/s). Show that the differential equation of motion is
$$\frac{dv}{dt} = 13.7 - 2.83v.$$
Using Newton's second law, $F = ma$,
$$F = ma = \frac{W}{g} \cdot \frac{dv}{dt} \text{ where } F = 19.4 - 4v$$
Thus, $19.4 - 4v = \frac{45.5\,\text{lb}}{32.2\,\text{ft/s}^2} \cdot \frac{dv}{dt}$

Solving for $\frac{dv}{dt}$: $\frac{dv}{dt} = 13.7 - 2.83v$

13. Find the displacement of the instrument package in Problem 11 after 1.00 s. Using the acceleration given in Problem 11, we will integrate twice to find the equation for displacement.
$$a = \frac{dv}{dt} = \frac{g(155-v)}{155}, \quad \frac{dv}{155-v} = \frac{g}{155}dt$$
Integrating: $\int \frac{dv}{155-v} = \frac{g}{155}\int dt \Rightarrow$
$$-\ln(155-v) = \frac{g}{155}t + C$$
At $t = 0$, $v = 0 \Rightarrow C = -\ln 155$
Therefore, $\ln(155-v) - \ln 155 = \frac{-g}{155}t$
$$\ln\frac{155-v}{155} = \frac{-g}{155}t$$
$$\frac{155-v}{155} = e^{-gt/155}$$
Solving for v: $v = 155 - 155e^{-gt/155}$
Substituting $v = \frac{ds}{dt}$ and integrating,
$$\int ds = \int 155dt + \frac{155^2}{g}\int e^{-gt/155}\left(\frac{-g}{155}\right)dt$$
$$s = 155t + \frac{155^2}{g}e^{-gt/155} + C_1$$
At $t = 0$, $s = 0 \Rightarrow C_1 = -\frac{155^2}{g}$
$$s(1) = 155 + \frac{155^2}{32.2}e^{-32.2/155} - \frac{155^2}{32.2} = 15.0\text{ ft}$$

17. Find the time at which the ball in the preceding problem is going at a speed of 70.0 ft/s.

$$V = 135 - 114e^{-t/4.19} = 70$$
$$114e^{-t/4.19} = 65$$

Dividing by 114 and taking the ln of both sides, we obtain,

$$\frac{-t}{4.19} = \ln\left(\frac{65}{114}\right)$$
$$t = -4.19\ln\left(\frac{65}{114}\right) = 2.36 \text{ s}$$

Exercise 9: Series RL and RC Circuits

1. If an inductor is discharged through a resistor, show that the current decays exponentially according to the function $i = \frac{E}{R}e^{-Rt/L}$.

Using Kirchhoff's voltage law gives:

$$Ri + L\frac{di}{dt} = 0 \Rightarrow \frac{di}{dt} + \frac{R}{L}i = 0$$

This is a first-order linear differential equation with

$$\int P(t)dt = \int \frac{R}{L}dt = \frac{R}{L}t \Rightarrow e^{\int Pdt} = e^{Rt/L}$$

Therefore $ie^{Rt/L} = \int e^{Rt/L} \cdot 0 dt = k$

At $t = 0$, $i = \frac{E}{R} \Rightarrow k = \frac{E}{R}$, therefore $ie^{Rt/L} = \frac{E}{R}$

$$i = \frac{E}{R}e^{-Rt/L}$$

5. Capacitor charging:

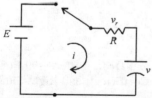

Summing voltages around the loop,

$$E - v_r = v \quad \text{or} \quad v_r = E - v$$

Since i is the same in the resistor and the capacitor,

$$i = \frac{v_r}{R} = C\frac{dv}{dt}$$

But $v_r = E - v$ so $\frac{E-v}{R} = C\frac{dv}{dt}$ or

$$\frac{E}{R} - \frac{v}{R} = C\frac{dv}{dt}$$
$$\frac{dv}{dt} + \frac{1}{RC}v = \frac{E}{RC} \quad \text{First-Order Linear}$$

I.F. $= e^{\int Pdx} = e^{\int \frac{1}{RC}dt} = e^{t/RC}$

$$ve^{t/RC} = \int \frac{E}{RC}e^{t/RC}dt = E\int e^{t/RC}\left(\frac{1}{RC}\right)dt$$
$$= Ee^{t/RC} + k$$

at $t = 0$, $v = 0$, so $0 = E + k$ or $k = -E$

then $ve^{t/RC} = Ee^{t/RC} - E$

The voltage across the capacitor is then

$$v = E - \frac{E}{e^{t/RC}} = E\left(1 - e^{-t/RC}\right)$$

9. For the circuit in Problem 8, find the current at $t = 1.55$ ms. At $t = 0.00155$ s

$$i = 10(0.00155 + 1) - \frac{10}{(0.00155+1)^2}$$
$$= 46.4 \text{ mA}$$

13. For the circuit in Problem 12, find the current at $t = 1.85$ ms.
When $t = 0.00185$ s

$$i = \frac{100(0.00185)\left[(0.00185)^2 + 12\right]}{\left[(0.00185)^2 + 4\right]^2}$$
$$= 139 \text{ mA}$$

17. For an RC circuit, the applied voltage is alternating, and is given by $e = E \sin \omega t$. If the switch is thrown from 2 to 1 (charging) when e is zero and increasing, show that the current is given by

$i = \frac{E}{Z}\left[\sin(\omega t + \phi) - e^{-t/RC}\sin\phi\right]$.

Summing the voltage drops around the loop gives

$$Ri + \frac{1}{C}\int i\,dt = E \sin \omega t$$

Taking the derivative,

$$R\frac{di}{dt} + \frac{i}{C} = \omega E \cos \omega t$$

since the derivative of $E \sin \omega t$ is $= \omega E \cos \omega t$. Dividing by R, our differential equation is then

$$\frac{di}{dt} + \frac{1}{RC}i = \frac{\omega E}{R}\cos \omega t$$

This is first-order linear, with an integrating factor $= e^{t/RC}$. The solution is then

$$ie^{t/RC} = \frac{\omega E}{R}\int e^{t/RC}\cos \omega t\,dt$$

Using integral 42, with $a = \frac{1}{RC}$ and $b = \omega$,

$$ie^{t/RC} = \frac{\omega E}{R}\left[\frac{e^{t/RC}}{\frac{1}{R^2C^2}+\omega^2}\left(\frac{1}{RC}\cos \omega t + \omega \sin \omega t\right)\right] + c = \frac{\omega E e^{t/RC}}{\frac{R}{R^2C^2}+R\omega^2}\left(\frac{1}{RC}\cos \omega t + \frac{\omega RC}{RC}\sin \omega t\right) + c$$

Dividing by $e^{t/RC}$,

$$i = \frac{E}{\left(\frac{1}{\omega RC^2}+R\omega\right)RC}(\cos \omega t + \omega RC \sin \omega t) + \frac{c}{e^{t/RC}} = \frac{E}{R^2\omega C+\frac{1}{\omega C}}(\cos \omega t + \omega RC \sin \omega t) + ce^{-t/RC}$$

$$= \frac{1}{\omega C}\frac{E}{\left(R^2+\frac{1}{\omega^2C^2}\right)}(\cos \omega t + \omega RC \sin \omega t) + ce^{-t/RC} = \frac{E}{Z}\left[\frac{X_c}{Z}\cos \omega t + \frac{R}{Z}\sin \omega t\right] + ce^{-t/RC}$$

since $X_c = \frac{1}{\omega C}$ and $Z^2 = R^2 + \frac{1}{\omega^2C^2}$. From the impedance triangle, $\frac{X_c}{Z} = \sin\phi$ and $\frac{R}{Z} = \cos\phi$, so

$$i = \frac{E}{Z}[\cos \omega t \sin\phi + \sin \omega t \cos\phi] + ce^{-t/RC} = \frac{E}{Z}\sin(\omega t + \phi) + ce^{-t/RC}$$

when $t = 0$, $i = \frac{E}{R}\sin \omega t = 0$, so

$$0 = \frac{E}{Z}\sin\phi + c, \quad \text{thus } c = -\frac{E}{Z}\sin\phi$$

Substituting back,

$$i = \frac{E}{Z}\left[\sin(\omega t + \phi) - e^{-t/RC}\sin\phi\right]$$
$$\quad\underset{\substack{\text{steady-state}\\\text{current}}}{}\quad\underset{\substack{\text{transient}\\\text{current}}}{}$$

Exercise 10: Second-Order Differential Equations

1. $y'' = 5$

Replacing y'' by $\frac{d}{dx}y'$ and separating variables,
$$dy' = 5dx$$

Integrating: $\frac{dy}{dx} = 5x + C_1 \Rightarrow dy = (5x + C_1)dx$

Integrating again, $y = \frac{5}{2}x^2 + C_1x + C_2$

5. $y'' - x^2 = 0$, where $y' = 1$ at $(0, 0)$

Replacing y'' by $\frac{d}{dx}y'$ and separating variables,
$$dy' = x^2\,dx$$

Integrating: $\frac{dy}{dx} = \frac{x^3}{3} + C_1$
$$y' = 1, \ x = 0 \Rightarrow C_1 = 1$$

Substituting for C_1 and separating the variables
$$dy = \left(\frac{x^3}{3}+1\right)dx$$

Integrating: $y = \frac{x^4}{12} + x + C_2$
$$x = 0, \ y = 0 \Rightarrow C_2 = 0$$

Therefore $y = \frac{x^4}{12} + x$

Exercise 11: Second-Order Differential Equations with Constant Coefficients and Right-Side Zero

1. The auxiliary equation is $m^2 - 6m + 5 = 0$
$$(m - 5)(m - 1) = 0 \text{ so } m = 5, \ m = 1$$
Therefore: $y = C_1e^x + C_2e^{5x}$

5. Auxiliary equation: $m^2 - m - 6 = 0$
$$(m - 3)(m + 2) = 0 \text{ so } m = 3, \ m = -2$$
Therefore: $y = C_1e^{3x} + C_2e^{-2x}$

9. Auxiliary equation: $6m^2 + 5m - 6 = 0$
$$(2m + 3)(3m - 2) = 0 \text{ so } m = \frac{2}{3}, \ m = \frac{-3}{2}$$
Therefore: $y = C_1e^{2x/3} + C_2e^{-3x/2}$

13. Auxiliary equation: $m^2 - 2m + 1 = 0$, or
$$(m - 1)(m - 1) = 0$$
so we get the equal roots, $m = 1$, $m = 1$.
Therefore: $y = C_1e^x + C_2xe^x$

17. Auxiliary equation: $m^2 + 2m + 1 = 0$
$$(m+1)(m+1) = 0$$
$m = -1$, $m = -1$ equal roots
Therefore: $\quad y = C_1 e^{-x} + C_2 x e^{-x}$

21. Auxiliary equation: $m^2 - 6m + 25 = 0$
$$m = \frac{6 \pm \sqrt{36 - 4(1)(25)}}{2} \text{ so } m = 3 \pm j4$$
Therefore: $\quad y = e^{3x}(C_1 \cos 4x + C_2 \sin 4x)$

25. Auxiliary equation: $m^2 - 4m + 5 = 0$
$$m = \frac{4 \pm \sqrt{16 - 4(1)(5)}}{2} \Rightarrow m = 2 \pm j$$
Therefore: $\quad y = e^{2x}(C_1 \cos x + C_2 \sin x)$

29. Auxiliary equation: $m^2 - 2m + 1 = 0$
$$(m-1)(m-1) = 0$$
$m = 1$, $m = 1$ equal roots
Therefore: $\quad y = C_1 e^x + C_2 e x^x$
At $x = 0$, $y = 5$, $C_1 + 0 = 5$.
Taking the derivative, $y' = 5e^x + C_2 x e^x + C_2 e x$
Substituting $C_1 = 5$, $y' = -9$, $x = 0$
$$-9 = 5e^0 + C_2, \text{ so } C_2 = -14$$
Therefore: $\quad y = 5e^x - 14x e^x$

33. Auxiliary equation: $m^2 - 4 = 0$
$$(m+2)(m-2) = 0$$
$$m = 2, \ m = -2$$
Therefore: $\quad y = C_1 e^{2x} + C_2 e^{-2x}$.
At $x = 0$, $y = 1$, $C_1 + C_2 = 1 \qquad (1)$
Calculating y' and substituting $x = 0$, $y' = -1$
$$y' = 2C_1 e^{2x} - 2C_2 e^{-2x} \Rightarrow$$
$$2C_1 - 2C_2 = -1 \qquad (2)$$
Solving equations (1) and (2) simultaneously
$$2C_1 + 2C_2 = 2$$
$$\underline{2C_1 - 2C_2 = -1}$$
$$4C_1 \qquad = 1$$
or $C_1 = \frac{1}{4}$
Then $C_2 = \frac{3}{4}$
Therefore: $\quad y = \frac{1}{4}e^{2x} + \frac{3}{4}e^{-2x}$

37. The auxiliary equation: $m^3 - 2m^2 - m + 2 = 0$
has the roots $m = 1$, $m = 2$, $m = -1$
Therefore: $\quad y = C_1 e^x + C_2 e^{-x} + C_3 e^{2x}$

41. The auxiliary equation: $m^3 - 3m^2 - m + 3 = 0$
has the roots $m = 1$, $m = 3$, $m = -1$
Therefore: $\quad y = C_1 e^x + C_2 e^{-x} + C_3 e^{3x}$

Exercise 12: Second-Order Differential Equations with Right-Side Not Zero

1. The auxiliary equation $m^2 - 4 = 0$, has roots $m = 2$, $m = -2$, so the complementary equation is
$$y_c = C_1 e^{2x} + C_2 e^{-2x}$$
The terms of $f(x)$ and its derivatives (less coefficients) are:
$$f(x) = C, \ f'(x) = 0, \ f''(x) = 0$$
So, $y_p = C$. Substituting into the DE, $y'' - 4y = 12$
$$0 - 4C = 12 \text{ from which } C = -3.$$
So, $y_p = -3$. Therefore: $y = y_c + y_p$, or
$$y = C_1 e^{2x} + C_2 e^{-2x} - 3$$

5. The auxiliary equation $m^2 - 4 = 0$, has roots $m = 2$, $m = -2$, so the complementary equation is
$y_c = C_1 e^{2x} + C_2 e^{-2x}$. The terms of $f(x)$ and its derivatives (less coefficients) are
$$(1) \ f(x) = Ax^3 + Bx \quad (2) \ f'(x) = Cx^2 + D \quad (3) \ f''(x) = Ex, \text{ a duplicate of (1)}.$$
The trial solution and its derivatives are then
$$y_p = Ax^3 + Cx^2 + Bx + D$$
$$y'_p = 3Ax^2 + 2Cx + B$$
$$y''_p = 6Ax + 2C$$
Substituting into the differential equation $y'' - 4y = x^3 + x$,
$$6Ax + C - 4(Ax^3 + Cx^2 + Bx + D) = x^3 + x$$
$$-4Ax^3 - 4Cx^2 + (6A - 4B)x + C - 4D = x^3 + x$$
Equating coefficients yields: $A = -\frac{1}{4}$, $B = -\frac{5}{8}$, and $C = D = 0$.
Thus $y_p = -\frac{1}{4}x^3 - \frac{5}{8}x$ and $y = C_1 e^{2x} + C_2 e^{-2x} - \frac{1}{4}x^3 - \frac{5}{8}x$

9. $y'' - 4y = 4x - 3e^x$. The auxiliary equation $m^2 - 4 = 0$ has roots $m = 2$, $m = -2$, so
$$y_c = C_1 e^{2x} + C_2 e^{-2x}$$
The terms of $f(x)$ and its derivatives are x, e^x, and a constant. The trial solution and its derivatives are then
$$y_p = Ax + Be^x + C, \quad y'_p = A + Be^x, \quad y''_p = Be^x$$
Substituting into $y'' - 4y = 4x - 3e^x$,
$$Be^x - 4Ax - 4Be^x - 4C = 4x - 3e^x$$
$$-4Ax - 3Be^x - 4C = 4x - 3e^x$$
Equating coefficients, $-4A = 4$, so $A = -1$. $-3B = -3$, so $B = 1$, and $C = 0$.
Thus $y_p = -x + e^x$ and $y = C_1 e^{2x} + C_2 e^{-2x} + e^x - x$

13. $y'' + 4y = \sin 2x$. The auxiliary equation $m^2 + 4 = 0$ has roots $2j$ and $-2j$, therefore, the complementary solution is
$$y_c = C_1 \cos 2x + C_2 \sin 2x$$
The terms of $f(x)$ and its derivatives (less coefficients) are
$$f(x) = \sin 2x \qquad f'(x) = \cos 2x \qquad f''(x) = \sin 2x$$
But $\sin 2x$ and $\cos 2x$ are duplicates of the terms in y_c, so we multiply each by x. The trial solution and its derivatives are then
$$y_p = Ax \sin 2x + Bx \cos 2x$$
$$y'_p = 2Ax \cos 2x + A \sin 2x - 2Bx \sin 2x + B \cos 2x$$
$$y''_p = -4Ax \sin 2x + 4A\cos 2x - 4Bx \cos 2x - 4B \sin 2x$$
Substituting into the differential equation:
$$-4Ax \sin 2x + 4A \cos 2x - 4Bx \cos 2x - 4B \sin 2x + 4Ax \sin 2x + 4B \cos 2x = \sin 2x$$
$$4A \cos 2x - 4B \sin 2x = \sin 2x$$
Equating coefficients: $A = 0$, $B = -\frac{1}{4}$
Thus $y_p = \dfrac{-x \cos 2x}{4}$ and $y = C_1 \cos 2x + C_2 \sin 2x - \dfrac{x \cos 2x}{4}$

17. $y'' + y = 2 \cos x - 3 \cos 2x$. The auxiliary equation $m^2 + 1 = 0$ has roots j and $-j$, so the complementary solution is
$$y_c = C_1 \cos x + C_2 \sin x$$
The terms of $f(x)$ and its derivatives (less coefficients) are $\sin x$, $\cos x$, $\sin 2x$, and $\cos 2x$.
We eliminate duplicates with y_c by multiplying $\sin x$ and $\cos x$ by x. The trial solution y_p and its derivatives is then
$$y_p = Ax \cos x + B \cos 2x + Cx \sin x + D \sin 2x$$
$$y'_p = -Ax \sin x + A \cos x - 2B \sin 2x + Cx\cos x + C \sin x + 2D \cos 2x$$
$$y''_p = -Ax \cos x - 2A \sin x - 4B \cos 2x - Cx \sin x + 2C \cos x - 4D \sin 2x$$
Substituting into the differential equation and simplifying:
$$-3B \cos 2x - 3D \sin 2x + 2C \cos x - 2A \sin x = 2 \cos x - 3 \cos 2x$$
Equating coefficients gives: $A = D = 0$, $B = 1$, and $C = 1$.
So $y_p = \cos 2x + x \sin x$ and $y = C_1 \cos x + C_2 \sin x + \cos 2x + x \sin x$

21. $y'' - 4y' + 5y = e^{2x}\sin x$. The auxiliary equation $m^2 - 4m + 5 = 0$ has roots $2 + j$ and $2 - j$, so the complementary solution is

$$y_c = e^{2x}(C_1\sin x + C_2\cos x)$$

The terms of $f(x)$ and its derivatives (less coefficients) are $e^{2x}\sin x$ and $e^{2x}\cos x$. To eliminate duplication, we multiply each by x. The trial solution y_p and its derivatives is

$$y_p = Axe^{2x}\sin x + Bxe^{2x}\cos x$$
$$y_p' = Axe^{2x}\cos x + 2Axe^{2x}\sin x + Ae^{2x}\sin x - Bxe^{2x}\sin x + 2Bxe^{2x}\cos x + Be^{2x}\cos x$$
$$= (A + 2B)xe^{2x}\cos x + (2A - B)xe^{2x}\sin x + Ae^{2x}\sin x + Be^{2x}\cos x$$
$$y_p'' = (A + 2B)[-xe^{2x}\sin x + 2xe^{2x}\cos x + e^{2x}\cos x] + (2A - B)[xe^{2x}\cos x + 2xe^{2x}\sin x + e^{2x}\sin x]$$
$$+ Ae^{2x}\cos x + 2Ae^{2x}\sin x - Be^{2x}\sin x + 2Be^{2x}\cos x$$

Substituting into the differential equation and simplifying:

$$e^{2x}\Big\{(A + 2B)[-x\sin x + 2x\cos x + \cos x] + (2A - B)[x\cos x + 2x\sin x + \sin x] + A\cos x + 2A\sin x$$

$$- B\sin x + 2B\cos x - 4[(A + 2B)x\cos x + (2A - B)x\sin x + A\sin x + B\cos x] + 5[Ax\sin x + Bx\cos x]\Big\}$$
$$= e^{2x}\sin x$$

Equating coefficients of $\sin x$ gives: $2A - B + 2A - B - 4A = 1$
from which $B = -\frac{1}{2}$.
Equating coefficients of $\cos x$ gives: $A + 2B + A + 2B - 4B = 0$
from which $A = 0$.
So $y_p = \left(-\frac{x}{2}\right)e^{2x}\cos x$ and $y = e^{2x}\left[C_1\sin x + C_2\cos x - \left(\frac{x}{2}\right)\cos x\right]$

25. $y'' + 4y = 2$; through $(0, 0)$ and $\left(\frac{\pi}{4}, \frac{1}{2}\right)$. The auxiliary equation $m^2 + 4 = 0$ has roots $2j$ and $-2j$, thus the complementary solution is:

$$y_c = C_1\cos 2x + C_2\sin 2x$$

The terms of $f(x)$ and its derivatives (less coefficients) are constants, thus

$$y_p = C, \quad y_p' = 0, \quad y_p'' = 0.$$

Substituting into the differential equation: $0 + 4C = 2$, so $C = \frac{1}{2}$ and $y_p = \frac{1}{2}$

Therefore $\quad y = C_1\cos 2x + C_2\sin 2x + \frac{1}{2}$

At $(0, 0)$, $\quad 0 = C_1 + \frac{1}{2}$ so $C_1 = -\frac{1}{2}$.

At $\left(\frac{\pi}{4}, \frac{1}{2}\right)$, $\quad C_2 = 0$

$y = -\frac{1}{2}\cos 2x + \frac{1}{2} = \frac{1 - \cos 2x}{2}$

29. $y'' + y = -2\sin x$, through $(0, 0)$ and $\left(\frac{\pi}{2}, 0\right)$. The auxiliary equation $m^2 + 1 = 0$ has roots j and $-j$, thus the complementary solution is

$$y_c = C_1\cos x + C_2\sin x$$

The terms of $f(x)$ and its derivatives (less coefficients) are $\sin x$ and $\cos x$, which duplicate the terms in y_c. The trial solution and its derivatives is thus:

$$y_p = Ax\sin x + Bx\cos x$$
$$y_p' = Ax\cos x + A\sin x + B\cos x - Bx\sin x$$
$$y_p'' = 2A\cos x - 2B\sin x - Ax\sin x - Bx\cos x$$

Substituting into the differential equation and simplifying,

$$2A\cos x - 2B\sin x = -2\sin x$$

so $A = 0$, $B = 1$. Therefore $y = y_c + y_p = C_1\cos x + C_2\sin x + x\cos x$

At $y = 0$, $x = 0 \Rightarrow C_1 = 0$ and at $y = 0$, $x = \frac{\pi}{4} \Rightarrow C_2 = 0$

Therefore $y = x\cos x$

Exercise 13: *RLC* Circuits

1. In an LC circuit, $C = 1.00$ microfarads, $L = 1.00$ H, and $E = 100$ V. At $t = 0$ the charge and current are both zero. Show that $i = 0.1 \sin 1000t$ A.

    Since $R = 0$, we obtain i using the formula:

    $$i = \frac{E}{w_n L} \sin w_n t = 100$$

    Computing $w_n = \frac{1}{\sqrt{LC}} = \frac{1}{\sqrt{1 \times 1 \times 10^{-6}}} = 10^3 = 1000$

    Therefore $i = \frac{100}{(1000)1} \sin 1000t = 0.1 \sin 1000t$ A

    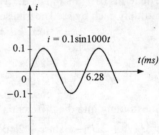

5. In an RLC circuit, $R = 1.55$ ohm, $C = 250$ microfarad, $L = 0.125$ H, and $E = 100$ V. The current and charge are 0 when $t = 0$. Show that a) the circuit is underdamped, and b) using Eq.A93, show that $i = 4.47e^{-6.2t} \sin 179t$ A.

    First we calculate $a = \frac{R}{2L}$ and $w_n = \frac{1}{\sqrt{LC}}$ to determine if the circuit is underdamped or overdamped.

    $$a = \frac{1.55}{2(0.125)} = 6.2 \qquad w_n = \frac{1}{\sqrt{(0.125)(250) \times 10^{-6}}} = 179$$

    Since $w_n > a$, we have an underdamped circuit and

    $$i = \frac{E}{w_d L} e^{-at} \sin w_d t$$

    where $w_d = \sqrt{w_n^2 - a^2} = \sqrt{179^2 - 6.2^2} = 179$

    $i = \frac{100}{179(0.125)} e^{-6.2t} \sin 179t = 4.47e^{-6.2t} \sin 179t$ A

9. For an RLC circuit, $R = 10.5$ ohms, $L = 0.125$ H, $C = 225$ microfarads, and $e = 100 \sin 175t$. Show that the steady-state current is $i = 9.03 \sin(175t + 18.5°)$.

    $$X = \omega L - \frac{1}{wC} = 175(0.125) - \frac{10^6}{175(225)} = -3.52$$
    $$Z^2 = R^2 + X^2 = 10.5^2 + (-3.52)^2$$
    $$Z = 11.07 \text{ ohms}$$
    $$\phi = \arctan \frac{X}{R} = \arctan\left(-\frac{3.52}{10.5}\right) = -18.5°$$
    $$i = \frac{E}{Z} \sin(\omega t - \phi)$$
    $$i = 9.03 \sin(175t + 18.5°)$$

13. Find the resonant frequency for the preceding circuit.

    $$\omega = \frac{1}{\sqrt{LC}} = \frac{1}{\sqrt{0.175(1.5 \times 10^{-3})}} = 61.7 \text{rad/s} = 9.84 \text{ Hz}$$

17. For the circuit in the preceding problem, find the amplitude of the steady-state current at resonance. At resonance the current will be maximum, thus

    $$i = \frac{E}{Z} \sin(\omega t - \phi) = \frac{E}{Z} = \frac{E}{R} = \frac{100}{10} = 10 \text{ A}$$

Chapter 35: Review Problems

1. $xy + y + xy' = e^x$

Substituting $\frac{dy}{dx}$ for y' and rearranging terms, yields the following first-order linear differential equation,

$$\frac{dy}{dx} + \frac{(x+1)}{x}y = \frac{e^x}{x}$$

where $P(x) = 1 + \frac{1}{x}$ and $Q(x) = \frac{e^x}{x}$

$$\int P(x)dx = \int\left(1 + \frac{1}{x}\right)dx = x + \ln x$$

$$\Rightarrow e^{x + \ln x} = xe^x$$

Substituting the integrating factor into our formula,

$$yxe^x = \int xe^x \cdot \frac{e^x}{x}dx = \frac{1}{2}\int e^{2x} \cdot 2dx = \frac{1}{2}e^{2x} + C$$

Therefore, $y = \frac{e^{2x}}{2xe^x} + \frac{C}{xe^x} \Rightarrow y = \frac{e^x}{2x} + \frac{C}{xe^x}$

5. $2y + 3x^2 + 2xy' = 0$

Substituting $\frac{dy}{dx}$ for y' and rearranging terms yields the following first-order linear differential equation,

$$\frac{dy}{dx} + \frac{y}{x} = \frac{-3x}{2}$$

where $P(x) = \frac{1}{x}$ and $Q(x) = \frac{-3x}{2}$

$$\int P(x)dx = \int \frac{1}{x}dx = \ln x \Rightarrow e^{\ln x} = x$$

Using the integrating factor with our formula,

$$yx = -3\int \frac{x^2}{2}dx = \frac{-x^3}{2} + C_1$$

$$y = \frac{-x^2}{2} + \frac{C_1}{x} \text{ or } x^3 + 2xy = C$$

9. $(1-x)\frac{dy}{dx} = y^2$

Separating variables, we obtain,

$$y^{-2}dy = \frac{dx}{1-x}$$

Integrating $\int y^{-2}dy = -\int \frac{-dx}{1-x}$

$$\Rightarrow y^{-1} = -\ln(1-x) + C_1$$

$$\frac{1}{y} = \ln|1-x| + C$$

$$y\ln|1-x| + Cy = 1$$

13. $y'\sin y = \cos x$, $(x = \frac{\pi}{4}$ when $y = 0)$

Separating variables, we obtain

$$\sin y\, dy = \cos x\, dx$$

Integrating: $\int \sin y\, dy = \int \cos x\, dx$

$$-\cos y = \sin x + C$$

Letting $x = \frac{\pi}{4}$, $y = 0 \Rightarrow C = -1.707$

Therefore: $\sin x + \cos y = 1.707$

17. A gear is rotating at 1550 rev/min. Its speed decreases exponentially at a rate of 9.50 percent per second, after the power is shut off. Find the gear's speed after 6.00 s.

This is exponential decay which obeys the formula $v = v_0 e^{-nt}$.

$$v_0 = 1550 \text{ rev/min} = 25.83 \text{ rev/s, } n = 0.0950 \text{ and } t = 6.00 \text{ s}$$

$$v = v_0 e^{-nt} = 25.83e^{-(0.095)6} = 25.83e^{-0.57} = 14.6 \text{ rev/s}$$

21. Find the equation of the curve which passes through the point $(1, 2)$ and whose slope is $y' = 2 + \frac{y}{x}$.

Substituting $\frac{dy}{dx}$ for y' and rearranging terms yields the following first-order linear differential equation,

$$\frac{dy}{dx} - \frac{y}{x} = 2 \text{, where } P(x) = -\frac{1}{x} \text{ and } Q(x) = 2$$

$$\int P(x)dx = -\int \frac{1}{x}dx = -\ln x \Rightarrow e^{-\ln x} = \frac{1}{x}$$

Using the integrating factor with our formula,

$$y \cdot \frac{1}{x} = \int \frac{1}{x}dx = 2\ln x + C \Rightarrow y = 2x\ln x + Cx$$

Letting $x = 1$, $y = 2$ gives $C = 2$.

Therefore: $y = 2x\ln x + 2x$

25. $y'' - y' - 2y = 5x$. The auxiliary equation $m^2 - m - 2 = 0$ has roots $m = 2$, $m = -1$, thus the complementary solution is:
$$y_c = C_1 e^{2x} + C_2 e^{-x}$$
The terms of $f(x)$ and their derivatives (less coefficients) are:
$$f(x) = Ax, \quad f'(x) = \text{constant}, \quad f''(x) = 0$$
So,
$$y_p = Ax + B, \quad y'_p = A, \quad y''_p = 0$$
Substituting in the DE: $y'' - y' - 2y = 5x$,
$$0 - A - 2(Ax + B) = 5x$$
$$-2Ax - (A + 2B) = 5x$$
Equating coefficients of x gives
$$-2A = 5 \quad \text{or} \quad A = \frac{-5}{2}$$
Equating constant terms gives
$$-A - 2B = 0 \quad \text{or} \quad B = \frac{5}{4}$$
So $y_p = \frac{-5}{2}x + \frac{5}{4}$
And $y = C_1 e^{2x} + C_2 e^{-x} - \frac{5}{2}x + \frac{5}{4}$

29. $y'' - 5y' = 6$; $y = y' = 1$ when $x = 0$
The auxiliary equation $m^2 - 5m = 0$ has roots $m = 0$, $m = 5$, thus the solution is $y_c = C_1 + C_2 e^{5x}$.
The right side is a constant, which duplicates a term in y_c, so
$$y_p = Ax, \quad y'_p = A, \quad y''_p = 0$$
Substituting gives $0 - 5A = 6$, or $A = -\frac{6}{5}$. Our equation is then
$$y = C_1 + C_2 e^{5x} - \left(\frac{6}{5}\right)x$$
and
$$y' = 5C_2 e^{5x} - \frac{6}{5}$$
At $x = 0$, $y = 1$, $1 = C_1 + C_2$ and $y' = 1$ when $x = 0$ so $1 = 5C_2 - \frac{6}{5}$ from which $C_2 = \frac{11}{25}$ and $C_1 = \frac{14}{25}$.
Then
$$y = \frac{14}{25} + \left(\frac{11}{25}\right)e^{5x} - \left(\frac{6}{5}\right)x$$
$$25y = 11e^{5x} - 30x + 14$$

33. $y''' - 6y'' + 11y' - 6y = 0$. The auxiliary solution: $m^3 - 6m^2 + 11m - 6 = 0$
$$(m - 2)(m - 3)(m - 1) = 0$$
$$m = 2, \quad m = 3, \quad m = 1$$
Therefore $y = C_1 e^{2x} + C_2 e^{3x} + C_3 e^x$

37. $y'' + 3y' + 2y = 0$; $y = 1$ and $y' = 2$ when $x = 0$.
The auxiliary equation $m^2 + 3m + 2 = 0$ has roots $m = -1$, $m = -2$.
Thus $y = C_1 e^{-x} + C_2 e^{-2x}$
Substituting $x = 0$, $y = 1$, yields $C_1 + C_2 = 1$ $\qquad$ (1)
Calculating y' and substituting $y' = 2$, and $x = 0$
$$y' = -C_1 e^{-x} - 2C_2 e^{-2x} \Rightarrow -C_1 - 2C_2 = 2 \qquad (2)$$
Adding equations (1) and (2)
$$3 = -C_2 \Rightarrow C_2 = -3 \text{ and } C_1 = 4$$
Therefore $y = 4e^{-x} - 3e^{-2x}$

41. $y'' + 5y' - y = 0$. The auxiliary equation $m^2 + 5m - 1 = 0$ has roots
$$m = \frac{-5 \pm \sqrt{25 + 4(1)(1)}}{2} = \frac{-5 \pm \sqrt{29}}{2} = -5.19 \text{ and } 0.192.$$
Therefore: $y = C_1 e^{-5.19x} + C_2 e^{0.192x}$

45. $y'' + 6y' + 9y = 0$. The auxiliary equation $m^2 + 6m + 9 = 0$ has roots $m = -3$, $m = -3$. Therefore
$$y = C_1 e^{-3x} + C_2 x e^{-3x}$$

49. For an RLC circuit, $R = 3.75$ ohms, $C = 15\bar{0}$ microfarad, $L = 0.100$ H, and $E = 12\bar{0}$ V. The current and charge are 0 when $t = 0$. Find i as a function of time.

First we calculate $a = \frac{R}{2L}$ and $\omega_n = \frac{1}{\sqrt{LC}}$ to determine if the circuit is underdamped or overdamped.

$$a = \frac{3.75}{2(0.1)} = 18.75 \qquad\qquad \omega_n = \frac{1}{\sqrt{(0.1)(150)\times 10^{-6}}} = 258$$

Since $w_n > a$, we have an underdamped circuit and

$$i = \frac{E}{\omega_d L}e^{-at}\sin \omega_d t$$

where $\omega_d = \sqrt{\omega_n^2 - a^2} = \sqrt{258^2 - 18.75^2} = 257.5$

$$i = \frac{120}{257.5(0.1)}e^{-18.75t}\sin 257.5t \Rightarrow i = 4.66e^{-18.75t}\sin 258t \text{ A}$$

CHAPTER 36: SOLVING DIFFERENTIAL EQUATIONS BY THE LAPLACE TRANSFORM AND BY NUMERICAL METHODS

Exercise 1: The Laplace Transform of a Function

1. By integral 8,
$$\mathcal{L}[6] = \int_0^\infty 6e^{-st}dt = -\frac{6}{s}\int_0^\infty -se^{-st}dt$$
$$= -\frac{6}{s}e^{-st}\Big|_0^\infty = \frac{6}{s}$$

5. $\mathcal{L}[\cos 5t] = \int_0^\infty \cos 5t \cdot e^{-st}dt$
$$\begin{pmatrix} u = e^{-st} & du = -se^{-st}dt \\ dv = \cos 5t & v = \frac{1}{5}\sin 5t \end{pmatrix}$$
$$= \frac{1}{5}e^{-st}\sin 5t\Big|_0^\infty + \frac{s}{5}\int_0^\infty \sin 5t \cdot e^{-st}dt$$
$$= 0 + \frac{s}{5}\int_0^\infty \sin 5t \cdot e^{-st}dt$$
$$\begin{pmatrix} u = e^{-st} & du = -se^{-st}dt \\ dv = \sin 5t & v = -\frac{1}{5}\cos 5t \end{pmatrix}$$
$$= \frac{s}{5}\left[-\frac{1}{5}e^{-st}\cos 5t\Big|_0^\infty - \frac{s}{5}\int_0^\infty \cos 5t \cdot e^{-st}dt\right]$$
$$= \frac{s}{25} - \frac{s^2}{25}\int_0^\infty \cos 5t \cdot e^{-st}dt$$
$$\left(1 + \frac{s^2}{25}\right)\int_0^\infty \cos 5t \cdot e^{-st}dt = \frac{s}{25}$$
Therefore, $\int_0^\infty \cos 5t \cdot e^{-st}dt = \frac{\frac{s}{25}}{\frac{(25+s^2)}{25}} = \frac{s}{s^2+25}$

9. By transform 9,
$$\mathcal{L}[3e^t + 2e^{-t}] = 3\mathcal{L}[e^t] + 2\mathcal{L}[e^{-t}]$$
$$= \frac{3}{s-1} + \frac{2}{s+1}$$

13. By transform 24,
$$\mathcal{L}[5e^{3t}\cos 5t] = 5\mathcal{L}[e^{3t}\cos 5t] = \frac{5(s-3)}{(s-3)^2+25}$$

17. By transforms 2 and 6,
$$\mathcal{L}[y'] + 2\mathcal{L}[y] = s\mathcal{L}[y] - y(0) + 2\mathcal{L}[y]$$
$$= s\mathcal{L}[y] - 1 + 2\mathcal{L}[y]$$

21. By transforms 2, 3, and 6,
$$\mathcal{L}[y''] + 3\mathcal{L}[y'] - \mathcal{L}[y]$$
$$= s^2\mathcal{L}[y] - sy(0) - y'(0)$$
$$\quad + 3\{s\mathcal{L}[y] - y(0)\} - \mathcal{L}[y]$$
$$= s^2\mathcal{L}[y] - s - 3 + 3s\mathcal{L}[y] - 3 - \mathcal{L}[y]$$
$$= s^2\mathcal{L}[y] + 3s\mathcal{L}[y] - \mathcal{L}[y] - s - 6$$

Exercise 2: Inverse Transforms

1. $\frac{1}{s}$: $f(t) = \mathcal{L}^{-1}[F(s)] = \mathcal{L}^{-1}\left[\frac{1}{s}\right] = 1$

5. $\frac{4}{4+s^2}$: $f(t) = \mathcal{L}^{-1}\left[\frac{4}{4+s^2}\right] = 2\mathcal{L}^{-1}\left[\frac{2}{2^2+s^2}\right]$
$$= 2\sin 2t$$

9. $\frac{(s+4)}{(s-9)^2}$: $f(t) = \mathcal{L}^{-1}\left[\frac{s}{[s-3^2]^2} + \frac{4}{[s-3^2]^2}\right]$
$$= e^{9t}(1 + 9t) + 4te^{9t}$$
$$= e^{9t}(1 + 13t)$$

13. $\frac{2s^2+1}{s(s^2+1)}$: $f(t) = \mathcal{L}^{-1}\left[\frac{2s^2}{s(s^2+1)}\right] + \mathcal{L}^{-1}\left[\frac{1}{s(s^2+1)}\right] = \mathcal{L}^{-1}\left[\frac{2s}{s^2+1}\right] + \mathcal{L}^{-1}\left[\frac{1}{s(s^2+1)}\right]$
$$= 2\cos t + 1 - \cos t = \cos t + 1$$

17. $\frac{2}{s^2+s-2}$
First we will separate using partial fractions: $\frac{2}{(s-1)(s+2)} = \frac{A}{s-1} + \frac{B}{s+2}$
$2 = A(s+2) + B(s-1)$, if $s = 1 \Rightarrow A = \frac{2}{3}$
$$\text{if } s = -2 \Rightarrow B = \frac{-2}{3}$$
Therefore: $F(s) = \frac{\frac{2}{3}}{s-1} + \frac{\frac{-2}{3}}{s+2}$
$$\mathcal{L}^{-1}[F(s)] = \frac{2}{3}\mathcal{L}^{-1}\left[\frac{1}{s-1}\right] - \frac{2}{3}\mathcal{L}^{-1}\left[\frac{1}{s+2}\right] = \frac{2}{3}e^t - \frac{2}{3}e^{-2t}$

Exercise 3: Solving Differential Equations by the Laplace Transform

1. $y' - 3y = 0, \qquad y(0) = 1$

   Taking the transform of both sides, we obtain,
   $$\mathcal{L}[y'] - 3\mathcal{L}[y] = \mathcal{L}[0]$$
   $$s\mathcal{L}[y] - y(0) - 3\mathcal{L}[y] = 0$$
   Substituting $y(0) = 1$ and solving for $\mathcal{L}[y]$
   $$(s - 3)\mathcal{L}[y] = 1 \Rightarrow \mathcal{L}[y] = \frac{1}{s-3}$$
   Taking the inverse transform: $y = e^{3t}$

5. $3y' - 2y = t^2, \qquad y(0) = 3$

   $$3\mathcal{L}[y'] - 2\mathcal{L}[y] = \mathcal{L}[t^2]$$
   $$3s\mathcal{L}[y] - 3y(0) - 2\mathcal{L}[y] = \frac{2}{s^2}$$
   Substituting 3 for $y(0)$ and solving for $\mathcal{L}[y]$
   $$(3s - 2)\mathcal{L}[y] = \frac{2}{s^3} + 9 \Rightarrow \mathcal{L}[y] = -\frac{9s^3 + 2}{s^3(3s-2)}$$
   Separating the right side of the last equation into two fractions.
   $$\frac{9s^3 + 2}{s^3(3s-2)} = \frac{9}{3s-2} + \frac{2}{s^3(3s-2)}$$
   Using the method of partial fractions on the last fraction
   $$\frac{2}{s^3(3s-2)} = \frac{A}{3s-2} + \frac{B}{s} + \frac{C}{s^2} + \frac{D}{s^3}$$
   $$2 = As^3 + Bs^2(3s - 2) + Cs(3s - 2) + D(3s - 2)$$
   $$s = \frac{2}{3} \Rightarrow A = \frac{27}{4}, \qquad s = 0 \Rightarrow D = -1$$
   Equating coefficients:
   $$2 = (A + 3B)s^3 + (-2B + 3C)s^2 + (3D - 2C)s - 2D$$
   $$\frac{27}{4} + 3B = 0 \Rightarrow B = \frac{-9}{4} \text{ and } -3 - 2C = 0 \Rightarrow C = \frac{-3}{2}$$
   Therefore:
   $$\mathcal{L}[y] = \frac{9s^3 + 2}{s^3(3s-2)} = \frac{9}{3s-2} + \frac{\frac{27}{4}}{3s-2} - \frac{\frac{9}{4}}{s} - \frac{\frac{3}{2}}{s^2} - \frac{1}{s^3} = \frac{\frac{21}{4}}{s - \frac{2}{3}} - \frac{\frac{9}{4}}{s} - \frac{\frac{3}{2}}{s^2} - \frac{1}{s^3}$$
   Taking the inverse transform of both sides, we get,
   $$y = \frac{21}{4}e^{2t/3} - \frac{9}{4} - \frac{3}{2}t - \frac{1}{2}t^2$$

9. $y'' + 2y' - 3y = 0, \quad y(0) = 0, \quad y'(0) = 2$

   $$\mathcal{L}[y''] + 2\mathcal{L}[y'] - 3\mathcal{L}[y] = \mathcal{L}[0]$$
   $$s^2\mathcal{L}[y] - sy(0) - y'(0) + 2s\mathcal{L}[y] - 2y(0) - 3\mathcal{L}[y] = 0$$
   Substituting $y(0) = 0$ and $y'(0) = 2$ and solving for $\mathcal{L}[y]$,
   $$(s^2 + 2s - 3)\mathcal{L}[y] = 2 \Rightarrow \mathcal{L}[y] = \frac{2}{(s+3)(s-1)}$$
   Using partial fractions,
   $$\frac{2}{(s+3)(s-1)} = \frac{A}{s+3} + \frac{B}{s-1} \Rightarrow 2 = A(s - 1) + B(s + 3)$$
   $$s = -3 \Rightarrow A = -\frac{1}{2} \text{ and } s = 1 \Rightarrow B = \frac{1}{2}$$
   Therefore, $\mathcal{L}[y] = \frac{-\frac{1}{2}}{s+3} + \frac{\frac{1}{2}}{s-1}$

   Taking the inverse transform:
   $$y = \frac{1}{2}e^t - \frac{1}{2}e^{-3t}$$

13. $2y'' + y = 4t,$ $y(0) = 3,$ $y'(0) = 0$

Taking the Laplace transform of both sides $2\mathcal{L}[y''] + \mathcal{L}[y] = 4\mathcal{L}[t]$

$$2s^2\mathcal{L}[y] - 2sy(0) - 2y'(0) + \mathcal{L}[y] = \frac{4}{s^2}$$

Substituting $y(0) = 3$ and $y'(0) = 0$ and solving for $\mathcal{L}[y]$

$$(2s^2 + 1)\mathcal{L}[y] = 6s + \frac{4}{s^2} \Rightarrow \mathcal{L}[y] = \frac{6s^3 + 4}{s^2(2s^2 + 1)}$$

Using method of partial fractions $\frac{6s^3 + 4}{s^2(2s^2 + 1)} = \frac{A}{s} + \frac{B}{s^2} + \frac{Cs + D}{2s^2 + 1}$

$6s^3 + 4 = s(2s^2 + 1)A + (2s^2 + 1)B + (Cs + D)s^2 = (2A + C)s^3 + (2B + D)s^2 + As + B$

Equating coefficients yields: $A = 0,\ B = 4,\ C = 6,\ D = -8$

Thus
$$\mathcal{L}[y] = \frac{4}{s^2} + \frac{6s - 8}{2s^2 + 1} = \frac{4}{s^2} + \frac{6s}{2s^2 + 1} - \frac{8}{2s^2 + 1}$$

Adjusting coefficients: $\mathcal{L}[y] = \frac{4}{s^2} + \frac{3s}{s^2 + \frac{1}{2}} - \frac{8}{\sqrt{2}}\left(\frac{\sqrt{2}/2}{(s^2 + \frac{1}{2})}\right)$

Taking the inverse transform

$$y = 4t + 3\cos\frac{\sqrt{2}}{2}t - \left(\frac{8}{\sqrt{2}}\right)\sin\frac{\sqrt{2}}{2}t$$

$$y = 4t + 3\cos 0.707t - 5.66\sin 0.707t$$

17. $3y'' + y' = \sin t,$ $y(0) = 2,$ $y'(0) = 3$

Taking the Laplace transform of both sides

$$3\mathcal{L}[y''] + \mathcal{L}[y'] = \mathcal{L}[\sin t]$$

$$3s^2\mathcal{L}[y] - 3sy(0) - 3y'(0) + s\mathcal{L}[y] - y(0) = \frac{1}{s^2 + 1}$$

Substituting $y(0) = 2$ and $y'(0) = 3$ and solving for $\mathcal{L}[y]$,

$$(3s^2 + s)\mathcal{L}[y] = \frac{1}{s^2 + 1} + 6s + 11$$

$$\mathcal{L}[y] = \frac{6s^3 + 11s^2 + 6s + 12}{s(3s + 1)(s^2 + 1)}$$

Using partial fractions and equating coefficients:

$$\frac{6s^3 + 11s^2 + 6s + 12}{s(3s + 1)(s^2 + 1)} = \frac{A}{s} + \frac{B}{3s + 1} + \frac{Cs + D}{s^2 + 1}$$

$$6s^3 + 11s^2 + 6s + 12 = A(3s + 1)(s^2 + 1) + Bs(s^2 + 1) + (Cs + D)s(3s + 1)$$

$$= (3A + B + 3C)s^3 + (A + C + 3D)s^2 + (3A + B + D)s + A$$

$A = 12,\ B = -\frac{297}{10},\ C = -\frac{1}{10}, D = -\frac{3}{10}$

Thus:
$$\mathcal{L}[y] = \frac{12}{s} - \frac{\frac{297}{10}}{3s + 1} - \frac{1}{10}\cdot\frac{s + 3}{s^2 + 1} = \frac{12}{s} - \frac{\frac{99}{10}}{s + \frac{1}{3}} - \frac{1}{10}\cdot\frac{s + 3}{s^2 + 1}$$

$$= \frac{12}{s} - 9.9\left(\frac{1}{s + \frac{1}{3}}\right) - 0.1\left(\frac{s}{s^2 + 1}\right) - 0.3\left(\frac{1}{s^2 + 1}\right)$$

$$y = 12 - 9.9e^{-t/3} - 0.1\cos t - 0.3\sin t$$

21. $y'' + 2y' + 3y = te^t$, $y(0) = 0$, $y'(0) = 0$

Taking the Laplace transform of both sides

$$\mathcal{L}[y''] + 2\mathcal{L}[y'] + 3\mathcal{L}[y] = \mathcal{L}[te^t]$$

$$s^2\mathcal{L}[y] - sy(0) - y'(0) + 2s\mathcal{L}[y] - 2y(0) + 3\mathcal{L}[y] = \frac{1}{(s-1)^2}$$

Substituting $y(0) = 0$ and $y'(0) = 0$, and solving for $\mathcal{L}[y]$

$$(s^2 + 2s + 3)\mathcal{L}[y] = \frac{1}{(s-1)^2} \Rightarrow \mathcal{L}[y] = \frac{1}{(s^2+2s+3)(s-1)^2}$$

Using partial fractions and equating coefficients:

$$\frac{1}{(s^2+2s+3)(s-1)^2} = \frac{As+B}{s^2+2s+3} + \frac{C}{s-1} + \frac{D}{(s-1)^2}$$

$$1 = (As + B)(s-1)^2 + C(s-1)(s^2 + 2s + 3) + D(s^2 + 2s + 3)$$

$$1 = (A + C)s^3 + (B + C - 2A)s^2 + (A - 2B + C + 2D)s + B - 3C + 3D$$

$A = \frac{1}{9}$, $B = \frac{1}{6}$, $C = -\frac{1}{9}$, $D = \frac{1}{6}$

Thus: $\mathcal{L}[y] = \frac{\frac{1}{9}s + \frac{1}{6}}{s^2+2s+3} - \frac{\frac{1}{9}}{s-1} + \frac{\frac{1}{6}}{(s-1)^2}$

Completing the square on 1st term, splitting the fractions and adjusting the coefficients:

$$\mathcal{L}[y] = \frac{1}{9} \cdot \frac{s + \frac{3}{2}}{(s+1)^2 + 2} - \frac{\frac{1}{9}}{s-1} + \frac{\frac{1}{6}}{(s-1)^2}$$

$$= \frac{1}{9} \cdot \frac{s+1}{(s+1)^2 + \left(\sqrt{2}\right)^2} + \frac{1}{18\sqrt{2}} \cdot \frac{\sqrt{2}}{(s+1)^2 + \left(\sqrt{2}\right)^2} - \frac{\frac{1}{9}}{s-1} + \frac{\frac{1}{6}}{(s-1)^2}$$

Taking the inverse transform:

$$y = 0.111e^{-t}\cos\sqrt{2}\,t + 0.0393e^{-t}\sin\sqrt{2}\,t - 0.111e^t + 0.167te^t$$

25. The rate of growth (bacteria/h) of a colony of bacteria is equal to 2.5 times the number present at any instant. How many bacteria are there after 24 h if there are 5000 at first?

The first statement of the problem written in mathematical form is

$$\frac{db}{dt} = 2.5b \quad \text{or} \quad b' - 2.5b = 0$$

Taking the Laplace transform of each term

$$\mathcal{L}[b'] - 2.5\mathcal{L}[b] = \mathcal{L}[0]$$

$$s\mathcal{L}[b] - b(0) - 2.5\mathcal{L}[b] = 0$$

Substituting $b(0) = 5000$ and solving for $\mathcal{L}[b]$

$$(s - 2.5)\mathcal{L}[b] = 5000 \Rightarrow \mathcal{L}[b] = \frac{5000}{s-2.5}, \text{ so } b = 5000e^{2.5t}$$

At $t = 24$ h, $b = 5000e^{2.5(24)} = 5.7 \times 10^{29}$ bacteria

Exercise 4: Electrical Applications

1. The current in a certain RC circuit satisfies the equation $172i' + 2750i = 115$. If i is zero at $t = 0$, show that $i = 41.8(1 - e^{-16t})$ mA.

Taking the Laplace transform of each term

$$172\mathcal{L}[i'] + 2750\mathcal{L}[i] = \mathcal{L}[115]$$

$$172s\mathcal{L}[i] - 172i(0) + 2750\mathcal{L}[i] = \frac{115}{s}$$

Solving for $\mathcal{L}[i]$ and using $i(0) = 0$, we obtain,

$$(172s + 2750)\mathcal{L}[i] = \frac{115}{s} \Rightarrow \mathcal{L}[i] = \frac{115}{s(172s+2750)}$$

Adjusting coefficients: $\mathcal{L}[i] = \frac{(7.19)(16)}{172s(s+16)} = 0.0418\left[\frac{16}{s(s+16)}\right]$

Taking the inverse transform:

$$i = 0.0418(1 - e^{-16t}) \text{ A} = 41.8(1 - e^{-16t}) \text{ mA}$$

5. The current in a certain RLC circuit satisfies the equation $3.15i + 0.0223i' + 1680\int_0^t i\,dt = 1$. If i is zero at $t = 0$, show that $i = 169e^{-70.5t}\sin 265t$ mA.

Transforming each term gives

$$3.15\mathcal{L}[i] + 0.0223\{s\mathcal{L}[i] - i(0)\} + \frac{1680\mathcal{L}[i]}{s} = \frac{1}{s}$$

Setting $i(0)$ to 0 and solving for $\mathcal{L}[i]$,

$$\mathcal{L}[i] = \frac{44.8}{s^2 + 141s + 75300}$$

Factoring the denominator,

$$s^2 + 141s + 75300 = (s^2 + 141s + 4970) + 75300 - 4970$$
$$= (s + 70.5)^2 - (-70300) = (s + 70.5)^2 - j^2(265)^2$$
$$= (s + 70.5 + j265)(s + 70.5 - j265)$$

after replacing -1 by j^2, and factoring the difference of two squares. So

$$\mathcal{L}[i] = \frac{44.8}{(s+70.5+j265)(s+70.5-j265)}$$
$$= \frac{A}{s+70.5+j265} + \frac{B}{s+70.5-j265}$$

Multiplying through by $s + 70.5 + j265$ gives

$$\frac{44.8}{s+70.5-j265} = A + \frac{B(s+70.5+j265)}{s+70.5-j265}$$

Setting $s = -70.5 - j265$ we get

$$A = \frac{44.8}{-70.5+j265+70.5-j265} = \frac{44.8}{-j530} = j0.0845$$

We find B by multiplying through by $s + 70.5 - j265$ and then setting $s = j265 - 70.5$, and we get $B = -j0.0845$ (work not shown). Substituting,

$$\mathcal{L}[i] = \frac{j0.0845}{s+70.5+j265} - \frac{j0.0845}{s+70.5-j265}$$

This still doesn't match any table entry. Let us now combine the two fractions over a common denominator.

$$\mathcal{L}[i] = \frac{j0.0845}{s+70.5+j265} - \frac{j0.0845}{s+70.5-j265}$$
$$= \frac{j0.0845(s+70.5-j265) - j0.0845(s+70.5+j265)}{(s+70.5+j265)(s+70.5-j265)}$$
$$= \frac{44.8}{(s+70.5)^2+(265)^2} = \left(\frac{44.8}{265}\right)\frac{265}{(s+70.5)^2+(265)^2}$$

Taking the inverse transform gives $i = 169e^{-70.5t}\sin 265t$ mA

Exercise 5: Numerical Solution of First-Order Differential Equations

NOTE: The problems in this section have been solved by using the modified Euler method. If you use a different method you may get slightly different answers.

1. $y' + xy = 3; \quad y(4) = 2.$
    Find $y(5)$.

X	Y
4.0	2.000
4.1	1.593
4.2	1.306
4.3	1.104
4.4	0.962
4.5	0.862
4.6	0.790
4.7	0.737
4.8	0.698
4.9	0.668
5.0	0.644

5. $1.94x^2y' + 8.23x \sin y - 2.99y = 0; \quad y(1) = 3$
    Find $y(2)$.

X	Y
1.0	3.000
1.1	3.468
1.2	4.104
1.3	4.851
1.4	5.548
1.5	6.093
1.6	6.489
1.7	6.775
1.8	6.985
1.9	7.142
2.0	7.261

9. $xy' + xy^2 = 8y \ln|xy|; \qquad y(1) = 3.$
Find $y(2)$.

X	Y
1.0	3.000
1.1	5.590
1.2	10.977
1.3	15.753
1.4	16.861
1.5	16.879
1.6	16.459
1.7	15.836
1.8	15.137
1.9	14.430
2.0	13.751

13.
```
10 '              RUNGE
20 '
30 ' THIS PROGRAM SOLVES A DIFFERENTIAL EQUATION
40 ' USING THE RUNGE-KUTTA METHOD
50 '
60   DEF FNM(X,Y) = (2.99*Y-8.23*X*SIN(Y))/1.94*X^2
70   W=.1:X0=1:   Y0=3
80   Y1=Y0
90   PRINT "X", "APPROX Y"
100  PRINT X0, Y0
110  FOR X1 = X0 TO 1 STEP W
120  X2 = X1+W
130  K1=FNM(X1,Y1)
140  K2=FNM(X1+W/2, Y1+K1*W/2)
150  K3=FNM(X1+W/2,Y1+K2*W/2)
160  K4=FNM(X1+W,Y1+K3*W)
170  Y2 = Y1 + W*(K1+2*K2+2*K3+K4)/6
180  PRINT X2 ,Y2
190  Y1=Y2
200  NEXT X1
```

Exercise 6: Numerical Solution of Second-Order Differential Equations

1. $y'' + y' + xy = 3; \qquad y'(1,1) = 2$

X	Y
1.00	1.0000
1.10	1.1995
1.20	1.3959
1.30	1.5863
1.40	1.7673
1.50	1.9361
1.60	2.0897
1.70	2.2254
1.80	2.3407
1.90	2.4334
2.00	2.5019

5. $xy'' - 2.4x^2y' + 5.3x \sin y - 7.4y = 0;$
$y'(1,1) = 0.1$

X	Y
1.00	1.0000
1.10	1.0262
1.20	1.0880
1.30	1.1941
1.40	1.3615
1.50	1.6202
1.60	2.0246
1.70	2.6765
1.80	3.7745
1.90	5.6952
2.00	9.0331

9. $y'' - xy' + 59.2xy = 74.1y \ln|xy|$;
   $y'(1,1) = 2$

X	Y
1.00	1.0000
1.10	0.9087
1.20	0.1899
1.30	-0.7845
1.40	-1.3143
1.50	-1.3809
1.60	-0.9255
1.70	0.2166
1.80	1.4101
1.90	2.2861
2.00	3.2673

Chapter 36: Review Problems

1. $f(t) = 2t: \; \mathcal{L}[f(t)] = 2\int_0^\infty te^{-st}dt \; \left(u = t, \; du = dt, \; dv = e^{-st}dt, \; v = -\frac{1}{s}e^{-st}\right)$

   $= 2\left[-\frac{t}{s}e^{-st}\Big|_0^\infty + \frac{1}{s}\int_0^\infty e^{-st}dt\right] = 2\left[-\frac{t}{s}e^{-st} - \frac{1}{s^2}e^{-s}t\Big|_0^\infty\right] = \frac{2}{s^2}$

5. $f(t) = 3te^{2t}: \; \mathcal{L}[f(t)] = \mathcal{L}[3te^{2t}] = \frac{3}{(s-2)^2}$

9. $y' + 3y, \qquad y(0) = 1: \; \mathcal{L}[y' + 3y] = \mathcal{L}[y'] + 3\mathcal{L}[y] = s\mathcal{L}[y] - y(0) + 3\mathcal{L}[y] = s\mathcal{L}[y] + 3\mathcal{L}[y] - 1$

   $= (s+3)\mathcal{L}[y] - 1$

13. $y'' + 3y' + 4y, \quad y(0) = 1, \quad y'(0) = 3:$

   $\mathcal{L}[y''] + 3\mathcal{L}[y'] + 4\mathcal{L}[y] = s^2\mathcal{L}[y] - y'(0) - sy(0) + 3s\mathcal{L}[y] - 3y(0) + 4\mathcal{L}[y] = (s^2 + 3s + 4)\mathcal{L}[y] - s - 6$

17. $F(s) = \frac{2s}{(s^2+5)^2} = \frac{1}{\sqrt{5}} \cdot \frac{2\sqrt{5}\,s}{\left(s^2 + \left(\sqrt{5}\right)^2\right)^2}$

   Taking the inverse transform, we obtain, $f(t) = \frac{1}{\sqrt{5}}t \sin \sqrt{5}\,t$

21. $F(s) = \frac{3s}{s^2 + 2s + 1} = \frac{3s}{(s+1)^2}$

   Taking the inverse transform of both sides of the equation, $f(t) = 3e^{-t}(1 - t)$

25. $y' - 3y = t^2,$ $\qquad y(0) = 2$
Taking the Laplace transform of each term,

$$\mathcal{L}[y'] - 3\mathcal{L}[y] = \mathcal{L}[t^2] \Rightarrow s\mathcal{L}[y] - y(0) - 3\mathcal{L}[y] = \tfrac{2}{s^3}$$

Substituting $y(0) = 2$ and solving for $\mathcal{L}[y]$

$$(s - 3)\mathcal{L}[y] = \tfrac{2}{s^3} + 2 = \tfrac{2(1+s^3)}{s^3}$$

$$\mathcal{L}[y] = \tfrac{2s^3+2}{s^3(s-3)}$$

Using partial fractions:

$$\tfrac{2s^3+2}{s^3(s-3)} = \tfrac{A}{s} + \tfrac{B}{s^2} + \tfrac{C}{s^3} + \tfrac{D}{s-3}$$

$$2s^3 + 2 = As^2(s - 3) + Bs(s - 3) + C(s - 3) + Ds^3$$

$$A = -\tfrac{2}{27}, \quad B = -\tfrac{2}{9}, \quad C = -\tfrac{2}{3}, \quad D = \tfrac{56}{27}$$

Thus $\qquad \mathcal{L}[y] = \dfrac{-\frac{2}{27}}{s} - \dfrac{\frac{2}{9}}{s^2} - \dfrac{\frac{2}{3}}{s^3} + \dfrac{\frac{56}{27}}{s-3}$

Taking the inverse transform of each term, weobtain,

$$y = -\tfrac{2}{27} - \tfrac{2}{9}t - \tfrac{1}{3}t^2 + \tfrac{56}{27}e^{3t}$$

29. $y'' + 2y = 4t,$ $\quad y(0) = 3,$ $\quad y'(0) = 0$
Taking the Laplace transform of each term,

$$\mathcal{L}[y''] + 2\mathcal{L}[y] = 4\mathcal{L}[t]$$

$$s^2\mathcal{L}[y] - y'(0) - sy(0) + 2\mathcal{L}[y] = \tfrac{4}{s^2}$$

Substituting $y(0) = 3$ and $y'(0) = 0$ and solving for $\mathcal{L}[y]$,

$$(s^2+2)\mathcal{L}[y] = \tfrac{4}{s^2} + 3s = \tfrac{4+3s^3}{s^2} \Rightarrow \mathcal{L}[y] = \tfrac{3s^3+4}{s^2(s^2+2)}$$

Using partial fractions,

$$\tfrac{3s^3+4}{s^2(s^2+2)} = \tfrac{A}{s} + \tfrac{B}{s^2} + \tfrac{Cs+D}{s^2+2}$$

$$3s^3 + 4 = As(s^2 + 2) + B(s^2 + 2) + (Cs + D)$$

$$= (A + C)s^3 + (B + D)s^2 + 2As + 2B$$

$$A = 0, \quad B = 2, \quad C = 3, \quad D = -2$$

$$\tfrac{3s^3+4}{s^2(s^2+2)} = \tfrac{2}{s^2} + \tfrac{3s-2}{s^2+2} = \tfrac{2}{s^2} + \tfrac{3s}{s^2+\left(\sqrt{2}\right)^2} - \tfrac{2}{\sqrt{2}}\left(\tfrac{\sqrt{2}}{s^2+\left(\sqrt{2}\right)^2}\right)$$

Taking the inverse transform of each term, we obtain,

$$y = 2t + 3\cos\sqrt{2}\,t - \sqrt{2}\sin\sqrt{2}\,t$$

33. The current in a certain RC circuit satisfies the equation $8.24i' + 149i = 100.$ If i is a zero at $t = 0$, write an equation for i. Taking the Laplace transform of each term

$$8.24\mathcal{L}[i'] + 149\mathcal{L}[i] = \mathcal{L}[100]$$

$$8.24s\mathcal{L}[i] - 8.24i(0) + 149\mathcal{L}[i] = \tfrac{100}{s}$$

Substituting $i(0) = 0$ and solving for $\mathcal{L}[i]$

$$(8.24s + 149)\mathcal{L}[i] = \tfrac{100}{s} \Rightarrow \mathcal{L}[i] = \tfrac{100}{s(8.24s+149)}$$

Using partial fractions, $\qquad \tfrac{100}{s(8.24s+149)} = \tfrac{A}{s} + \tfrac{B}{8.24s+149}$

$$100 = A(8.24s + 149) + Bs \Rightarrow A = \tfrac{100}{149}, B = \tfrac{-824}{149}$$

Thus, $\qquad \mathcal{L}[i] = \dfrac{\frac{100}{149}}{s} - \dfrac{\frac{824}{149}}{8.24s+149} = \tfrac{0.671}{s} - \tfrac{0.671}{s+18.08}$

Taking the inverse transform of each term,

$$i = 0.671 - 0.671e^{-18.1t} \text{ A} = 0.671(1 - e^{-18.1t}) \text{ A}$$

37. $y' + xy = 4y;$ $\qquad y(2) = 2.$
Find $y(3)$.

X	Y
2.0	2.000
2.1	2.428
2.2	2.919
2.3	3.474
2.4	4.095
2.5	4.778
2.6	5.521
2.7	6.317
2.8	7.156
2.9	8.026
3.0	8.913

41. $y'' + y'\ln y + 5.27(x - 2y)^2 = 2.84x;$
$y'(1,1) = 3.$ $\qquad$ Find $y(2)$.

X	Y
1.00	1.0000
1.10	1.2774
1.20	1.4701
1.30	1.5363
1.40	1.4779
1.50	1.3371
1.60	1.1698
1.70	1.0199
1.80	0.9111
1.90	0.8520
2.00	0.8444

45. $7.35y' + 2.85x \sin y - 7.34x = 0;$
$y(2) = 4.3.$ $\qquad$ Find $y(3)$.

X	Y
2.0	4.300
2.1	4.581
2.2	4.878
2.3	5.184
2.4	5.492
2.5	5.792
2.6	6.079
2.7	6.350
2.8	6.604
2.9	6.841
3.0	7.065

CHAPTER 37: INFINITE SERIES

Exercise 1: Convergence and Divergence of Infinite Series

1. The geometric series $9 + 3 + 1 + \cdots$
   Here, $a = 9$, $r = \frac{1}{3}$. The sum to infinity is
   $$S = \frac{a}{1-r} = \frac{9}{1-\frac{1}{3}} = \frac{9}{\frac{2}{3}} = \frac{27}{2}$$
   Since the limit of the sum is finite, the series converges.

5. The geometric series $16 + 8 + 4 + 2 + \cdots$
   The general term is $u_n = 16\left(\frac{1}{2}\right)^{n-1}$. Taking the limit as n approaches infinity,
   $$\lim_{n\to\infty} u_n = \lim_{n\to\infty} 16\left(\frac{1}{2}\right)^{n-1} = 0$$
   Since the limit of the nth term approaches zero, we cannot tell if the series converges or diverges.

9. $\frac{1}{2!} + \frac{2}{3!} + \frac{3}{4!} + \cdots$ The general term is $\frac{n}{(n+1)!}$.
   Using the ratio test, we obtain,
   $$\lim_{n\to\infty}\left|\frac{u_{n+1}}{u}\right| = \lim_{n\to\infty}\left|\frac{n+1}{(n+2)!} \cdot \frac{(n+1)!}{n}\right|$$
   $$= \lim_{n\to\infty}\left|\frac{n+1}{n^2+2n}\right| = \lim_{n\to\infty}\left|\frac{\frac{1}{n}+\frac{1}{n^2}}{1+\frac{2}{n}}\right| = 0$$
   A limit less than 1 tells that the series converges.

13. $1 + \frac{1}{2} + \frac{1}{3} + \frac{1}{4} + \cdots$ The general term is $\frac{1}{n}$.
    Using the ratio test, we obtain,
    $$\lim_{n\to\infty}\left|\frac{u_{n+1}}{u_n}\right| = \lim_{n\to\infty}\left|\frac{1}{n+1} \cdot \frac{n}{1}\right|$$
    $$= \lim_{n\to\infty}\left|\frac{1}{1+\frac{1}{n}}\right| = 1$$
    Test fails.

Exercise 2: Maclaurin Series

1. $1 + x + x^2 + \cdots + x^{n-1} + \cdots$
   $$\lim_{n\to\infty}\left|\frac{u_{n+1}}{u_n}\right| = \lim_{n\to\infty}\left|\frac{x^n}{x^{n-1}}\right| = \lim_{n\to\infty}|x| = |x|$$
   $|x| < 1$ implies convergence for x's in the range $-1 < x < 1$.

5. $x - \frac{x^3}{3} + \frac{x^5}{5} - \frac{x^7}{7} + \cdots$
   Using the ratio test on the general term of
   $$\frac{(-1)^{n+1}x^{2n-1}}{2n-1},$$
   $$\lim_{n\to\infty}\left|\frac{u_{n+1}}{u_n}\right| = \lim_{n\to\infty}\left|\frac{x^{2n+1}}{2n+1} \cdot \frac{2n-1}{x^{2n-1}}\right|$$
   $$= \lim_{n\to\infty}\left|\frac{x^2(2n-1)}{2n+1}\right|$$
   $$= \lim_{n\to\infty}\left|\frac{x^2\left(2-\frac{1}{n}\right)}{2+\frac{1}{n}}\right| = |x^2|$$
   $x^2 < 1$ indicates convergence for x's in the range $-1 < x < 1$.

9. $e^x = 1 + x + \frac{x^2}{2!} + \frac{x^3}{3!}\cdots$
   Taking successive derivatives and evaluating each at $x = 0$,
   $$\begin{array}{ll} f(x) = e^x & f(0) = e^0 = 1 \\ f'(x) = e^x & f'(0) = e^0 = 1 \\ f''(x) = e^x & f''(0) = e^0 = 1 \\ f'''(x) = e^x & f'''(0) = e^0 = 1 \end{array}$$
   Substituting into the Maclaurin Series gives
   $$f(x) = 1 + x + \frac{x^2}{2!} + \frac{x^3}{3!} + \cdots + \frac{x^{n-1}}{(n-1)!} + \cdots$$

13. $\ln(1+x) = x - \frac{x^2}{2} + \frac{x^3}{3} - \frac{x^4}{4} + \cdots$
    Taking successive derivatives and evaluating at $x = 0$,
    $$\begin{array}{ll} f(x) = \ln(1+x) & f(0) = 0 \\ f'(x) = \frac{1}{(1+x)} & f'(0) = 1 \\ f''(x) = -\frac{1}{(1+x)^2} & f''(0) = -1 \\ f'''(x) = \frac{2}{(1+x)^3} & f'''(0) = 2 \\ f^{(IV)}(x) = -\frac{6}{(1+x)^4} & f^{(IV)}(0) = -6 \end{array}$$
    Substituting into the Maclaurin Series,
    $$\ln(1+x) = x - \frac{x^2}{2} + \frac{x^3}{3} - \frac{x^4}{4} + \cdots$$
    $$+ \frac{(-1)^{n-1}x^n}{n} + \cdots$$

17. $\sec x = 1 + \frac{x^2}{2} + \frac{5x^4}{24} + \cdots$

Taking successive derivatives and evaluating at $x = 0$,

$f(x) = \sec x$ $f(0) = 1$

$f'(x) = \sec x \cdot \tan x$ $f'(0) = 0$

$f''(x) = \sec x(1 + 2\tan^2 x)$ $f''(0) = 1$

$f'''(x) = \sec x \cdot \tan x(5 + 6\tan^2 x)$ $f'''(0) = 0$

$f^{(IV)}(x) = \sec x(28\tan^2 x + 24\tan^4 x + 5)$

                         $f^{(IV)}(0) = 5$

Substituting into the Maclaurin series gives

$$\sec x = 1 + \frac{x^2}{2} + \frac{5x^4}{24} + \cdots$$

21. $\sqrt{e}$ Using three terms of the series,

$e^x = 1 + x + \frac{x^2}{2!} + \frac{x^3}{3!} + \cdots$, with $x = \frac{1}{2}$,

we obtain,

$\sqrt{e} = e^{1/2} \simeq 1 + \frac{1}{2} + \frac{1/4}{2} = 1 + \frac{1}{2} + \frac{1}{8}$

    $= \frac{13}{8} = 1.625$

25. $\sqrt{1.1}$ Using three terms of the series,

$\sqrt{1 + x} = 1 + \frac{x}{2} - \frac{x^2}{8} + \frac{x^3}{16} + \cdots$, with $x = 0.1$,

we obtain,

$\sqrt{1.1} \simeq 1 + \frac{0.1}{2} - \frac{0.01}{8} = \frac{8 + 0.4 - 0.01}{8}$

    $= \frac{8.39}{8} = 1.049$

Exercise 3: Taylor Series

1. For $a = 1$ $\frac{1}{x} = 1 - (x - 1) + (x - 1)^2$

        $- (x - 1)^3 + \cdots$

Taking successive derivatives and evaluating each at $x = 1$

$f(x) = \frac{1}{x}$ $f'(1) = 1$

$f'(x) = -\frac{1}{x^2}$ $f'(1) = -1$

$f''(x) = \frac{2}{x^3}$ $f''(1) = 2$

$f'''(x) = -\frac{6}{x^4}$ $f'''(1) = -6$

Substituting into the Taylor Series gives

$f(x) = \frac{1}{x} = 1 - (x - 1) + (x - 1)^2$

        $- (x - 1)^3 + \cdots$

5. For $a = \frac{\pi}{3}$ $\sec x = 2 + 2\sqrt{3}\left(x - \frac{\pi}{3}\right)$

        $+ 7\left(x - \frac{\pi}{3}\right)^2 + \cdots$

Taking successive derivatives and evaluating each at $x = \frac{\pi}{3}$

$f(x) = \sec x$ $f\left(\frac{\pi}{3}\right) = 2$

$f'(x) = \sec x \cdot \tan x$ $f'\left(\frac{\pi}{3}\right) = 2\sqrt{3}$

$f''(x) = \sec x(1 + 2\tan^2 x)$ $f''\left(\frac{\pi}{3}\right) = 14$

Substituting into the Taylor Series,

$\sec x = 2 + 2\sqrt{3}\left(x - \frac{\pi}{3}\right) + 7\left(x - \frac{\pi}{3}\right)^2 \cdots$

9. For $a = -1$

$$e^{-x} = e\left[1 - (x + 1) + \frac{(x+1)^2}{2!} + \cdots\right]$$

Taking successive derivatives and evaluating each at $x = -1$

$f(x) = e^{-x}$ $f(-1) = e$

$f'(x) = -e^{-x}$ $f'(-1) = -e$

$f''(x) = e^{-x}$ $f''(-1) = e$

Substituting into the Taylor Series,

$$e^{-x} = e\left[1 - (x - 1) + \frac{(x-1)^2}{2!} + \cdots\right]$$

13. $\cos 46.5°$ (Use $\cos 45° = 0.707107$)

Using three terms of the series,

$$\cos x = \frac{\sqrt{2}}{2}\left[1 - \left(x - \frac{\pi}{4}\right) - \frac{\left(x - \frac{\pi}{4}\right)^2}{2} + \cdots\right]$$

with $x = 46.5° \cdot \frac{\pi}{180} = 0.811578$, we obtain,

$\cos 46.5° \cong \frac{\sqrt{2}}{2}\left[1 - \left(0.811578 - \frac{\pi}{4}\right)\right.$

        $\left. - \frac{(0.811578 - \frac{\pi}{4})^2}{2}\right] \cong 0.688353$

17. $\sqrt[3]{-0.8}$ $\sqrt[3]{x} = -1 + \frac{x+1}{3} + \frac{(x+1)^2}{9} + \cdots$

with $x = -0.8$, we obtain,

$\sqrt[3]{-0.8} \cong -1 + \frac{-0.8+1}{3} + \frac{(-0.8+1)^2}{9}$

    $= -0.92889$

Exercise 4: Operations with Power Series

1. $e^{-x} = 1 - x + \frac{x^2}{2!} - \frac{x^3}{3!} + \cdots$

     $e^x = 1 + x + \frac{x^2}{2!} + \frac{x^3}{3!} + \cdots$

Replacing x with $-x$, we obtain,

$$e^{-x} = 1 + (-x) + \frac{(-x)^2}{2!} + \frac{(-x)^3}{3!} + \cdots$$

or

$$e^{-x} = 1 - x + \frac{x^2}{2!} - \frac{x^3}{3} + \cdots$$

5. $\sin x^2 = x^2 - \frac{x^6}{3!} + \frac{x^{10}}{5!} - \cdots$

Replacing the x with x^2, we obtain,

$$\sin x^2 = x^2 - \frac{x^6}{3!} + \frac{x^{10}}{5!} + \cdots$$

9. $\cosh x = \frac{(e^x + e^{-x})}{2} = 1 + \frac{x^2}{2!} + \frac{x^4}{4!} + \frac{x^6}{6!} + \cdots$

Since $\frac{e^x}{2} = \frac{1}{2} + \frac{x}{2} + \frac{x^2}{2\cdot2} + \frac{x^3}{2\cdot3} + \cdots$ and

$\frac{e^{-x}}{2} = \frac{1}{2} - \frac{x}{2} + \frac{x^2}{2\cdot2!} - \frac{x^3}{2\cdot3!} + \cdots$

we obtain, by adding, the following series,

$\cosh x = \frac{e^x}{2} + \frac{e^{-x}}{2} = 1 + \frac{x^2}{2!} + \frac{x^4}{4!} + \cdots$

13. $xe^{-2x} = x - 2x^2 + 2x^3 - \cdots$ The series for e^x is given by,

$$e^x = 1 + x + \frac{x^2}{2!} + \frac{x^3}{3!} + \cdots$$

Replacing x with $-2x$, we obtain,

$$e^{-2x} = 1 - 2x + 2x^2 - \frac{4x^3}{3} + \cdots$$

Multiplying by x yields,

$$xe^{-2x} = x - 2x^2 + 2x^3 - \frac{4x^4}{3} + \cdots$$

17. $\frac{e^x - 1}{x} = 1 + \frac{x}{2!} + \frac{x^2}{3!} + \frac{x^3}{4!} + \cdots$

The series for e^x is given by,

$$e^x = 1 + x + \frac{x^2}{2!} + \frac{x^3}{3!} + \cdots$$

Therefore, $e^x - 1 = x + \frac{x^2}{2!} + \frac{x^3}{3!} + \cdots$

Dividing each term by x,

$$\frac{e^x - 1}{x} = 1 + \frac{x}{2!} + \frac{x^2}{3!} + \frac{x^3}{4!} + \cdots$$

21. Find a series for $\cos x$ by differentiating the series for $\sin x$. The series for $\sin x$ is given by,

$$\sin x = x - \frac{x^3}{3!} + \frac{x^5}{5!} - \frac{x^7}{7!} + \cdots$$

Taking the derivative of both sides,

$$\cos x = 1 - \frac{x^2}{2!} + \frac{x^4}{4!} - \frac{x^6}{6!} + \cdots$$

25. Find the series for $\ln(1-x)$ by integrating the series for $\frac{1}{(1-x)}$. The series for $\frac{1}{(1-x)}$ is given by,

$\frac{1}{1-x} = 1 + x + x^2 + x^3 + \cdots$ (long division)

Integrating each term,

$\ln(1-x) = \int \frac{1}{1-x}dx = -\int \frac{-1}{1-x}dx$

$= -\left[x + \frac{x^2}{2} + \frac{x^3}{3} + \frac{x^4}{4} + \cdots\right]$

29. $\int_0^{1/4} e^x \ln(x+1)dx$

The series for e^x and $\ln(x+1)$ are

$$e^x = 1 + x + \frac{x^2}{2!} + \cdots$$

and $\ln(x+1) = x - \frac{x^2}{2} + \frac{x^3}{3} - \cdots$

Multiplying the two series together term by term, $e_x \ln(x+1) = x + \frac{x^2}{2} + \frac{x^3}{3} + \frac{x^4}{12} \cdots$

Integrating the first three terms from $x = 0$ to $x = \frac{1}{4}$

$$\int_0^{1/4} e^x \ln(x+1) = \frac{x^2}{2} + \frac{x^3}{6} + \frac{x^4}{12}\Big|_0^{1/4}$$

$$= \frac{\frac{1}{16}}{2} + \frac{\frac{1}{64}}{6} + \frac{\frac{1}{256}}{12} = 0.0342$$

Exercise 5: Fourier Series

1. Write seven terms of the Fourier series given.

$a_0 = 4$	$a_1 = 3$	$a_2 = 2$	$a_3 = 1$
	$b_1 = 4$	$b_2 = 3$	$b_3 = 2$

A Fourier series is of the form:

$$f(x) = \frac{a_0}{2} + a_1 \cos x + a_2 \cos 2x$$
$$+ \cdots + a_n \cos nx + \cdots$$
$$+ b_1 \sin x + b_2 \sin 2x$$
$$+ \cdots + b_n \sin nx + \cdots$$

Substituting the given coefficients, we obtain,

$$f(x) = 2 + 3\cos x + 2\cos 2x + \cos 3x$$
$$+ 4\sin x + 3\sin 2x + 2\sin 3x$$

5. Verify series Number 6. The given function is $f(x) = \frac{x}{\pi}$ for $0 \le x \le 2\pi$. Solving for a_0:

$$a_0 = \frac{1}{\pi}\int_0^{2\pi} \frac{1}{\pi}x\,dx = \frac{1}{\pi^2}\cdot\frac{x^2}{2}\Big|_0^{2\pi} = \frac{4\pi^2}{2\pi^2} = 2$$

Solving for a_n:

$$a_n = \frac{1}{\pi}\int_0^{2\pi} f(x)\cos nx\,dx = \frac{1}{\pi}\int_0^{2\pi} \frac{x}{\pi}\cos nx\,dx$$

$$= \frac{1}{\pi^2}\int_0^{2\pi} x\cos nx\,dx$$

Integrating by parts with:

$$u = x, \qquad\qquad dv = \cos nx\,dx$$
$$du = dx, \qquad\qquad v = \frac{1}{n}\sin nx$$

$$a_n = \frac{1}{\pi^2}\left[\frac{x}{n}\sin nx + \frac{1}{n^2}\cos nx\right]_0^{2\pi} = 0$$

Thus we have no cosine terms. Solving for b_n:

$$b_n = \frac{1}{\pi}\int_0^{2\pi} f(x)\sin nx\,dx = \frac{1}{\pi}\int_0^{2\pi} \frac{x}{\pi}\sin nx\,dx$$

$$= \frac{1}{\pi^2}\int_0^{2\pi} x\sin nx\,dx$$

Integrating by parts with:

$$u = x, \qquad\qquad dv = \sin(nx)\,dx$$
$$du = dx, \qquad\qquad v = -\frac{1}{n}\cos nx$$

$$b_n = \frac{1}{\pi^2}\left[-\frac{x}{n}\cos nx + \frac{1}{n^2}\sin nx\right]\Big|_0^{2\pi}$$

Substituting $n = 1, 2, \cdots$ gives

$$b_1 = -\frac{2}{\pi} \qquad b_2 = -\frac{2}{2\pi} \qquad b_3 = -\frac{2}{3\pi}$$

Substituting into the Fourier Series,

$$f(x) = 1 - \frac{2}{\pi}\left[\sin x + \frac{1}{2}\sin 2x + \frac{1}{3}\sin 3x + \cdots\right]$$

9. Verify series Number 10.

The half-wave function over one cycle is given by,

$$f(x) = \sin x \qquad 0 \le x \le \pi,$$
$$f(x) = 0, \qquad \pi \le x \le 2\pi$$

Solving for a_0:

$$a_0 = \frac{1}{\pi}\int_0^{2\pi} f(x)dx = \frac{1}{\pi}\int_0^{\pi}\sin x\,dx + \frac{1}{\pi}\int_{\pi}^{2\pi} 0\,dx = -\frac{1}{\pi}\cos x\Big|_0^{\pi} = \frac{2}{\pi}$$

Solving for a_1:

$$a_1 = \frac{1}{\pi}\int_0^{\pi}\sin x\cos x\,dx = \frac{1}{2\pi}\int_0^{\pi}\sin 2x\,dx = \frac{1}{4\pi}(-\cos 2x)\Big|_0^{\pi} = 0$$

Solving for a_n: $(n > 1)$

$$a_n = \frac{1}{\pi}\int_0^{2\pi} f(x)\cos(nx)dx = \frac{1}{\pi}\int_0^{\pi}\sin x\cos nx\,dx = \frac{1}{2\pi}\int_0^{\pi}(\sin(1+n)x + \sin(1-n)x)dx$$

$$= \frac{1}{2\pi}\left[\frac{-\cos(1+n)x}{1+n} - \frac{\cos(1-n)x}{1-n}\right]_0^{\pi} = -\frac{1}{2\pi}\left[\frac{\cos(1+n)x}{1+n} - \frac{\cos(n-1)x}{n-1}\right]_0^{\pi}$$

For $n = 3, 5, 7, \cdots, a_n = 0$ and $a_2 = -\frac{2}{3\pi}$, $a_4 = -\frac{2}{15\pi}$, $a_6 = -\frac{2}{35\pi}$.

Solving for b_1 :

$$b_1 = \frac{1}{\pi}\int \sin x\ \sin x\,dx = \frac{1}{2\pi}\int(1 - \cos 2x)dx$$
$$= \frac{1}{2\pi}\left[x - \frac{1}{2}\sin 2x\right]_0^{\pi} = \frac{1}{2}$$

Solving for b_n: $(n > 1)$

$$b_n = \frac{1}{\pi}\int_0^{2\pi} f(x)\sin(nx)dx = \frac{1}{\pi}\int_0^{\pi}\sin x\sin(nx)dx = \frac{1}{2\pi}\int_0^{\pi}[\cos(1-n)x - \cos(1+n)x]dx$$

$$= \frac{1}{2\pi}\left[\frac{\sin(1-n)x}{1-n} - \frac{\sin(1+n)x}{1+n}\right]\Big|_0^{\pi} = 0$$

Substituting into the Fourier series, we obtain,

$$f(x) = \frac{1}{\pi} + \frac{1}{2}\sin x - \frac{2}{3\pi}\cos 2x - \frac{2}{15\pi}\cos 4x - \frac{2}{35\pi}\cos 6x + \cdots$$
$$f(x) = \frac{1}{\pi}\left(1 + \frac{\pi}{2}\sin x - \frac{2}{3}\cos 2x - \frac{2}{15}\cos 4x - \frac{2}{35}\cos 6x + \cdots\right)$$

Exercise 6: Waveform Symmetries

1. Fig. 37-9   a. Neither
5. Fig. 37-9   e. Neither
9. Fig. 37-9   c. no*

*but could have half-wave symmetry if x axis were shifted upward.

13. Verify series Number 2. From the picture we determine that the square wave is an even function, has half-wave symmetry, and has no constant terms. Thus, we are looking for a function that has only odd harmonic cosine terms. The function is defined below,

$$f(x) = 1 \ \text{ for } -\frac{\pi}{2} \le x \le \frac{\pi}{2}$$
$$f(x) = -1 \ \text{ for } \frac{\pi}{2} \le x \le \frac{3\pi}{2}$$

Solving for a_n:

$$a_n = \frac{1}{\pi}\int_{-\pi}^{\pi} f(x)\cos nx\,dx = \frac{1}{\pi}\int_{-\pi/2}^{\pi/2}\cos nx\,dx$$
$$- \frac{1}{\pi}\int_{\pi/2}^{3\pi/2}\cos nx\,dx$$

$$= \frac{1}{n\pi}\sin(nx)\Big|_{-\pi/2}^{\pi/2} - \frac{1}{n\pi}\sin(nx)\Big|_{\pi/2}^{3\pi/2}$$

$$= \frac{3}{n\pi}\sin\left(\frac{\pi}{2}n\right) - \frac{1}{n\pi}\sin\left(\frac{3\pi}{2}n\right)$$

From which $a_1 = \frac{4}{\pi}$, $a_3 = -\frac{4}{3\pi}$, $a_5 = \frac{4}{5\pi}$,
$a_7 = -\frac{4}{7\pi}$, $\cdots$

Thus the Fourier Series is

$$f(x) = \frac{4}{\pi}\left(\cos x - \frac{1}{3}\cos 3x + \frac{1}{5}\cos 5x\right.$$
$$\left. - \frac{1}{7}\cos 7x + \cdots\right)$$

Exercise 7: Waveforms with Period of 2L

1. Verify waveform (a), Figure 37-12. If the x axis is shifted up $\frac{1}{2}$ units, we see that the function is odd and has half-wave symmetry. Thus, we will have only sine terms with odd harmonics. The function $f(x)$ is defined as follows:

$$f(x) = -\tfrac{1}{2} \text{ for } -1 \le x \le 0$$
$$f(x) = \tfrac{1}{2} \text{ for } 0 \le x \le 1$$

Solving for b_n:

$$b_n = \tfrac{1}{L}\int_{-L}^{L} f(x)\sin\left(\tfrac{n\pi x}{L}\right)dx = -\tfrac{1}{2}\int_{-1}^{0}\sin(n\pi x)dx + \tfrac{1}{2}\int_{0}^{1}\sin(n\pi x)dx$$
$$= \left.\tfrac{\cos n\pi x}{2n\pi}\right|_{-1}^{0} - \left.\tfrac{\cos n\pi x}{2n\pi}\right|_{0}^{1} = \tfrac{1}{n\pi} - \tfrac{\cos n\pi}{n\pi}$$
$$n = 1, b_1 = \tfrac{2}{\pi}; \; n = 3, b_3 = \tfrac{2}{3\pi};$$
$$n = 5, b_5 = \tfrac{2}{5\pi}$$

Substituting into the formula for waveforms with period $2L$, remembering to add a constant term of $\frac{1}{2}$, we obtain,

$$f(x) = \tfrac{1}{2} + \tfrac{2}{\pi}\sin \pi x + \tfrac{2}{3\pi}\sin 3\pi x + \tfrac{2}{5\pi}\sin 5\pi x + \cdots$$
$$= \tfrac{1}{2} + \tfrac{2}{\pi}\left(\sin \pi x + \tfrac{1}{3}\sin 3\pi x + \tfrac{1}{5}\sin 5\pi x + \cdots\right)$$

5. Verify series (e) in Figure 37-12. From the picture we see that the function is even and therefore has only cosine terms. The equation of the waveform from -1 to 1 is $f(x) = x^2$. Solving for a_0:

$$a_0 = \int_{-1}^{1} x^2 dx = \left.\tfrac{x^3}{3}\right|_{-1}^{1} = \tfrac{2}{3}$$

Integrating by parts twice, we solve for a_n:

$$a_n = \int_{-1}^{1} x^2\cos n\pi x\, dx = \left.\tfrac{x^2}{n\pi}\sin(n\pi x)\right|_{-1}^{1} - \tfrac{2}{n\pi}\int_{-1}^{1} x\sin(n\pi x)dx$$

$$u = x^2 \qquad\qquad dv = \cos n\pi x\, dx$$
$$du = 2x\, dx \qquad v = \left(\tfrac{1}{n\pi}\right)\sin(n\pi x)$$
$$u = x \qquad\qquad dv = \sin(n\pi x)dx$$
$$du = dx \qquad\qquad v = \left(-\tfrac{1}{n\pi}\right)\cos n\pi x$$

$$a_n = \left.\tfrac{x^2}{n\pi}\sin(n\pi x)\right|_{-1}^{1} - \tfrac{2}{n\pi}\left[\left.-\tfrac{x}{n\pi}\cos n\pi x\right|_{-1}^{1} + \tfrac{1}{n\pi}\int_{-1}^{1}\cos n\pi x\, dx\right]$$
$$= \left[\tfrac{x^2}{n\pi}\sin(n\pi x) + \tfrac{2x}{n^2\pi^2}\cos(n\pi x) - \tfrac{2}{n^3\pi^3}\sin(n\pi x)\right]_{-1}^{1} = \tfrac{4}{n^2\pi^2}\cos n\pi$$

From which we determine:

$$a_1 = -\tfrac{4}{\pi^2}, \; a_2 = \tfrac{4}{4\pi^2}, \; a_3 = -\tfrac{4}{9\pi^2}, \; a_4 = \tfrac{4}{16\pi^2}$$

Substituting into the formula for waveforms with period $2L$,

$$f(x) = \tfrac{1}{3} - \tfrac{4}{\pi^2}\cos \pi x + \tfrac{4}{4\pi^2}\cos 2\pi x - \tfrac{4}{9\pi^2}\cos 3\pi x + \tfrac{4}{16\pi^2}\cos 4\pi x + \cdots$$
$$= \tfrac{1}{3} - \tfrac{4}{\pi^2}\left(\cos \pi x - \tfrac{1}{4}\cos 2\pi x + \tfrac{1}{9}\cos3\pi x - \tfrac{1}{16}\cos 4\pi x + \cdots\right)$$

Exercise 8: A Numerical Method for Finding Fourier Series

Note: The calculations in this section were done by computer.

1.

x	0°	20°	40°	60°	80°	100°	120°	140°	160°	180°
y	0	2.1	4.2	4.5	7.0	10.1	14.3	14.8	13.9	0

x	y	$y\sin x$	$y\cos x$	$y\sin 3x$	$y\cos 3x$	$y\sin 5x$	$y\cos 5x$
0	0.0	0.00	0.00	0.00	0.00	0.00	0.00
20	2.1	0.72	1.97	1.82	1.05	2.07	−0.36
40	4.2	2.70	3.22	3.64	−2.10	−1.43	−3.95
60	4.5	3.90	2.25	0.01	−4.50	−3.90	2.24
80	7.0	6.89	1.22	−6.05	−3.51	4.48	5.38
100	10.1	9.95	−1.75	−8.76	5.03	6.53	−7.71
120	14.3	12.39	−7.14	−0.05	14.30	−12.35	−7.22
140	14.8	9.53	−11.33	12.79	7.45	−5.15	13.88
160	13.9	4.77	−13.05	12.07	−6.90	13.67	2.51
180	0.0	0.00	0.00	0.00	0.00	0.00	0.00
SUMS		50.84	−24.60	15.46	10.82	3.92	4.77
y-Avg		5.08	−2.46	1.55	1.08	0.39	0.48

Fourier Coefficients					
B1	A1	B3	A3	B5	A5
10.17	−4.92	3.09	2.16	0.78	0.95

Thus, $y = -4.92\cos x + 2.16\cos 3x + 0.95\cos 5x$
$\qquad + 10.17\sin x + 3.09\sin 3x + 0.78\sin 5x$

5.
```
100    DIM X(20),Y(20),XD(20)
110    ' THIS PROGRAM WILL COMPUTE THE COEFFICIENTS OF THE
120    ' FIRST 6 TERMS OF A FOURIER SERIES FOR A GIVEN SET
130    ' OF EXPERIMENTAL DATA.
140    '
150    ' THE PROGRAM ASSUMES HALF-WAVE SYMMETRY, A PERIOD
160    ' OF TWO PI, STARTING AT X = 0
170    '
180    INPUT "THE STEP SIZE BETWEEN X VALUES";D
200    '
210    ' YOU WILL NOW BE ASKED TO INPUT THE Y COORDINATES
220    '
230    FOR Z = 0 TO 180/D
240    INPUT Y(Z+1)
250    K=K+1
260    NEXT Z
270    '
280    X(1) = 0: XD(1) = 0: D1 = D*3.14/180
290    '
300    FOR Z = 1 TO K
310    XD(Z+1) = XD(Z) + D
320    X(Z+1) = X(Z) + D1
330    NEXT Z
340    PRINT "X Y YSINX YCOSX YSIN3X YCOS3X YSIN5X";
350    PRINT " YCOS5X"
360    '
370    FOR Z = 1 TO K
380         S = Y(Z)*SIN(X(Z))
390             STOTAL = STOTAL + S          (cont'd)
```

```
400              C = Y(Z)*COS(X(Z))
410                  CTOTAL = CTOTAL + c
420              S3 = Y(Z)*SIN(3*X(Z))
430                  S3TOTAL = S3TOTAL + S3
440              C3 = Y(Z)*COS(3*X(Z))
450                  C3TOTAL = CSTOTAL + C3
460              S7 = Y(Z)*SIN(7*X(Z))
480                  S7TOTAL = S7TOTAL + S7
490              C7 = Y(Z)*COS(7*X(Z))
500                  C7TOTAL = C7TOTAL + C7
510
520              PRINT USING "### ##.# ###.##";XD(Z),Y(Z),S;
530           PRINT USING " ###.## ###.## ###.##";C,S3,C3;
540           PRINT USING " ###.## ###.##";S5,C5
560      NEXT Z
570      '
580      PRINT "-------------------------"
590      '
600      PRINT "SUMS";
610      PRINT USING "     ###.## ###.##";STOTAL,CTOTAL;
620      PRINT USING " ###.## ###.##";S3TOTAL,C3TOTAL;
630      PRINT USING " ###.## ###.##";S5TOTAL,C5TOTAL
640      PRINT
650      PRINT " Y-AVG";
660      PRINT USING " ###.## ###.##";STOTAL/K,CTOTAL/K;
670      PRINT USING " ###.## ###.##";S3TOTAL/K,C3TOTAL/K;
680      PRINT USING " ###.## ###.##";S5TOTAL/K,C5TOTAL/K
690      PRINT:PRINT
700      PRINT "  A(N) AND B(N) COEFFICIENTS"
710      PRINT   "---------------------------"
720      PRINT "  B1 A1 B3 A3 B5 A5"
730      PRINT
740      PRINT USING "    ###.## ###.##";2*STOTAL/K,2*CTOTAL/K;
750      PRINT USING "  ###.## ###.##;2*S3TOTAL/K,2*C3TOTAL/K;
760      PRINT USING "  ###.## ###.##;2*S5TOTAL/K,2*C3TOTAL/K
770      END
```

Chapter 37: Review Problems

1. e^3:   The series for e^x is
$$e^x = 1 + x + \frac{x^2}{2!} + \frac{x^3}{3!} + \cdots$$
Substituting $x = 3$, we obtain,
$$e^3 = 1 + 3 + \frac{9}{2} = 8.5 \quad \text{(By calculator, } e^3 = 20.1\text{)}$$

5. Use the ratio test to find the interval of convergence of the power series
$$1 + x + x^2 + x^3 + \cdots + x^n + \cdots$$
$$\lim_{n\to\infty} \left| \frac{u_n+1}{u_n} \right| = \lim_{n\to\infty} \left| \frac{x^{n+1}}{x^n} \right| = \lim_{n\to\infty} |x| = |x|$$
$|x| < 1$ implies convergence for x's in the range $-1 < x < 1$.

9. Multiply earlier series to verify the series
$$x^2 e^{2x} = x^2 + 2x^3 + 2x^4 + \cdots$$
The series for e^{2x}, from Exercise 4 Problem 2, is
$$e^{2x} = 1 + 2x + 2x^2 + \frac{4x^3}{3} + \cdots$$
Multiplying each term by x^2,
$$x^2 e^{2x} = x^2 + 2x^3 + 2x^4 + \frac{4x^5}{3} + \cdots$$

13. $\int_0^1 \frac{e^x-1}{x} dx$
From Exercise 4 Problem 17,
$$\frac{e^x-1}{x} = 1 + \frac{x}{2!} + \frac{x^2}{3!} + \frac{x^3}{4!} + \cdots$$
Integrating from $x = 0$ to $x = 1$,
$$\int_0^1 \frac{e^x-1}{x} dx = x + \frac{x^2}{4} + \frac{x^3}{18} + \frac{x^4}{96}\cdots \Big|_0^1$$
$$= 1 + \frac{1}{4} + \frac{1}{18} + \frac{1}{96}\cdots \simeq 1.316$$

17. Verify the Maclaurin expansion
$$(1+x)^3 = 1 + 3x + 3x^2 + x^3$$
Taking successive derivatives and evaluating each at $x = 0$.

$$f(x) = (1+x)^3 \qquad f(0) = 1$$
$$f'(x) = 3(1+x)^2 \qquad f'(0) = 3$$
$$f''(x) = 6(1+x) \qquad f''(0) = 6$$
$$f'''(x) = 6 \qquad f'''(0) = 6$$

Substituting into Maclaurin Series,
$$(1+x)^3 = 1 + 3x + 3x^2 + x^3$$

21. Fig. 37-15   c. Odd

25. Fig. 37-15   c. Yes

29. Write a series for (c), Figure 37-15. From the picture we determine that the square wave is an odd function with half-wave symmetry. Thus, we are looking for a function that has only odd harmonic sine terms. The function $f(x)$ is defined as follows:
$$f(x) = 1 \text{ for } -1 \le x \le 0,$$
$$f(x) = -1 \text{ for } 0 \le x \le 1$$

Solving for b_n:
$$b_n = \frac{1}{L}\int_{-L}^{L} f(x)\sin\frac{n\pi x}{L}\,dx = \int_{-1}^{0}\sin(n\pi x)\,dx$$
$$+ \int_{0}^{1}\sin(n\pi x)\,dx$$
$$= \left.\frac{-\cos n\pi x}{n\pi}\right|_{-1}^{0} + \left.\frac{\cos n\pi x}{n\pi}\right|_{0}^{1} = \frac{2(-1+\cos n\pi)}{n\pi}$$
$$b_1 = \frac{-4}{\pi}, \qquad b_3 = \frac{-4}{3\pi}, \qquad b_5 = -\frac{4}{5\pi}, \cdots$$

Substituting into the formula for waveforms with period $2L$,
$$f(x) = -\frac{4}{\pi}\sin \pi x - \frac{4}{3\pi}\sin 3\pi x - \frac{4}{5\pi}\sin 5\pi x + \cdots$$
$$= -\frac{4}{\pi}\left(\sin \pi x + \frac{1}{3}\sin 3\pi x + \frac{1}{5}\sin 5\pi x + \cdots\right)$$